Self-Powered Cyber Physical Systems

Scrivener Publishing
100 Cummings Center, Suite 541J
Beverly, MA 01915-6106

Publishers at Scrivener
Martin Scrivener (martin@scrivenerpublishing.com)
Phillip Carmical (pcarmical@scrivenerpublishing.com)

Self-Powered Cyber Physical Systems

Edited by
Rathishchandra R. Gatti
Chandra Singh
Rajeev Agrawal
and
Felcy Jyothi Serrao

Scrivener Publishing

WILEY

This edition first published 2023 by John Wiley & Sons, Inc., 111 River Street, Hoboken, NJ 07030, USA
and Scrivener Publishing LLC, 100 Cummings Center, Suite 541J, Beverly, MA 01915, USA

For more information about Scrivener publications please visit www.scrivenerpublishing.com.

Wiley Global Headquarters
111 River Street, Hoboken, NJ 07030, USA

For details of our global editorial offices, customer services, and more information about Wiley products visit us at www.wiley.com.

Library of Congress Cataloging-in-Publication Data

ISBN 9781119841883

Cover image: Engineer controlling Robotic Arm, BiancoBlue | Dreamstime.com
Cover design by Kris Hackerott

Set in size of 11pt and Minion Pro by Manila Typesetting Company, Makati, Philippines

Printed in the USA

10 9 8 7 6 5 4 3 2 1

Contents

Preface

This book is an attempt at a very futuristic vision of realizing self-powered Cyber-Physical Systems (CPS) by applying a multitude of current technologies such as ULP electronics, thin film electronics, ULP transducers, autonomous wireless sensor networks using energy harvesters at the component level, and efficient, clean energy for powering data centers and machines at the system level.

This is the need of the hour for cyber-physical systems, since data requires energy when it is stored, transmitted or converted to other forms. Since the verge is towards ubiquitous computing with massive deployment of sensors and actuators, cyber-physical systems will become energy hungry. Hence, there is a need for research to understand energy patterns and distribution in cyber-physical systems and to adopt new technologies to transcend to self-powered cyber-physical systems.

This book will explore the recent trends in energy management and self-powered devices and methods in the cyber-physical world. Self-powered devices include autonomous sensors, self-powered actuators, robots, and renewable-powered data centers. This book will also explore several advances in cyber physical systems, which may not be directly linked to self-powered autonomous systems but are enabling technologies that trend towards self–powered autonomy in the design and deployment of CPS.

Chapter 1 summarizes the low-cost, self-powered sensory transducers used in different application areas. The research aims at the different ways to generate/recharge power sources and their uses in different applications. It also focuses on recent challenges and the future scope of the device. Moreover, the work enlightens on the inter-relation of self-powered devices to the evolution of Green IoT.

Chapter 2 focuses on harvesting ambient energy sources to power wireless sensor networks. The two popular energy-harvesting devices that transform mechanical energy into electrical energy are piezoelectric nanogenerators and triboelectric nanogenerators. Utilizing these devices, self-powered sensors can be built. This chapter gives an overview of how the utilization of nanogenerators can lead to the development of self-powered sensors and their applications.

Chapter 3 focuses on various conventional actuator methodologies with the recent non-contact sensing equipment suitable for soft robots due to the low modulus factor. The inadequate nature of control over the soft robots in terms of conventional methods is overcome with the present changes in material fabrication, and design structure is taken into account to make flexible and reliable soft robots.

Chapter 4 focuses on various IDBOATS schemes, a multi-objective optimization method designed for DGDCs targeting reducing the likelihood of mean task loss and improving the profit of DGDC providers by partitioning the tasks among different ISPs and rate of task services associated with each GDC. It adopts the merits of a Mutation-based Local Search Algorithm (MLSA) for improving the diversity of solutions and preventing the issue of local optima by intelligently scheduling tasks of diversified applications and allocating the available resources within the bounds of response time.

Chapter 5 focuses on an overview of wireless power transmission techniques & also focuses on current issues and potential growth in the field.

Chapter 6 elaborates on work on computing machine learning algorithms that have to be proposed for CPS to be carried out, and application areas in the Internet of Things are ascertained.

Chapter 7 focuses on predominant concepts such as Grid to vehicles (G2V), Vehicle to Grid (V2G), and Virtual Power plant (VPP) will be discussed to evaluate the feasibility of a super system of smart Grid where renewables, EVs and the Grid are seamlessly integrated. During such integration, a smart grid will have multiple stakeholders, such as EV aggregators, Utility Service Providers, electrical transmission system operators (ETSOs) and electrical distribution system operators (EDSO) discussed in this chapter.

Chapter 8 discusses the recent technological trends in systems and subsystems of RES microgrids and EVs that aid in seamless integration, considering the optimal load matching of power demand and supply.

Chapter 9 focuses on different challenges that are discussed along with the state-of-art solutions and possible future solutions. Finally, it also focuses on the impact and leverages of fast charging technologies on self-powered automotive cyber-physical systems.

Chapter 10 covers the anonymous routing schemes, which help to hide the vehicular identity from others. Various intrusion detection systems are also discussed. Briefly, the survey covers most information about VANET and its security systems.

Chapter 11 discussed the different novel techniques presented in this chapter aimed at predicting the distress on pavement using a data-driven methodology. The main focus of this study is to manage the difficulties and ensure safe and comfortable roadway use. Diagnosing deterioration types and using proper maintenance techniques is critical, especially in the early phases.

Chapter 12 portrays the fundamental ideas of an autonomous vehicle. The car business is one of the main ventures in the globe today. The ascent of driverless vehicles will significantly affect organizations and experts. Driverless vehicles could trade corporate task forces for transport, or moving delegates and employees would obtain valuable time in the day by functioning rather than driving during everyday drives. Advancements in the present field are additionally guaranteed to modify the vehicle insurance market by diminishing mishaps. The link and synchronization of radar and ultrasonic sensors and optical cameras permit driverless driving.

Chapter 13 presents the basics of Intelligent Transportation Systems and their protection worries throughout their applications. Intelligent Transportation Systems (ITS) focus on incorporating detecting, investigating, and executing correspondence into movement, well-being, and comfort. All vehicle producers continuously make more brilliant and futuristic vehicles, changing the feel of movement.

Chapter 14 discusses different types of digital twins, such as digital twin prototypes, product digital twin, production digital twins, and performance digital twins and their applications in the automotive domain. It also explains the different levels of digital twins, such as descriptive twin, informative twin, predictive twin, comprehensive twin, and autonomous twin, which are essential during the product's lifecycle. State-of-the-art digital twin applications in the automotive domain are discussed, highlighting the benefits and challenges in adoption.

Chapter 15 discusses the state-of-the-art and the future applications of farming robotics that will be deployed to farming CPS. In the present and future era of automation being implemented into industrial technologies, the capabilities of the Robot Operating System are advancing swiftly. Even though this technology is relatively new to the robotics world, it is available openly to any early adopters. The fundamentals of ROS are presented in Chapter 16.

Chapter 17 dicusses the novel techniques of industry 5.0 of dividing work between human workers and collaborative robots, as well as safety technology, which were the most pressing development needs. There needed to be a better response from representatives from various industries.

Chapter 18 elaborates about the powered exoskeleton and how it impacts our current technologies and associated applications across different industries, such as the military, healthcare, and consumer electronics. Several state-of-the-art developed exoskeletons for critical applications such as the rehabilitation of a patient or boosting the physical attributes of a soldier are discussed.

Chapter 19 discusses the state-of-the-art development of consumer robots, different types of consumer robots that are commercialized and the future trends of consumer robotics.

Chapter 20 discusses soft robotics as a novel technique. Soft robots are made of natural polymers known for their flexibility and durability, allowing for new and innovative implementations. Soft robotics is now at the forefront of its field. Soft robotics' progress in waste management might point the way to a more sustainable future.

Chapter 21 mainly focuses on the scope of the use of robots in the agricultural industry. We also look into the current methods for preparing the ground before growing, seeding, reaping, and measuring yield.

Acknowledgements

First of all, we would like to thank the authors for their valuable contributions to the book chapters and their patient cooperation. We would also like to thank the publisher Scrivener-Wiley for providing us with the opportunity to publish this book and being very friendly and supportive during our publishing process. We would also like to thank our reviewers for reviewing the chapters. Lastly, we would like to thank our respective organisations for their continued support of our publications.

Editors –SPCPS
Dr. Rathishchandra R. Gatti
Prof. Chandra Singh
Dr. Rajeev Agrawal
Dr. Jyothi F. Serrao

1

Self-Powered Sensory Transducers: A Way Toward Green Internet of Things

Rajeev Ranjan

School of Computer Science and Applications, REVA University, Bangalore, Karnataka, India

Abstract

A sensory transducer is a device that can convert environmental energy or event into a signal/energy that can be stored for further processing. For these conversions, small and low-cost devices are preferred by many engineers. Most of the devices used in such ubiquitous applications work in a hostile environment, where changing or replacing the power module is almost impossible. To overcome these problems, energy scavenging may be the solution. Researchers focus more on such devices because they regenerate power from the environment and can be used for a more extended period with less maintenance monitoring systems with the Internet of Things (IoT).

This work summarizes an overview of the low-cost, self-powered sensory transducers used in different areas of applications. The research aims at the different ways to generate/recharge the power sources and their uses on different applications. It also focuses on recent challenges and the future scope of the device. Moreover, the work enlightens on the interrelation of self-powered devices to the evolution of Green IoT.

***Keywords*:** Self-powered transducers, sensors, WSAN, Internet of Things, green computing, energy harvesting, wireless sensor networks

1.1 Introduction

On May 19, 2021, Cyclone Tauktae struck the coast of India with a speed of 80 kmph. Likewise, two cyclones hit Indian states in the era of COVID-19 pandemic. However, the early detection of such kinds of natural calamities,

Email: rajeev.ranjan@reva.edu.in

Rathishchandra R. Gatti, Chandra Singh, Rajeev Agrawal and Felcy Jyothi Serrao (eds.) Self-Powered Cyber Physical Systems, (1–40) © 2023 Scrivener Publishing LLC

for example, could reduce damage and life causalities. This early detection and monitoring of events are possible with advanced devices like sensors and actuators. A sensor works on the principle of transduction, which can convert environmental parameters in a readable signal form [1, 2]. The actuator performs necessary actions to the environment based on the sensed data [3].

The invention of sensors and actuators makes human life more manageable. The goal of the invention is to monitor objects and events in the area where human intervention is almost impossible [4]. In the present era, sensors can be used in almost every part of life. These sensors and actuators form Wireless Sensors and Actuator Networks (WSANs) to communicate wirelessly in a single-hop or multi-hop manner [5], and hence, they experience all the challenges faced by any wireless device. Furthermore, the device needs to work in a hostile environment with constrained resources [6–8] (Figure 1.1).

The sensory transducers require power to operate. A typical sensor required 50 microwatts (mW) to 150 mW of power in an active state [19]. As a result of the hostile working environment and complex applications, changing the power module is a bit tedious. Therefore, to overcome the problem, there are two ways: (i) to minimize the energy demand by the sensor devices and network, (ii) self-powered devices by generating energy from the environment. In the early days of sensor research, scientists proposed many techniques to optimize energy consumption. However, it lowers the Quality-of-Service (QoS) and affects the environment. Therefore, scientists have shown the path for the second option. Self-powered devices will help the system run for a more extended period with high QoS [9].

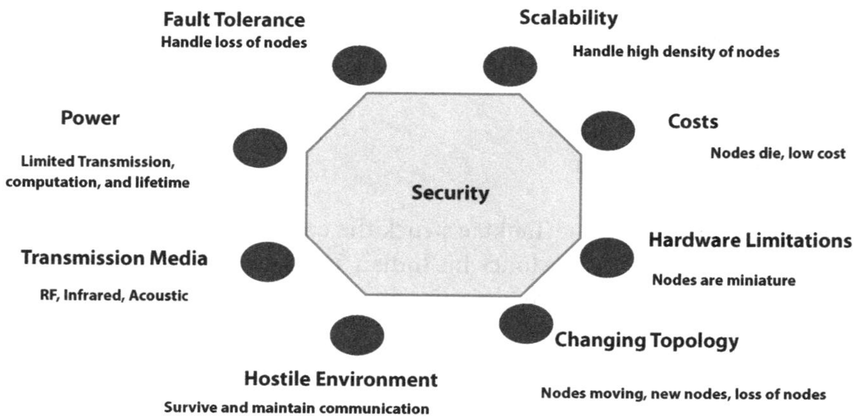

Figure 1.1 Constraints and challenges in WSAN.

Sensor is the primary element of a typical IoT application because it creates data in the form of signals [10–12]. The data created by sensor networks are saved, processed, and analyzed to gain information. The term IoT, coined by Kevin Ashton in 1999 [13], is associated with a heterogeneous set of such sensors, communicating devices, and networks working together with the Internet to achieve the goal(s).

IoT plays a significant role in information and communication technology (ICT) to automate and monitor remote events. Various hardware devices and software modules are working together in the ICT sector to achieve a goal in almost all human life fields. The hardware module includes electronic chips, antenna, processing units, and I/O devices. These devices need to work with different communication protocols, scheduling algorithms, and data processing in a few cases. In this process, most 2% of total greenhouse emissions are caused by ICT and related areas [14]. To minimize these emissions, self-powered sensors and actuators are one of the technologies leading to Green IoT [15]. Green IoT portraits the concept of reduced energy consumption of IoT devices and WSNs to make the environment safe.

In this work, the need and development of self-powered devices have been studied. The significant contribution of the work focuses on the following:

- The necessity of self-powered systems in IoT
- Finding a connection of the devices with green computation
- Parameter requirement and scope for development of such devices in different applications
- Future scope and challenges of the devices in the current scenario

1.2 Need of the Work

Self-powered systems play an important role in IoT applications because of high energy demand. However, developing and work with these systems is challenging in nature. A few reasons are:

- The sensors or actuators used in IoT are smaller in size and cost. These networks are always resource-constrained networks.
- These devices have to work in a hostile environment with different sets of hardware and software.

- The IoT applications use different transmission media, for example, RF, infrared, etc., for communication.
- The survivability and maintenance of the sensor mote working for ICT sectors are very challenging. However, applicability and users are enormous in numbers.

Moreover, generating power from the environment also requires advanced techniques and processes, which require more resources. Therefore, extensive research is the need of the hour, which concentrates on the process, techniques, requirements, applicability, challenges, and future scope of the work. It is also required exploring its ways toward green ICT Technology.

1.3 Energy Scavenging Schemes in WSAN

Energy scavenging from the environment is required to fill the gap between energy demand and supply in WSAN. As nonrechargeable primary batteries cannot increase the network lifetime and thus lower the QoS, researchers have proposed techniques for power generation. As a result of this, self-powered systems and devices become possible [16]. In this section, a few of such techniques and methods have been discussed.

1.3.1 Photovoltaic or Solar Cell

Sun is a significant source of energy in our universe. The average solar power radiation arrives at the earth's atmosphere annually is approximately 1361 W/m^2 [17]. This gives a huge potential for the power-hungry device to work with a longer lifetime. The use of PV cell in WSN gives more advantages over another energy-generating mechanism.

Solar is a continuous and huge source of energy. It provides highest power density among other energy scavenging techniques [19]. Table 1.1 highlights the energy generated by PV cell.

Table 1.1 Energy harvesting by PV cell.

Outdoors	direct sun	15 mW/cm^2
	cloudy day	0.15 mW/cm^2
Indoor	standard office desk	0.006 mW/cm^2
	<60 W desk lamp	0.57 mW/cm^2

It can be used for power constrained, low cost, small size devices, as well as for heavy machines in an industry.

A typical solar cell consists of a semiconductor of NP junction [18] (Figure 1.2). When the light energy emits the semiconductor's surface, it converts the light energy into electrical energy. This ability of the conversion is called energy conversion efficiency. The energy conversion efficiency is calculated as the ratio of energy output from the solar cell to input energy from the sun. The typical conversion efficiency of a solar cell is approximately 10% to 15%. Some researchers have proposed a maximum power point tracker (MPPT) algorithm based photovoltaic power supply for WSN [20–23]. It is a prevalent technique used in PV cells to maximize power extraction. It has been proven that the algorithm improves the utilization by 85-90% with little effect from environmental disturbance and noise. Much earlier research focuses on solar power-based energy harvesting.

In the one hand, many researchers have focused on a solar-powered cell in many applications; on the other hand, there are certain limitations, which restrict the use of PV cell. The PV cell can be used only for terrestrial applications; however, many underwater and underground applications exist where PV cell usage is restricted. As the sensor and actuator network is a highly densely deployed and low-cost network, equipping each sensor with a solar panel will be a tedious task, increasing the deployment cost. The charging rate of a PV cell depends on the size of the solar surface and the application area. There is always a need to periodically clean the solar cell surface to ensure the required efficiency of the cell. As per the knowledge, the energy generation prediction is very general and prone to substantial prediction error.

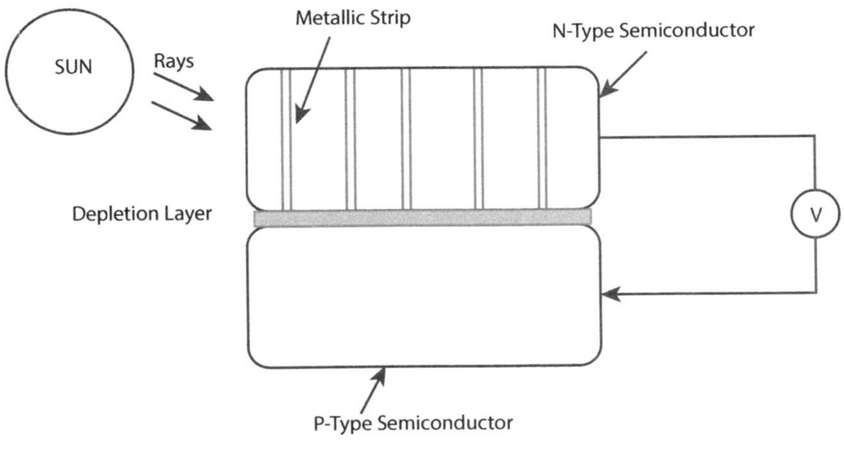

Figure 1.2 Solar/PV cell.

1.3.2 Temperature Gradient

Microthermoelectric generator works on the principle of the Seebeck effect [24], where the temperature difference between two thermocouples can be converted into electric voltage (Figure 1.3). The generator comprises two heat-conducting ceramic plates and a series of p-type and n-type semiconductor material-made thermocouples. The generator can deliver power in the range of mW. For example, as per the empirical study, it can generate the power of 2.7 mW and 20°C with a load of 10Ω. Therefore, it can be used for applications and devices where energy demand is less or nonperiodic based.

The use of a microthermoelectric generator has unique advantages. It uses only heat as input to convert into electric voltage [115]. It can generate power for more than one hundred thousand hours continuously, and it can be applied the heat as a source of any size. The thermocouple materials and other modules consist entirely of solid, fixed materials, and hence are more durable. It is also an environment-friendly option for self-powered devices [117–120].

However, the uses of thermoelectric generators in the WSN may lead to a few limitations. It cannot be used for applications with high energy demand. Therefore, it can hardly be used for periodic monitoring of events or objects. These generators only have an efficiency rate between 5% and 10%. The setup cost of these microgenerators for each sensor and actuator nodes are very costly, and it is not feasible. It may have poor thermal characteristics and high output resistance.

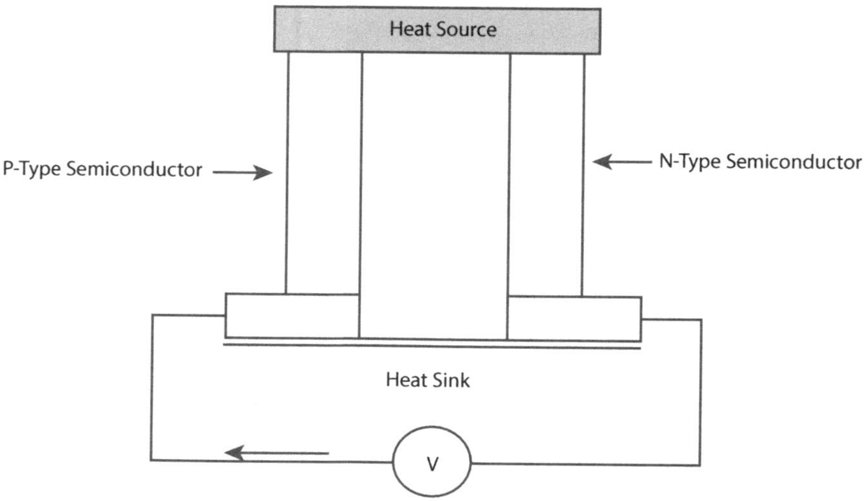

Figure 1.3 Thermoelectric cell [24].

1.3.3 Pressure Variations

The piezoelectric generator works on the pressure variations and converts mechanical stresses in the material to electrical energy [24]. This piezoelectric transducer produces very high direct current output impedance, and it is advantageous in WSN applications. The well-known examples of these generators are in the heel of shoes to recharge the pacemaker inside the heart. The piezoelectric generators are made of a piezoelectric crystal, placed between two metal plates (Figure 1.4). Mechanical strain is applied on the material using the plates, which results in the excess negative and positive charges appear on opposite sides of the crystal faces. The metal plates collect these charges to produce a voltage. A microgenerator can produce 13 to 40 V of voltage.

The use of a pressure variation based energy generator has many advantages [25]. The communication between sensors may be affected by temperature variations, but the piezoelectric generator can work in a range of temperature conditions. The carbon emissions in the generator are significantly less, and that is why it is environmentally friendly. There is very little or no effect of external electromagnetic fields on the generators.

On the other hand, a piezoelectric generator possesses a few limitations. It is not suitable for large-scale energy scavenging because they are expensive. It is used for dynamic measurement only, not suitable for static conditions. Some piezoelectric crystals are soluble in water and therefore it is not suitable for a moist environment. If the temperature of the crystal increases above a particular threshold value, it stops working.

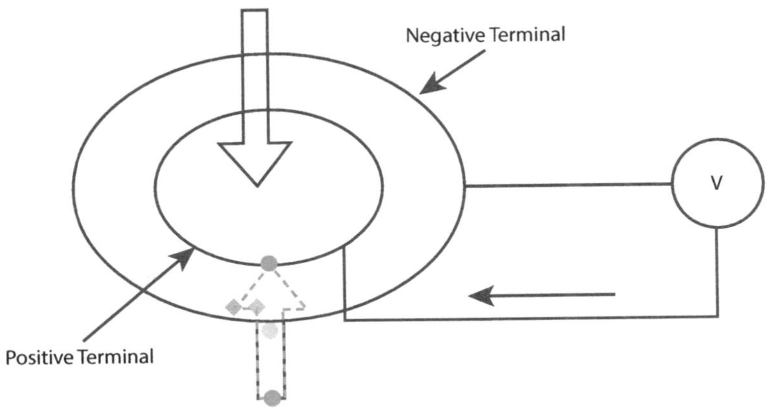

Figure 1.4 Piezoelectric cell [25].

1.3.4 Plant Microbial Fuel

Plant microbial fuel helps indoor and outdoor applications in WSN and IoT [26, 27]. It generates bioelectric current through the plant-microbe metabolisms at the rhizodiposits part of the plants. It generates a very low current of 100 $\mu W/cm^2$ and thus, it can be used in a selected set of WSN & IoT applications. For example, Schamphelaire *et al.* [28] have generated biofuel from the rhizodeposition of rice. Typical plant-microbe fuel systems consist of an anode and cathode, separated by the plant membrane. Plant bacteria grow at the cathode and convert the bio species into carbon dioxide, protons, and electrons.

There are many advantages of microbial fuel in WSN and IoT applications. Biofuels are beneficial to the environment as it reduces the pollution in the environment. A wide range of different renewable substrates can generate energy, i.e., pee power, plant-microbe, desalination cells, and many more. The fuel cell performance is based on the size of the plant.

However, plant microbial fuel has a few disadvantages. It can generate very low electric current and the efficiency of the cell will not be based on plant height. The cost of biofuel is very high. It emits sulphur dioxide which causes acid rain.

1.3.5 Wind/Liquid Flow

Wind and liquid flow cause the generation of nontraditional energy [29]. Wind or liquid turbine rotates propeller-like blades around a rotor. The rotor turns the shaft to generate an electric current (Figure 1.5). Thus, the kinetic energy will be converted into mechanical energy and further, the

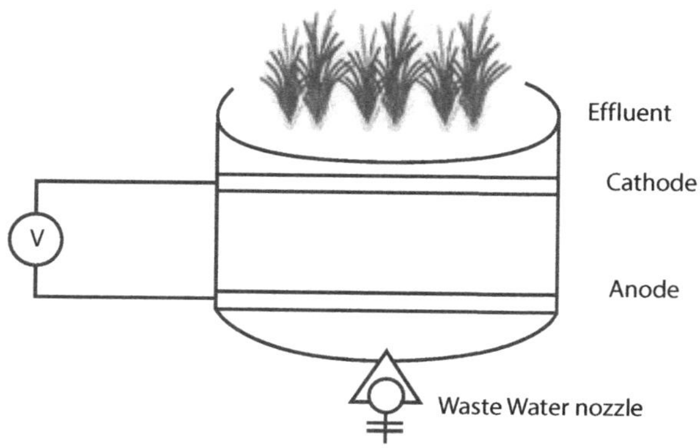

Figure 1.5 Plant microbial fuel [27].

mechanical energy is converted into electric energy. It can generate unlimited huge power, and thus, the applicability of the energy is very high. The main advantages of these power sources are it is highly inexpensive power generators and can produce high power. The land used for wind for liquid turbines can often also be used for other purposes, such as farming and thus it saves space. Wind turbines can be easily implementable on sensor devices working in the terrestrial environment whereas fluid turbines can be implemented in underwater applications. However, the main disadvantage of these types of electric sources is the cost of hydroelectricity is high and is always not suitable for low-cost network devices.

1.3.6 Vibrations

The vibration based electric generators work on the principle of Faraday's law of converting kinetic energy into electric energy. A small scale vibration-electricity generator generates 10 μW to 100 mW of power [30], which is enough for the sensor nodes and small power constraint devices working for monitoring purpose. Vibration based electric generators use a type of resonator which will vibrate with a certain frequency. A transducer mechanism is used to convert the kinetic energy to electric energy.

The major advantages of these generators are in industrial applications where the source of vibrations continuously exists as long as the machines are in operation. There are various types of vibration sources available in the environment, like, heartbeat and sinusoidal, which can be used as the input to the generators. However, theoretically, all types of vibrations can be converted into electric current, but practically, it is not always possible. There are certain types of vibrations are preferred in sensor applications. Scientists are working to identify the alternate sources of energy generation to reduce the load on primary nonrechargeable batteries so that there is a continuous supply to power-hungry devices. The energy scavenging techniques can increase the network lifetime of the WSN and IoT applications.

1.3.7 Friction

Friction-based electric generation is called as triboelectric effect where electrification can be done by the triboelectric materials like polymers, metals, and inorganic materials (Figure 1.6). The major advantage of such nanogenerators is there is no or less interference of the electrostatic scanning and thus it provides high efficiency. Energy density produced by the generator is proportional to pressing frequency. It requires low cost and maintenance.

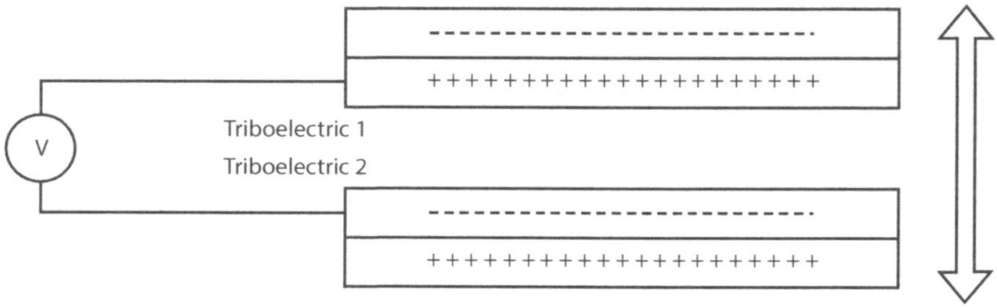

Figure 1.6 Triboelectric effect [30].

However, it has high internal resistance between layers. The energy density of triboelectric generators is low compared to pyroelectric generators. Adverse environmental conditions may reduce its electrical output value.

1.4 Self Powered Systems and Green IoT (G-IoT)

It is estimated that nearly 50 billion wireless and wired devices have been connected to IoT by 2050 [31, 32]. These huge numbers of connected devices are working with the Internet and other related networks for many applications. This became possible due to the technological advancement of low-cost, low-size miniature, and portable devices. As per the statistics, it is estimated that IoT devices generate 79.4 Zettabytes by 2025 and 50,000 Zettabytes by 2050. Storage and processing of this vast data would require high bandwidth and energy. In the process, it emits a large number of greenhouse gases, which may reach 10-12% in five years, according to the study by U.S. Environmental Protection Agency. In this direction, scientists have focused on technologies to reduce greenhouse emission. Green IoT (G-IoT) is an energy-efficient approach to IoT to reduce the greenhouse effect caused by existing applications.

G-IoT aims at three types of IoT approaches:

i) To develop smart low power hardware.
ii) To reduce the energy consumption in the devices and
iii) To harvest the power from the environment.

Several earlier kinds of research focus on the first two options of saving or reducing energy consumption. The first option targets smart scheduling of tasks on cores [33], minimized processing power on devices [34], follows the layers of devices [35], and the use of passive low power sensors [36].

Researchers have also proposed many "energy–efficient" protocols and techniques to achieve the goals. The works include an optimized power model [37], dynamic modulation techniques [38], scheduling schemes [39], sleep-awake techniques [40], finding the least congested optimized route [41], topology control [42], and many more [43–46]. However, these works require thinking over the "trade-off" between resource optimization and QoS requirements. Therefore, in recent work, scientists explored the third option. The generations of new energy solutions have reduced the problem of "trade-offs" and aim to reduce the greenhouse effects.

G-IoT focuses on every design process of both hardware and software to reduce the greenhouse effect. The evolution of such power generating and wireless charging capabilities has shown how energy constrained devices work for a longer period in a more controlled and reliable way. Various Researchers have surveys and proposed energy charging to the sensor nodes deployed in the area [47–50]. This is a new and unique way to explore more, as it creates an environment in which sensor nodes can share energy between neighbours.

1.5 Application Area and Scope of Self-Powered System in G-IoT

The applicability of sensor-assisted networks has gained much focus in the recent era. In this section, different applications running on the self-powered system have been discussed briefly. Today, we have various self-powered systems and techniques ranging from agriculture to medical applications and from underground mineral exploration to Unmanned Arial Vehicles. Different WSN and IoT applications can be categorized into three parts based on the area of deployment.

1.5.1 Terrestrial Applications

The applications above the earth's surface are listed under terrestrial applications. The deployment of the nodes in this category will be random or fixed deployment. In this, the major focused areas of applications are:

1.5.1.1 Agriculture

WSN and IoT applications in agriculture are huge. Considerable research work has been conducted to increase the agricultural yield in various parts of the world [51–60]. Among these, solar powered systems gain more mileage than other wireless charging systems [51–54]. Udutalapally *et al.* [51] have proposed a machine learning model to monitor the health of rice, tomato, potato, and corn plants. Results indicate that the PV cells can generate power 8.33 V to 16.3 V when the solar cell has inserted up to 30 cm in the ground. The generated power was enough to work micro controlled water motor and soil sensors. Sharma *et al.* [52] have proposed a solar powered system to monitor the plant growth. The result shows that the lifetime of the solar powered system increased by approximately 20 times with network throughput increased by 33%. Na & Isaac [53] implemented a solar powered system with the dimension of 170 x 170 x 2 mm, the solar panel was providing an output of 6 V at 3.65 W. Similarly, Heble *et al.* [54] installed the solar panel for maize crop monitoring and claimed the network lifetime increased by 83%. The model has been developed with Zigbee with LoRa module.

Another area of deployment of IoT and WSNs is in monitoring the agricultural yield. Ikeda *et al.* [55] have developed a microthermoelectric generator powered soil sensor. Researchers claim that the proposed method can harvest 100 μW–370 μW on average, which is enough for the microcontroller to perform. Researchers in [56, 57] have developed a biofuel powered WSN system for land and flora health monitoring. The biofuel system [56] generates average output power in the range from 200 μW to 300 μW. Reference [57] proposed the plant microbial fuel powered system for land monitoring. Keswani *et al.* [58], Nigussie *et al.* [59] and Benyezza *et al.* [60] suggested the energy charging of the sensor nodes using a fluid turbine. The fluid turbine helps in smart irrigation based on weather conditions in different zones and terrains. Akhund *et al.* [61] used the fluid turbine to develop an intelligent poultry farm. Terteil *et al.* [62] have proposed a real-time, low-cost precision agriculture monitoring system which consumes less power and emits less greenhouse gas. It is a LoRa based security system that would also detect the chemical composition of the soil [115]. Table 1.2 features the self energy systems in agricultural applications.

1.5.1.2 Smart Home and Cities

Development and design of smart cities focus more on self-powered G-IoT to optimize the power need. Researchers have proposed various models to fulfill the needs. Kaur *et al.* [63] have designed a five-layered architecture for G-IoT to reduce energy consumption. Researchers in [64] proposed G-IoT–assisted intelligent parking system. It is well known that the boundaries of smart cities and home applications are expanding very fast

Table 1.2 Self-powered system in agricultural applications.

Work	Energy harvester	Parameter sensed	Power generated
[51]	Solar	Soil moisture, humidity, temperature	8.33 V to 16.3 V (Daytime)
[52]	Solar	Humidity, temperature, pressure	0.75V/cm^2
[53]	Solar	Soil moisture, humidity, temperature	6 V
[54]	Solar	Soil moisture, soil temp, humidity, light intensity, temperature	-
[55]	Thermoelectric	Soil Temp	100 µW–370 µW
[56]	Bio - Fuel	Biosensor	200 µW to 300 µW
[57]	Bio - Fuel	pH value, temperature, humidity, type of soil	200 mV
[58]	Fluid Turbine & Solar	Soil moisture, soil Temperature, humidity, CO_2 sensor, light intensity	-
[59]	Fluid Turbine	Temperature, moisture, pH, microclimate	15 mW/cm^2
[60]	Fluid Turbine	Soil humidity and temperature	-

and widespread and large amount of techniques/algorithms have been proposed in various areas. Solar-powered self-charging points for network devices gain mileage in these applications also [65–68]. Researchers in [69] have deployed the PV-equipped WSN in the forest area to evaluate the environmental effects of a forest fire. The study proves that the network is reliable even two months after the fire. However, the network lifetime decreases in a harsh environment where fire propagation speed is higher. Research in [70–72] proposed the PV-enabled WSN for data-centric applications. The researchers have proposed secure data access control using encryption methods to maintain the balance among security, energy efficiency, and interoperability. Similarly, PV-equipped WSNs have been proposed for time synchronization [73–76], urban computing [77, 78], and so on. Solar power-based energy generation and distribution has attracted much attention from Industry and academics.

Furthermore, these areas have been conducted to generate power from sources other than solar plants. Yun *et al.* [79] have investigated the use of thermoelectricity to charge the nanosensors used in home applications. Imperial study shows that the temperature difference across an insulation panel harvests ~100 μW of energy. Likewise, researchers in [80–82] have studied extensively the use and scope of piezoelectric generators to charge and supply power to the devices in smart city and smart home applications. Thyagaraj Naidu [83] proposed to generate the vibration power from human movement.

Wind and fluid turbine is a hot topic in the area of the self-powered system for smart applications. Smart grid-based energy generation and distribution is the need of the hour for every big city. Various works, for example, generation and storage of such energy [84–86], forecast of wind energy [87, 88], distribution [89], have been proposed. Carmo *et al.* [90] have used the energy to develop vapor-compression heat pumps in residential use. They also proposed the technique to maintain and monitor the health of such turbines to work longer.

1.5.1.3 Industry

Sensor-enabled Industrial monitoring was started during the fourth Industrial revolution [91], initiated in the 1990s with increased connectivity and information exchange. This era is advanced in product development and monitoring automation. The exchange of information which makes the revolution was possible due to WSAN and IoT. Due to many such industrial applications, starting from raw material to final product quality, quantity, and process monitoring, it has gained much interest among scientists, engineers, and policymakers [92]. Energy consumption in any Industrial Internet of Things (IIoT) system depends on the device usages, the complexity of the algorithm or protocols, the size of the packet, the size of the data, total transmissions and retransmissions, and retransmissions mobility of the nodes. Moreover, most of the WSAN-IoT system uses a web interface to monitor and control the components. These web interfaces increase carbon emission.

In order to provide the G-IoT solutions in Industrial applications, pressure, vibration, heat, and motion are the primary sources of renewable energy [14]. Table 1.3 underlines the self energy systems in industrial applications. Firmansah *et al.* [93] have used piezoelectric power to monitor the Induction motor. The results obtained showed that the vibration generated by the sensor was 0.53V at a frequency of 50Hz. Plessis *et al.* [94] have proved that WSAN can monitor the machine with piezoelectric generated power and cantilever architecture. Similarly, works with references [95–98] generated power with piezoelectric generators to monitor and diagnose machines. In 2018, Hou *et al.* [99] have used the temperature

Table 1.3 Self-powered system in industrial applications.

Work	Energy harvester	Parameter sensed	Power generated
[93]	Piezoelectric	Vibration of motor	0.53V at 50Hz
[94]	Piezoelectric	Temperature, voltage, vibration	2.8 mW at 100 Hz
[95]	Piezoelectric	Accelerometer, temperature, voltage	0.04W at 1 Khz
[96]	Piezoelectric & Electromagnetic	Temperature, vibration	0.003 to 0.015 mw at 100hz
[97]	Magneto-Piezoelectric	Temperature, movement, accelerometer, magnetometer, gyroscope, light, humidity, barometer	39.2 mw (max)
[98]	Piezoelectric And Electromagnetic	Temperature, humidity, light, motion	25.45 mw at 60 Hz
[99]	Temperature Gradient	Temperature	32.7 mw and 34.7 mw (Two Thermoelectric Generator)
[100]	Temperature Gradient	-	-
[101]	Solar And Radio Frequency	Temperature	-

gradient-based self-powered WSN and monitored the temperature generated by the machines. The paper concludes that the energy conversion rate is ~27%, and it can recharge the WSN nodes with an active period of 0.9s. McNinch *et al.* [100] have discussed the use of thermocouples in the mining industry. Williams *et al.* [101] have proposed a hybrid approach of the solar and radiofrequency powered system in data centers monitoring. They have applied the technique with a sleep-wake scheduling scheme to optimize the power usages. Pozo *et al.* [102] have suggested using another hybrid method with wind turbine and PV cell in IIoT. The requirement of these hybrid methods balances the considerable energy requirement in Industry to work continuously with different devices.

Moreover, various IoT architectures and layered approaches have also been proposed to reduce the greenhouse effects. Shekhar Rao *et al.* [103] have designed architecture with a 5G network to manage large data generation by IIoT devices. Ghader *et al.* [104] have proposed a data management layer to cache data called proxy nodes to optimize network energy consumption. Chiarotti *et al.* [121] have proposed a thermoelectric generator for industrial plant.

1.5.1.4 Medicines

Wearable sensory transducers are one of the critical technologies in personalized applications. With other communicating devices, biosensors form Body Area Networks (BAN), which can measure blood glucose, temperature, heartbeat, oxygen level and help doctors and health personnel monitor the patient remotely and effectively. The low-powered sensors help in early-stage disease detection, diagnosis, and management. It can be used to monitor quarantined people and severe patients during the COVID-19 pandemic [105]. Numerous solutions have been proposed with low power biosensors; however, energy consumption and optimization are significant hurdles in timely reporting and identifying disease. Lou *et al.* [106] have done a detailed survey on self–powered biosensors for medical applications. As per the literature, biosensors can be charged by the electric current generated by mechanical forces, such as vibration and friction, temperature gradient, and solar energy. Similarly, researchers in [107, 108] have also studied various self-powered sensors for medicine and health applications.

Different researchers have also implemented self-powered biosensors in different health applications. Mao *et al.* [109] have devised an auto-powered sensor system to monitor the Maximal Lactate Steady State (MLSS) during the exercise of sportsmen. The biosensors were getting charged by piezoelectric signals generated by the motion of the sportsman. The researchers in [110] proposed biosensors attached with the diaper to measure the glucose level in the urine of a diabetic patient. The system was powered by an enzymatic biofuel cell, which used the glucose in urine as fuel to generate electricity. Zhang *et al.* [111] has used motion powered system for respiratory monitoring.

1.5.1.5 Environment Monitoring

Environment monitoring is a way to keep track and predict climate change across the globe. It includes (i) air quality monitoring [116, 131–134], (ii) water quality monitoring [149, 150], (iii) waste monitoring, and (iv) disaster monitoring [122–124]. As per UNICEF reports, the UNECE Ministers endorsed recommendations for strengthening environmental monitoring and information systems in European and Central Asian countries. It was prepared by the UNECE Working Group on Environmental Monitoring at the Fifth Ministerial Conference on Environment Europe held at Kyiv in 2003 [112]. Moreover, G-IoT focuses on environmental and climate clean initiatives. However, the traditional power supply through battery affects the climate mainly because it has a limited lifetime and increased e-waste [113, 114]. In this work [113], researchers have proposed a wind–hybrid generated power to charge the wireless node for climate monitoring (temperature, humidity, and pressure). The system uses three nanogenerators, namely, triboelectric piezoelectric nanogenerator and electromagnetic generator, to produce power.

Water quality and contamination will be discussed in next section in details.

1.5.1.6 Structural Monitoring

Structural monitoring includes damage and load over a structure and prediction of the lifetime of the same [125]. In the recent era, it can be monitored and analyzed remotely by WSN and IoT. Aono *et al.* [126] has proposed a pressure powered system to detect cracks in the steel frame. Zheng *et al.* has generated power through windmill to recharge the sensor cell used in subway tunnel monitoring. In reference [126], researchers have generated energy using the magnetic levitation, which can be used to charge a lithium battery. The battery gives power to the rail side Zigbee module. The researchers claim that the device can produce up to 2.3V with the rail vehicle running speed of 130 kmph. Researchers in [127] have powered the wireless sensor system using piezoelectric material to monitor large-scale bridges remotely. Based on the empirical studies conducted, it was found that the energy harvested were enough to detect the lifetime of the bridge by finding the cracks, damage and load of the bridge. Salehi *et al.* [127] recently studied in details about the self-powered system used in structural monitoring. The literature highlighted that majority of the structural monitoring and civil engineering applications use strain and triboelectric voltage for energy harvesting.

1.5.1.7 Indoor Applications

Applications related to smart homes and indoor applications are interlinked to each other. In this era of smart devices, advanced IoT and sensor devices can be used in indoor applications to make our homes smart where different devices communicate. The indoor applications include intruder detection, appliances monitoring and control, indoor air quality, floor sterilization, dust removal and many more. Major advantage of such applications is the deployment of the nodes may be deterministic rather than random deployment [128]. Therefore, the energy consumption required for deploying the node is negligible. However, self-sustainable sensor nodes and other communicating devices are required to gain the QoS in the device to device communication. Researchers in [129, 130] have proposed different methods of self-sustainable sensor nodes applications to detect intruders. Dong *et al.* [129] have used four Electromagnetic harvesters in series to recharge a 470-μF storage capacitor through a rectifier. Sultana *et al.* [130] have used a piezoelectric generator to provide charge to the microcontroller. References [131–133] have discussed self-powered systems to monitor the air quality in smart homes. Tran *et al.* [131] have generated power from radio frequency operated at 915 MHz to operate 0.5 mW microcontroller units. The power generated by RF can be used to monitor volatile organic compounds, temperature, humidity, and atmospheric pressure inside the building. Researchers in [132] have tried indoor PV cells to provide power for the same application. Jiang *et al.* [133] have proposed the use of plant microbial fuel to operate the sensors deployed for air quality monitoring. Ma *et al.*, in 2017 [134], developed a system to monitor and sterilization of floors with technology-enabled devices powered by triboelectric nanogenerators. The nanogenerator has been designed by polyvinyl chloride (PVC) flooring with a thin copper film, which can generate power-density of 0.5 $\mu A/cm^2$.

1.5.1.8 Arial Vehicles

Unmanned arial vehicles (UAVs), airplanes and aircraft monitoring are one of the promising applications of WSNs and IoT. It includes monitoring of machineries, tracking of the vehicles, load and fuel monitoring, etc. Various researches show various self-powered system designs to control and monitor remotely. Khoshnoud *et al.* [135] and Shearwood *et al.* [136] have shown the use of solar-powered multi-rotor in UAVs. Shearwood [136] and his team have proposed the use of radio frequency generated power to localize and track bees using UAV. Elahi [137] and Somov [138] have proposed the use of piezoelectric power for airplane structural monitoring. Zelenika *et al.* [139] in 2020 and Le *et al.* [140] in 2005 have studied extensively regarding the energy harvesting methods and applications in the aeronautical engineering field. Both the work explained the different schemes for energy harvesting in different parts of the airplane, i.e., solar, heat, vibration, RF, airflow, and acoustic energy.

1.5.1.9 Military Applications

WSNs were initially designed for military related applications [138, 141, 142]. The applications include border surveillance, war-zone classification, vehicle tracking, army mental and physical health monitoring, aircraft structural health monitoring, damage assessment, and many more. This is the one application area where secure communication between different parties plays a significant role. In this scenario, self-powered WSN and IoT system helps in building intelligent applications. For example, solar powered UAVs and aircraft for border surveillance [136], motion charged military vehicles [143] and many more. Among military applications, a strong focus is on mobile data aggregation and dissemination to analyze different conditions.

1.5.1.10 Underwater Applications

In the recent era, underwater wireless sensor networks (UWSNs) have attracted both the industry and academia due to its various applications. However, the basic requirements of underwater applications are different from terrestrial applications. Signal interference, noise, multipath propagation, absorption in signaling in these conditions will be more. Therefore, most of the system uses either acoustic signal or Ultra Wide Band (UWB) [144] signal for the purpose. Moreover, if the applications require working in both above the surface and underwater, then refraction, reflection, scattering and Doppler effects in the signal may increase, which causes more noise in the output signal. However, the applications pertaining to this category is huge, for example, water quality, submarine monitoring, ground water availability, and so on. Therefore, self-powered system plays a major role in this set of applications. The main source of energy harvesting in UWSN are piezoelectric, water turbines, triboelectric nanogenerators, and plant-based microfuel cell generators [148].

1.5.1.11 Submarine and Event Localization

Location and tracking are the process to identify the location of an event and node over the space (special) and over the time (temporal) respectively. Researchers in [145, 146] have presented self powered system to localize event in the UWSNs. Saeed *et al.* [145] have proposed a hybrid model of energy generation using hybrid acoustic wave and optical wave. As optical wave travels very faster and acoustic wave travels slower, average received signal strength could be used for high accuracy of localization. The combined use of the signal can harvest power to recharge the sensors. Zhang *et al.* [146] have discussed about nonmyopic energy required to provide power to the underwater sensor devices.

1.5.1.12 Water Contamination

Water quality is the major concern in today's society. Out of 71% water available in world, only ~2.5% of the water is available for drinking purpose [150]. Moreover, 70% to 80% of the water comes from underground sources. References in [147–149], researchers have proposed many methods to generate power in order to recharge sensor nodes in UWSNs. Chen *et al.* [147] have proposed plant microfuel-powered system for energy harvesting to monitor water quality. Researchers proved empirically that the microbio#fuel can produce 100 times more than the steam water, which can be used to charge the temperature, pH and dissolved oxygen sensors. Researchers in [148] presented wireless power transfer through wave propagation method, which is out of the scope of the current work. Carminati *et al.* [149] has presented a water turbine generated power to check the water contamination level. Experiment proves the turbine can generate a maximum current of 110 mA with a flow rate of 20 L/min.

1.5.1.13 Intelligent Water Distribution and Smart Meter

Water distribution in urban and rural areas requires more attention now days to reduce wastage. Moreover, smart meters with remote maintenance facility are needed to cope up increasing water supply demand. Therefore, sensor and valve actuator enabled IoT system has been proposed and implemented by various researchers. Again, having concern with energy consumption and greenhouse effect, various self powered systems have/had been proposed [151–158]. Li & Chong [151] and Tasic *et al.* [152] have proposed water turbine powered smart meter for water distribution. Researchers in [151] have chosen a 12 V 10 W water turbine which can generate a power range of 32-37 W which is enough to recharge the meter battery. Reference [152] have used the meter for hot water consumption in Swiss and proved that each household can save on average per day 0.6 kWh of heat energy. Pimenta & Chaves [153] have proposed mechanical vibration power can increase the lifespan of 20 Wh rechargeable lithium battery up to 11 years. Similarly, Chen *et al.* [154] have shown triboelectric nanogenerator can generate an output power of 143 V at 0.75 Hz and at a water depth of 60 m which is more than enough to supply power to the 12-15 temperature sensors and LED lights. Ordoñez *et al.* [155], Gould in 2020 [156], Zhong *et al.* [158] have proposed vibration, temperature gradient and hybrid electromagnetic-triboelectric generated power respectively for the purpose.

1.5.1.14 Underground Applications

Sensor and IoT applications in underground have their own significance. Applications like petroleum and mineral mining and exploration, structural health of the underground earth, lava and tectonic plate's movement to predict and prevent natural calamities are now much more important for modern human civilization. However, energy harvesting and distribution among devices in such applications is a tedious task. Due to large interference, non line of sight, and unavailability of energy sources like solar power adds more complexity to the application domain. Therefore vibration and pressure based energy sources would be preferred by most of the scientists [159, 160].

1.5.1.15 Coal and Petroleum Mining Application

Mineral exploration and extraction like coal, petroleum product, iron, mica etc is a part of human life and thus its demand aggregates with an increased population [161]. However, energy generation and recharging the Underground WSN is more challenging. Molaei *et al.* [162] in 2014 proved that the Zigbee module consumes lesser energy than Bluetooth, UWB and Wi-Fi in Underground space. Singh [163] and He *et al.* [164] have suggested the implement the Piezo-electric energy harvesting technique for the application.

1.5.1.16 Underground Structural Monitoring

Monitoring underground train tunnels, roadways, underpass etc. has become easier with the economically low cost, high power technological advancement in the field. In order to provide power to such applications, air flow [165], vibrations [166, 167] and Radio frequency [168] have been used. Cammarano *et al.* [165] have implanted the idea of vibration energy as energy harvesting for underground train tunnels. The work was carried out with six temperature and humidity sensor motes which were getting charged by the wind turbine. The wind turbine harvests 100 mA at a voltage of 7.2 V. Researchers in [168] used the radio frequency signal to charge the nodes for underground pavement monitoring.

1.6 Challenges and Future Scope of the Self-Powered G-IoT

G-IoT focuses on monitoring devices or events or both in such a way that it will reduce environmental hazard. Energy harvesting in WSNs and IoT applications has provided extra mileage in the technological world. It reduced the dependency on the primary source of power like alkaline batteries and it helps in reducing the greenhouse effect in the environment. However, there are a few technological challenges exist in the way. These challenges can be categorized into two parts: i) challenges pertain to energy efficient design and protocols ii) challenges with focus to the energy harvesting techniques (Table 1.4).

1.6.1 Challenges Pertain to Energy Efficient Design and Protocols

In this subsection, few of the challenges and future extensions of the energy efficient hardware and software design techniques will be discussed.

1.6.2 Size and Cost of the Harvester

Sensor network is a densely deployed network and hence the number of sensor motes in the given area will be higher. These sensor nodes may be heterogeneous on homogeneous. Therefore, there is a requirement for a large amount of economically feasible energy harvesters which can generate sufficient power for the nodes, and hence, scientists and engineers are working to make such mini-size miniature devices [169].

Table 1.4 Challenges and future of G-IoT.

Energy Efficient Design and Protocols	Size and cost of the Energy harvester
	Dynamic Modulation Scaling
	Energy efficient routing and scheduling protocols
	Application specific passive wake up receivers
	Redefined Protocol with application specific goals
	Embedded operating systems
	AI and cloud assisted Lifetime prediction
	Service-oriented Architecture
	Intelligent Web Interface
	Cross Layer Exploitations
	Security aspects and need of Standardization
Energy Harvesting Techniques	Generic energy generator
	Hybrid Energy Sources
	Cooperation among different energy sources
	Energy storage
	Intelligent prediction model
	Focus on Energy harvesting in Underwater and Underground applications

1.6.3 Energy-Efficient Routing and Scheduling Protocols

To reduce greenhouse emission, the implementation of energy efficient routing protocols is an alternate way to increase battery life. The researchers reported so far suggest the following methods to reduce energy consumption:

- Single hop and multi-hop routing technique
- Designing dynamic routing rather than static routing with a balanced load over all the nodes
- Packet size and channel condition
- Energy consumption by the node in the handshaking process

- Routing in topology controlled networks
- Intelligent power control based on inter–node distance and residual energy
- Network aware sleep-awake scheduling mechanism
- Energy aware packet forwarding and routing procedures

Approaches to obtain the goal of reduction of energy consumption in WSNs are categorized along with energy harvesting techniques. For example, the routing policies may focus on the maximum hop distance rather than the shorter or optimized hop. The data-centric network may focus on the information centric or knowledge centric.

1.6.4 Design of Application-Specific Passive Wake-Up Receivers

Bello *et al.* [170] have studied extensively on RF based energy harvesting through passive wake-up radios. The article point out the need for application specific design of such wake-up receives to meet the different requirements of varieties of applications. It also focuses on the requirements of wake-up receivers which were suitable for mobile nodes as most of such receivers have been designed for stationary nodes.

1.6.5 Redefined Protocol with Application-Specific Goals

The conventional IoT protocols focus mainly on energy efficiency. However, the significance of QoS parameters depends on the type of application. For example, in multimedia applications, the requirement of reduced latency is more significant than the reliability factor. The choices of QoS in certain applications are still challenging and it requires a tradeoff between the available resource and the required QoS. Moreover, in the energy harvested WSNs, the aim should be more on the type and amount of information gathered based on the amount of energy harvested.

1.6.6 Embedded Operating Systems

A general purpose operating system works as a middleware between the hardware and the applications. It restricts and controls the access of different resources to users and supports the simultaneous execution of different processes. Embedded OS on the other hand works in more constrained environment and is designed to work for specific tasks.

These systems should be supported by energy management to execute the energy efficient WSN application [19]. The component of sensor mote like sensors, microcontrollers, transceivers, timers etc. should be handled easily and efficiently. All this requires an appropriate programming model to support energy management without imposing a huge burden on system resources.

1.6.7 AI and Cloud-Assisted Lifetime Prediction Techniques

The lifetime of the network – consists of primary or secondary batteries-can be predicted generally without any model or a specific technique, which may lead to error. Artificial intelligence and cloud based algorithms can be used to predict the battery lifetime with reduced error. Data processing and prediction of other WSAN parameters could be developed at cloud platform.

1.6.8 Design of Energy-Efficient/Harvested Service-Oriented Architecture

Service-oriented architecture (SoA) is an approach in form of architecture to provide a set of application specific services to the users over the network. Several WSN architectures are available with IoT in the literature for specific applications. Now, scientists should devote their attention on such SoA with the aim of G-IoT.

1.6.9 Smart Web Interfaces for Monitoring

As discussed in Section 1.4, the majority of the monitoring and controlling of devices are possible by web interfaces and these interfaces add more greenhouse emission in the environment. With the increased number of devices and users, the number of such interfaces also will increase. Therefore, designing of intellect web interfaces should be focused to cut down the CO_2 emission and support G-IoT mission.

1.6.10 Cross Layer Exploitations with Energy Harvesting

The tradition techniques focus on the layered approach, where the energy optimization takes place at each layers separately. Cross-layered approach must be exploited to optimize energy and other resource parameters for IoT applications.

1.6.11 Security Aspects and Need of Standardization

Data and network privacy is a major aspect in wireless communication. To prevent the sensitive information and network nodes from possible security attacks, it is necessary to implement security mechanism to the network. However, public and private keys used in encryption/decryption increases the network overhead, and hence, increase the energy consumption. Therefore, energy-efficient security protocols must be developed to secure the communication and data.

1.6.12 Challenges Related to Energy Harvesting Techniques

In this subsection, challenges and future scope related to energy harvesting techniques will be discussed.

1.6.13 Generic Energy Generator

There are multiple ways to generate power which requires harvester specific hardware and interface to manage and distribute the power. A general purpose mini size energy harvester with single interface can be designed to reduce the complexities of each energy generators.

1.6.14 Hybrid Energy Sources

Energy scavenging by different energy sources need to be explored more in a set of applications, to reduce the cost and to increase the power generation. For example, energy harvested by solar power in a rainy season can be compensated by fluid turbine. In the recent times, possibilities of many hybrid nanogenerators have been proposed, but, many other possibilities could be thought-of [50].

1.6.15 Cooperation Among Different Energy Sources

In large-scale systems and hybrid energy generators, where two or more energy scavenging techniques generate power, mutual cooperation between such generators are need to be strengthened. Cloud based third party monitoring among these generators can be a possible solution.

1.6.16 Energy Storage

Udutalapally *et al.* [51] have shown that the solar installations generates surplus power during sunny day and scarce during night time. Similarly, in many long-lasting applications the energy generation by the nanogenerators are affected by self-discharge, environmental parameter, and hardware impurity etc. Therefore, scientists are researchers should focus on the storage and distribution of surplus energy in the network, whenever required.

1.6.17 Intelligent Prediction Model for Amount of Harvested Energy

The energy harvesting error by a given generator can be calculated as:

$$HE = \frac{\hat{E} - \underline{E}}{\hat{E}}$$

where $\hat{E}$ is the predicted energy harvested per unit time, and $\underline{E}$ is the actual energy harvested in the unit time. The error in the harvesting is mainly due to the material/hardware quality, application area, environmental parameter, self discharge of the battery, capacity of battery under load etc. The prediction model of the energy harvesting can be controlled and monitored by cloud platform along with AI and ML technique.

1.6.18 Focus on Energy Generator for Underwater and Underground Applications

Industry and academia should devote more research on finding more nanomaterials and nanogenerator options which can generate adequate energy for the devices working for underground applications. The possible solutions may be bio-fuel and integrated circuit.

1.7 Conclusion

In this chapter, self-powered WSN and IoT systems have been discussed in detail. The chapter discusses different options of energy harvesting techniques with their advantages, disadvantages and possible working domains. It also gives the overview of the proposed techniques in different set of applications. It shows that there are much sufficient options available in terrestrial applications, but more options need to be discovered in the underground applications. Finally, it explores the existing technical challenges and future research scope in the given field. In the future work, different techniques of wireless power transfer in WSN will be discussing in detail.

References

1. Fahmy, H.M.A., Energy harvesting in WSNs, in: *Wireless Sensor Networks*, pp. 41–99, Springer, Cham, 2020.
2. Fain, G. L. *Sensory transduction*. Sinauer, Sunderland, MA, 2019.
3. Chaudhary, S., Johari, R., Bhatia, R., Gupta, K., Bhatnagar, A., Craiot: Concept, review and application(s) of IoT, in: *2019 4th International Conference on Internet of Things: Smart Innovation and Usages (IoT-SIU)*, 2019, April, IEEE, pp. 1–4.
4. Culler, D., Estrin, D., Srivastava, M., Guest editors' introduction: Overview of sensor networks. *Computer*, 37, 8, 41–49, Aug, 2004.
5. Pandeeswari, G., Suganthi, M., Asokan, R., Performance of single hop and multi hop relaying protocols in cognitive radio networks over Weibull fading channel. *J. Ambient Intell. Humaniz. Comput.*, 12, 3, 3921–3927, 2021.
6. Sharma, R., Prakash, S., Roy, P., Methodology, applications, and challenges of WSN-IoT, in: *2020 International Conference on Electrical and Electronics Engineering (ICE3)*, 2020, February, IEEE, pp. 502–507.
7. Hashish, S. and Tawalbeh, H., Quality of service requirements and challenges in generic WSN infrastructures. *Proc. Comput. Sci.*, 109, 1116–1121, 2017.
8. Maitra, T. and Roy, S., Research challenges in BAN due to the mixed WSN features: Some perspectives and future directions. *IEEE Sens. J.*, 17, 17, 5759–5766, 2017.
9. Singh, J., Kaur, R., Singh, D., Energy harvesting in wireless sensor networks: A taxonomic survey. *Int. J. Energy Res.*, 45, 1, 118–140, 2021.
10. Astill, J., Dara, R.A., Fraser, E.D., Roberts, B., Sharif, S., Smart poultry management: Smart sensors, big data, and the Internet of Things. *Comput. Electron. Agric.*, 170, 105291, 2020.
11. Liu, D., Chen, B., An, J., Li, C., Liu, G., Shao, J., Wang, Z.L., Wind-driven self-powered wireless environmental sensors for Internet of Things at long distance. *Nano Energy*, 73, 104819, 2020.
12. Zhao, X., Kang, Z., Liao, Q., Zhang, Z., Ma, M., Zhang, Q., Zhang, Y., Ultralight, self-powered and self-adaptive motion sensor based on triboelectric nanogenerator for perceptual layer application in Internet of things. *Nano Energy*, 48, 312–319, 2018.
13. Aslam, F., Aimin, W., Li, M., Ur Rehman, K., Innovation in the era of IoT and industry 5.0: Absolute innovation management (AIM) framework. *Information*, 11, 2, 124, 2020 Feb.

14. Tang, X., Wang, X., Cattley, R., Gu, F., Ball, A.D., Energy harvesting technologies for achieving self-powered wireless sensor networks in machine condition monitoring: A review. *Sensors*, 18, 12, 4113, 2018.
15. Nandyala, C.S. and Kim, H.K., Green IoT agriculture and healthcare application (GAHA). *Int. J. Smart Home*, 10, 4, 289–300, 2016.
16. Knight, C., Davidson, J., Behrens, S., Energy options for wireless sensor nodes. *Sensors*, 8, 12, 8037–66, 2008 Dec.
17. Kovácik, J., Mináriková, N., Emmer, Š., Šugár, P., Šugárová, J., Ludrovcová, B., Al–Qutub, A., Al–Athel, K., Rodriguez, J., Cañadas, I., Titanium solar metallurgy–earth and space, in: *MATEC Web of Conferences 2019*, vol. 304, EDP Sciences, p. 07005.
18. Sharma, H., Haque, A., Jaffery, Z.A., Solar energy harvesting wireless sensor network nodes: A survey. *J. Renew. Sustain. Energy*, 10, 2, 023704, 2018 Mar 13.
19. Karl, H. and Willig, A., *Protocols and architectures for wireless sensor networks*, John Wiley & Sons, 2007 Oct 8.
20. Zhang, L., Yu, Y., Ma, H., Zhang, Y., Cao, P., Design of photovoltaic power supply MPPT circuit for WSN node based on current observation. *Int. J. Online Eng.*, 14, 7, 45–61, 2018 Jul 1.
21. Li, K., Yang, X., Shu, S., Ma, C., Jia, H., MPPT algorithm for wireless sensor network solar power system. *J. Tsinghua Univ. (Sci. Technol.)*, 51, 5, 2011.
22. Sharma, S., Piezoelectric energy harvesting and management in WSN using MPPT algorithm, in: *2016 International Conference on Wireless Communications, Signal Processing and Networking (WiSPNET)*, 2016 Mar 23, IEEE, pp. 2228–2232.
23. López-Lapeña, O., Penella, M.T., Gasulla, M., A new MPPT method for low-power solar energy harvesting. *IEEE Trans. Ind. Electron.*, 57, 9, 3129–38, 2009 Dec 4.
24. Edvinsson, N., Energy harvesting power supply for wireless sensor networks: Investigation of piezo-and thermoelectric micro generators, 2013.
25. Bridgeless, L.P., Vibration based piezoelectric energy harvesting utilizing bridgeless rectifier circuit. *Jurnal Kejuruteraan.*, 28, 87–94, 2016.
26. Osorio de la Rosa, E., Vázquez Castillo, J., Carmona Campos, M., Barbosa Pool, G.R., Becerra Nuñez, G., Castillo Atoche, A., Ortegón Aguilar, J., Plant microbial fuel cells–based energy harvester system for self-powered IoT applications. *Sensors*, 19, 6, 1378, 2019 Jan.
27. Osorio-de-la-Rosa, E., Vázquez-Castillo, J., Castillo-Atoche, A., Heredia-Lozano, J., Castillo-Atoche, A., Becerra-Nuñez, G., Barbosa, R., Arrays of plant microbial fuel cells for implementing self-sustainable wireless sensor networks. *IEEE Sens. J.*, 21, 2, 1965–74, 2020 Aug 28.
28. Schamphelaire, L.D., Bossche, L.V., Dang, H.S., Höfte, M., Boon, N., Rabaey, K., Verstraete, W., Microbial fuel cells generating electricity from rhizodeposits of rice plants. *Environ. Sci. Technol.*, 42, 8, 3053–8, 2008 Apr 15.

29. Jushi, A., Pegatoquet, A., Le, T.N., Wind energy harvesting for autonomous wireless sensor networks, in: *2016 Euromicro Conference on Digital System Design (DSD)*, 2016 Aug 31, IEEE, pp. 301–308.
30. Iqbal, M., Nauman, M.M., Khan, F.U., Abas, P.E., Cheok, Q., Iqbal, A., Aissa, B., Vibration-based piezoelectric, electromagnetic, and hybrid energy harvesters for microsystems applications: A contributed review. *Int. J. Energy Res.*, 45, 1, 65–102, 2021 Jan.
31. Pawar, P. and Trivedi, A., Device-to-device communication based IoT system: Benefits and challenges. *IETE Tech. Rev.*, 36, 4, 362–74, 2019 Jul 4.
32. Omotosho, A., Ayemlo Haruna, B., Mikail Olaniyi, O., Threat modeling of internet of things health devices. *J. Appl. Secur. Res.*, 14, 1, 106–21, 2019 Jan 2.
33. Suresh, A., Reyana, A., Varatharajan, R., CE Multi-core architecture for optimization of energy over heterogeneous environment with high performance smart sensor devices. *Wirel. Pers. Commun.*, 103, 2, 1239–52, 2018 Nov.
34. García, G.J., Jara, C.A., Pomares, J., Alabdo, A., Poggi, L.M., Torres, F., A survey on FPGA-based sensor systems: Towards intelligent and reconfigurable low-power sensors for computer vision, control and signal processing. *Sensors*, 14, 4, 6247–78, 2014 Apr.
35. Foteinos, V., Kelaidonis, D., Poulios, G., Vlacheas, P., Stavroulaki, V., Demestichas, P., Cognitive management for the internet of things: A framework for enabling autonomous applications. *IEEE Veh. Technol. Mag.*, 8, 4, 90–9, 2013 Nov 12.
36. Reindl, L., Scholl, G., Ostertag, T., Ruppel, C.C., Bulst, W.E., Seifert, F., SAW devices as wireless passive sensors, in: *1996 IEEE Ultrasonics Symposium. Proceedings*, 1996 Nov 3, vol. 1, IEEE, pp. 363–367.
37. Kumar, S., Kaiwartya, O., Rathee, M., Kumar, N., Lloret, J., Toward energy-oriented optimization for green communication in sensor enabled IoT environments. *IEEE Syst. J.*, 14, 4, 4663–73, 2020 Apr 27.
38. Migabo, E., Djouani, K., Kurien, A., An energy-efficient and adaptive channel coding approach for narrowband internet of things (NB-IoT) systems. *Sensors*, 20, 12, 3465, 2020 Jan.
39. Firouzi, R., Rahmani, R., Kanter, T., Distributed-reasoning for task scheduling through distributed internet of things controller. *Proc. Comput. Sci.*, 184, 24–32, 2021 Jan 1.
40. Haimour, J. and Abu-Sharkh, O., Energy efficient sleep/wake-up techniques for IOT: A survey, in: *2019 IEEE Jordan International Joint Conference on Electrical Engineering and Information Technology (JEEIT)*, 2019 Apr 9, IEEE, pp. 478–484.
41. Srinidhi, N.N., Kumar, S.D., Venugopal, K.R., Network optimizations in the Internet of Things: A review. *Eng. Sci. Technol., Int. J.*, 22, 1, 1–21, 2019 Feb 1.
42. Rai, S., Ananad, N., Varma, S., Scrutinizing localized topology control in wsn using rigid graphs, in: *2015 2nd International Conference on Computing*

for Sustainable Global Development (INDIACom), 2015 Mar 11, IEEE, pp. 1712–1715.
43. Ranjan, R. and Varma, S., Cross layer approach for energy minimization in Wireless Sensor Networks, in: *IMPACT-2013*, 2013 Nov 23, IEEE, pp. 6–10.
44. Ranjan, R. and Varma, S., Collision-free time synchronization for multi-hop wireless sensor networks. *J. Comput. Intell. Electron. Syst.*, 1, 2, 200–6, 2012 Dec 1.
45. Sahni, Y., Cao, J., Liu, X., MidSHM: A middleware for WSN-based SHM application using service-oriented architecture. *Future Gener. Comput. Syst.*, 80, 263–74, 2018 Mar 1.
46. Sharma, P.K., Kumar, N., Park, J.H., Blockchain technology toward Green IoT: Opportunities and challenges. *IEEE Netw.*, 34, 4, 263–9, 2020 Mar 3.
47. Alsharif, M.H., Kim, S., Kuruoğlu, N., Energy harvesting techniques for wireless sensor networks/radio-frequency identification: A review. *Symmetry*, 11, 7, 865, 2019 Jul.
48. Ahmed, S., Khan, M.A., Ishtiaq, A., Khan, Z.A., Ali, M.T., Energy harvesting techniques for routing issues in wireless sensor networks. *Int. J. Grid Util. Comput.*, 10, 1, 10–21, 2019.
49. Basagni, S., Naderi, M.Y., Petrioli, C., Spenza, D., Conti, M., Giordano, S., Stojmenovic, I., Wireless sensor networks with energy harvesting. *Mobile Ad Hoc Netw.*, 1, 701–36, 2013 Mar 5.
50. Sah, D.K. and Amgoth, T., Renewable energy harvesting schemes in wireless sensor networks: A survey. *Inf. Fusion*, 63, 223–47, 2020 Nov 1.
51. Udutalapally, V., Mohanty, S.P., Pallagani, V., Khandelwal, V., sCrop: A Internet-of-Agro-Things (IoAT) enabled solar powered smart device for automatic plant disease prediction. *arXiv preprint* arXiv:2005.06342, 2020 May 9.
52. Sharma, H., Haque, A., Jaffery, Z.A., Maximization of wireless sensor network lifetime using solar energy harvesting for smart agriculture monitoring. *Ad Hoc Networks*, 94, 101966, 2019 Nov 1.
53. Sadowski, S. and Spachos, P., Solar-powered smart agricultural monitoring system using internet of things devices, in: *2018 IEEE 9th Annual Information Technology, Electronics and Mobile Communication Conference (IEMCON)*, 2018 Nov 1, IEEE, pp. 18–23.
54. Heble, S., Kumar, A., Prasad, K.V., Samirana, S., Rajalakshmi, P., Desai, U.B., A low power IoT network for smart agriculture, in: *2018 IEEE 4th World Forum on Internet of Things (WF-IoT)*, 2018 Feb 5, IEEE, pp. 609–614.
55. Ikeda, N., Shigeta, R., Shiomi, J., Kawahara, Y., Soil-monitoring sensor powered by temperature difference between air and shallow underground soil. *Proceedings of the ACM on Interactive, Mobile, Wearable and Ubiquitous Technologies*, 2020 Mar 18, vol. 4(1), pp. 1–22.
56. Brunelli, D., Tosato, P., Rossi, M., Microbial fuel cell as a biosensor and a power source for flora health monitoring, in: *2016 IEEE Sensors*, 2016 Oct 30, IEEE, pp. 1–3.

57. Pietrelli, A., Micangeli, A., Ferrara, V., Raffi, A., Wireless sensor network powered by a terrestrial microbial fuel cell as a sustainable land monitoring energy system. *Sustainability*, 6, 10, 7263–75, 2014 Oct.
58. Keswani, B., Mohapatra, A.G., Mohanty, A., Khanna, A., Rodrigues, J.J., Gupta, D., De Albuquerque, V.H., Adapting weather conditions based IoT enabled smart irrigation technique in precision agriculture mechanisms. *Neural Comput. Appl.*, 31, 1, 277–92, 2019 Jan.
59. Nigussie, E., Olwal, T., Musumba, G., Tegegne, T., Lemma, A., Mekuria, F., IoT-based irrigation management for smallholder farmers in rural sub-Saharan Africa. *Proc. Comput. Sci.*, 177, 86–93, 2020 Jan 1.
60. Benyezza, H., Bouhedda, M., Rebouh, S., Zoning irrigation smart system based on fuzzy control technology and IoT for water and energy saving. *J. Cleaner Prod.*, 302, 127001, 2021 Jun 15.
61. Akhund, T.M., Snigdha, S.R., Reza, M.S., Newaz, N.T., Saifuzzaman, M., Rashel, M.R., Self-powered IoT-based design for multi-purpose smart poultry farm, in: *International Conference on Information and Communication Technology for Intelligent Systems*, 2020 May 15, Springer, Singapore, pp. 43–51.
62. Ali, T.A., Choksi, V., Potdar, M.B., Precision agriculture monitoring system using green internet of things (g-iot), in: *2018 2nd International Conference on Trends in Electronics and Informatics (ICOEI)*, 2018 May 11, IEEE, pp. 481–487.
63. Kaur, G., Tomar, P., Singh, P., Design of cloud-based green IoT architecture for smart cities, in: *Internet of Things and Big Data Analytics Toward Next-Generation Intelligence*, pp. 315–333, Springer, Cham, 2018.
64. Sant, A., Garg, L., Xuereb, P., Chakraborty, C., A novel green IoT-based pay-as-you-go smart parking system. *CMC Comput. Mater. Contin.*, 67, 3, 3523–44, 2021 Jan 1.
65. Wang, S., Wang, X., Wang, Z.L., Yang, Y., Efficient scavenging of solar and wind energies in a smart city. *ACS Nano*, 10, 6, 5696–700, 2016 Jun 28.
66. Dispenza, G., Sergi, F., Napoli, G., Randazzo, N., Di Novo, S., Micari, S., Antonucci, V., Andaloro, L., Development of a solar powered hydrogen fueling station in smart cities applications. *Int. J. Hydrog. Energy*, 42, 46, 27884–93, 2017 Nov 16.
67. Bezplyug, E.A., Afonina, V.S., Grosheva, A.S., Smart city and heritage preservation with the help of modern technologies, in: *Russian Economy: Goals, Challenges and Achievements*, pp. 30–33, 2018.
68. Kabir, M.H., Roy, S., Ahmed, M.T., Alam, M., IoT based solar powered smart waste management system with real time monitoring-an advancement for smart city planning. *Glob. J. Comput. Sci. Technol. (GJCST)*, 20, G5, 11–20, 2020.
69. Kassan, R., Châtelet, E., Soukieh, J., Reliability assessment of photovoltaic wireless sensor networks for forest fire propagation detection. *Int. J. Model. Simul.*, 38, 1, 50–65, 2018 Jan 2.

70. Bianchi, G., Capossele, A.T., Petrioli, C., Spenza, D., AGREE: Exploiting energy harvesting to support data-centric access control in WSNs. *Ad Hoc Networks*, 11, 8, 2625–36, 2013 Nov 1.
71. Kim, J.M., Lee, H.S., Yi, J., Park, M., Power adaptive data encryption for energy-efficient and secure communication in solar-powered wireless sensor networks. *J. Sens.*, 2016, 1–9, 2016 Feb.
72. Sudarsono, A., Huda, S., Fahmi, N., Al-Rasyid, M.U., Kristalina, P., Secure data exchange in environmental health monitoring system through wireless sensor network. *Int. J. Eng. Technol. Innov.*, 6, 2, 103, 2016 Apr 1.
73. Sun, X., Su, Y., Huang, Y., Tan, J., Yi, J., Hu, T., Zhu, L., Photovoltaic modules monitoring based on WSN with improved time synchronization. *IEEE Access*, 7, 132406–12, 2019 Sep 17.
74. Li, Z., Chen, W., Li, M., Lei, J., Incorporating energy heterogeneity into sensor network time synchronization. *IEEE Trans. Parallel Distrib. Syst.*, 26, 1, 163–73, 2014 Feb 24.
75. Kawagoe, H. and Sugano, M., Implementation of time synchronization for energy harvesting wireless sensor network, in: *Proceedings of the 2017 VI International Conference on Network, Communication and Computing*, 2017 Dec 8, pp. 175–178.
76. Wu, T., Liu, Y., Li, H., Xue, C.J., Lee, H.G., Yang, H., SATS: An ultra-low power time synchronization for solar energy harvesting WSNs, in: *Proceedings of the 2016 International Symposium on Low Power Electronics and Design*, 2016 Aug 8, pp. 106–111.
77. Dehwah, A.H., Mousa, M., Claudel, C.G., Lessons learned on solar powered wireless sensor network deployments in urban, desert environments. *Ad Hoc Networks*, 28, 52–67, 2015 May 1.
78. Ju, J., Park, I., Lee, Y., Cho, J., Cho, H., Yoe, H., Shin, C., An implementation of the salt-farm monitoring system using wireless sensor network, in: *International Conference on Security-Enriched Urban Computing and Smart Grid*, 2010 Sep 15, Springer, Berlin, Heidelberg, pp. 647–655.
79. Yun, M., Ustun, E., Nadeau, P., Chandrakasan, A., Thermal energy harvesting for self-powered smart home sensors, in: *2016 IEEE MIT Undergraduate Research Technology Conference (URTC)*, 2016 Nov 4, IEEE, pp. 1–4.
80. Al Ahmad, M. and Allataifeh, A., Piezoelectric-based energy harvesting for smart city application, in: *Information Innovation Technology in Smart Cities*, pp. 343–356, Springer, Singapore, 2018.
81. Kalyazina, S. and Lepekhin, A., Implementing an energy harvesting technology within the concept of Smart city, in: *MATEC Web of Conferences*, vol. 193, EDP Sciences, p. 01030, 2018.
82. Toh, C.K., Sanguesa, J.A., Cano, J.C., Martinez, F.J., Advances in smart roads for future smart cities. *Proc. R. Soc. A*, 476, 2233, 20190439, 2020 Jan 29.
83. Thyagaraj Naidu, S., Energy harvesting from human motion in smart cities: An alternative renewable energy source. Bachelor's Thesis, Tampere University of Applied Sciences, Pirkanmaa, Finland, 2021.

84. Kashem, S.B., Chowdhury, M.E., Ahmed, J., Ashraf, A., Shabrin, N., Wind power integration with smart grid and storage system: Prospects and limitations. *Int. J. Adv. Comput. Sci. Appl.*, 11, 5, 552–69, 2020 May.
85. Pinnangudi, B., Kuykendal, M., Bhadra, S., Smart grid energy storage, in: *The Power Grid 2017*, pp. 93–135, Academic Press, Cambridge, Massachusetts, USA, Jan 1.
86. Cheng, M., Zhang, Y.Q., Zhang, J.Z., Development and research progress of wind power generators. *J. Electric Power Sci. Technol.*, 3, 2–9, 2009.
87. Ioakimidis, C.S., Oliveira, L.J., Genikomsakis, K.N., Wind power forecasting in a residential location as part of the energy box management decision tool. *IEEE Trans. Industr. Inform.*, 10, 4, 2103–11, 2014 Jun 30.
88. Burlando, M. and Meissner, C., Evaluation of two ANN approaches for the wind power forecast in a mountainous site. *Int. J. Renew. Energy Res. (IJRER)*, 7, 4, 1629–38, 2017 Dec 30.
89. Mosetti, G.P., Poloni, C., Diviacco, B., Optimization of wind turbine positioning in large wind farms by means of a genetic algorithm. *J. Wind Eng. Ind. Aerodyn.*, 51, 1, 105–16, 1994 Jan 1.
90. Carmo, C., Detlefsen, N., Nielsen, M., Smart Grid enabled heat pumps: An empirical platform for investigating how residential heat pumps can support large-scale integration of intermittent renewables. *Energy Proc.*, 61, 1695–8, 2014 Jan 1.
91. Xu, L.D., Xu, E.L., Li, L., Industry 4.0: State of the art and future trends. *Int. J. Prod. Res.*, 56, 8, 2941–62, 2018 Apr 18.
92. Levina, A., Kalyazina, S., Ershova, A., Schuur, P.C., IIOT within the architecture of the manufacturing company, in: *Proceedings of the International Scientific Conference-Digital Transformation on Manufacturing, Infrastructure and Service*, 2020 Nov 18, pp. 1–6).
93. Firmansah, A., Mufti, N., Affandi, A.N., Zaeni, I.A., Self-powered IoT based vibration monitoring of induction motor for diagnostic and prediction failure. *IOP Conf. Ser.: Mater. Sci. Eng.*, 588, 1, 012016, 2019 Aug 1, IOP Publishing.
94. du Plessis, A.J., Huigsloot, M.J., Discenzo, F.D., Resonant packaged piezoelectric power harvester for machinery health monitoring, in: *Smart Structures and Materials 2005: Industrial and Commercial Applications of Smart Structures Technologies*, 2005 May 5, vol. 5762, International Society for Optics and Photonics, pp. 224–235.
95. Discenzo, F.M., Chung, D., Automation, R., Heights, M., Loparo, K.A., Pump condition monitoring using self-powered wireless sensors. *Sound Vib.*, 40, 12–15, 2006.
96. Pekosławski, B., Pietrzak, P., Makowski, M., Szafoni, Ł., Napieralski, A., Study on application of piezoelectric vibration energy harvesters for powering of wireless sensor nodes in large rotary machine diagnostic systems, in: *Proceedings of the 2009 NSTI Nanotechnology Conference and Expo*, Houston, TX, USA, 3–7 May 2009, vol. 1, pp. 530–533.

97. Cho, J.Y., Kim, J., Kim, K.B., Ryu, C.H., Hwang, W., Lee, T.H., Sung, T.H., Significant power enhancement method of magneto-piezoelectric energy harvester through directional optimization of magnetization for autonomous IIoT platform. *Appl. Energy*, 254, 113710, 2019 Nov 15.
98. Jung, I., Choi, J., Park, H.J., Lee, T.G., Nahm, S., Song, H.C., Kim, S., Kang, C.Y., Design principles for coupled piezoelectric and electromagnetic hybrid energy harvesters for autonomous sensor systems. *Nano Energy*, 75, 104921, 2020 Sep 1.
99. Hou, L., Tan, S., Zhang, Z., Bergmann, N.W., Thermal energy harvesting WSNs node for temperature monitoring in IIoT. *IEEE Access*, 6, 35243–9, 2018 Jun 28.
100. McNinch, M., Parks, D., Jacksha, R., Miller, A., Leveraging IIoT to improve machine safety in the mining industry. *Min. Metall. Explor.*, 36, 4, 675–81, 2019 Aug.
101. Williams, J.M., Gao, F., Qian, Y., Song, C., Khanna, R., Liu, H., Solar and RF energy harvesting design model for sustainable wireless sensor tags, in: *2020 IEEE Topical Conference on Wireless Sensors and Sensor Networks (WiSNeT)*, 2020 Jan 26, IEEE, pp. 1–4.
102. Pozo, B., Garate, J.I., Ferreiro, S., Double smart power management with wind mini-reactor and photovoltaic cell energy harvester for Industry 4.0 IIoT devices, in: *Energy Harvesting and Storage: Materials, Devices, and Applications VIII*, vol. 10663, p. 106630L, International Society for Optics and Photonics, Bellingham, Washington, USA, 2018 May 15.
103. Rao, V.C., Kumarswamy, P., Phridviraj, M.S., Venkatramulu, S., Rao, V.S., 5G enabled industrial Internet of Things (IIoT) architecture for smart manufacturing, in: *Data Engineering and Communication Technology 2021*, pp. 193–201, Springer, Singapore, 2021.
104. Ghaderi, A., Qavany, Z., Movahedi, Z., A green data management layer in industrial IoT. *IOP Conf. Ser.: Mater. Sci. Eng.*, 1067, 1, 012131, 2021 Feb 1. IOP Publishing.
105. Mujawar, M.A., Gohel, H., Bhardwaj, S.K., Srinivasan, S., Hickman, N., Kaushik, A., Aspects of nano-enabling biosensing systems for intelligent healthcare; towards COVID-19 management. *Mater. Today Chem.*, 5, 100306, 2020 Jun.
106. Lou, Z., Li, L., Wang, L., Shen, G., Recent progress of self-powered sensing systems for wearable electronics. *Small*, 13, 45, 1701791, 2017 Dec.
107. Reid, R.C. and Mahbub, I., Wearable self-powered biosensors. *Curr. Opin. Electrochem.*, 19, 55–62, 2020 Feb 1.
108. Gao, W., Self-powered wearable biosensors, in: *Energy Harvesting and Storage: Materials, Devices, and Applications XI*, vol. 11722, p. 117220O, International Society for Optics and Photonics, Bellingham, Washington, USA, 2021 Apr 12.

109. Mao, Y., Yue, W., Zhao, T., Shen, M., Liu, B., Chen, S., A self-powered biosensor for monitoring maximal lactate steady state in sport training. *Biosensors*, 10, 7, 75, 2020 Jul.
110. Zhang, J., Liu, J., Su, H., Sun, F., Lu, Z., Su, A., A wearable self-powered biosensor system integrated with diaper for detecting the urine glucose of diabetic patients. *Sens. Actuators B: Chem.*, 341, 130046, 2021 Aug 15.
111. Zhang, X., Ai, J., Zou, R., Su, B., Compressible and stretchable magnetoelectric sensors based on liquid metals for highly sensitive, self-powered respiratory monitoring. *ACS Appl. Mater. Interfaces*, 13, 13, 15727–37, 2021 Mar 29.
112. World Health Organization, Protecting children's health in a changing environment: Report of the fifth ministerial conference on environment and health, World Health Organization. Regional Office for Europe, 2010.
113. Rahman, M.T., Salauddin, M., Maharjan, P., Rasel, M.S., Cho, H., Park, J.Y., Natural wind-driven ultra-compact and highly efficient hybridized nanogenerator for self-sustained wireless environmental monitoring system. *Nano Energy*, 57, 256–68, 2019 Mar 1.
114. Touati, F., Galli, A., Crescini, D., Crescini, P., Mnaouer, A.B., Feasibility of air quality monitoring systems based on environmental energy harvesting, in: *2015 IEEE International Instrumentation and Measurement Technology Conference (I2MTC) Proceedings*, 2015 May 11, IEEE, pp. 266–271.
115. Chottirapong, K., Manatrinon, S., Dangsakul, P., Kwankeow, N., Design of energy harvesting thermoelectric generator with wireless sensors in organic fertilizer plant, in: *2015 6th International Conference of Information and Communication Technology for Embedded Systems (IC-ICTES)*, 2015 Mar 22, IEEE, pp. 1–6.
116. Wang, C.H., Huang, Y.K., Zheng, X.Y., Lin, T.S., Chuang, C.L., Jiang, J.A., A self sustainable air quality monitoring system using WSN, in: *2012 Fifth IEEE International Conference on Service-Oriented Computing and Applications (SOCA)*, 2012 Dec 17, IEEE, pp. 1–6.
117. Lee, F.Y., Navid, A., Pilon, L., Pyroelectric waste heat energy harvesting using heat conduction. *Appl. Therm. Eng.*, 37, 30–7, 2012 May 1.
118. Pandya, S., Velarde, G., Zhang, L., Wilbur, J.D., Smith, A., Hanrahan, B., Dames, C., Martin, L.W., New approach to waste-heat energy harvesting: Pyroelectric energy conversion. *NPG Asia Mater.*, 11, 1, 1–5, 2019 Jun 7.
119. Zabek, D. and Morini, F., Solid state generators and energy harvesters for waste heat recovery and thermal energy harvesting. *Therm. Sci. Eng. Prog.*, 9, 235–47, 2019 Mar 1.
120. Rattner, A.S. and Garimella, S., Energy harvesting, reuse and upgrade to reduce primary energy usage in the USA. *Energy*, 36, 10, 6172–83, 2011 Oct 1.
121. Chiarotti, U., Moroli, V., Menchetti, F., Piancaldini, R., Bianco, L., Viotto, A., Baracchini, G., Gaspardo, D., Nazzi, F., Curti, M., Gabriele, M., Development of a small thermoelectric generators prototype for energy harvesting from low temperature waste heat at industrial plant. *J. Nanosci. Nanotechnol.*, 17, 3, 1586–91, 2017 Mar 1.

122. Qian, J. and Jing, X., Wind-driven hybridized triboelectric-electromagnetic nanogenerator and solar cell as a sustainable power unit for self-powered natural disaster monitoring sensor networks. *Nano Energy*, 52, 78–87, 2018 Oct 1.
123. Mihajlovic, Z., Joza, A., Milosavljevic, V., Rajs, V., Zivanov, M., Energy harvesting wireless sensor node for monitoring of surface water, in: *2015 21st International Conference on Automation and Computing (ICAC)*, 2015 Sep 11, IEEE, pp. 1–6.
124. Hande, A., Bridgelall, R., Zoghi, B., Vibration energy harvesting for disaster asset monitoring using active RFID tags. *Proc. IEEE.*, 98, 9, 1620–8, 2010 Jul 1.
125. Gao, M., Wang, P., Wang, Y., Yao, L., Self-powered ZigBee wireless sensor nodes for railway condition monitoring. *IEEE Trans. Intell. Transp. Syst.*, 19, 3, 900–9, 2017 Jun 16.
126. Aono, K., Hasni, H., Pochettino, O., Lajnef, N., Chakrabartty, S., Quasi-self-powered piezo-floating-gate sensing technology for continuous monitoring of large-scale bridges. *Front. Built Environ.*, 5, 29, 2019 Mar 26.
127. Salehi, H., Burgueño, R., Chakrabartty, S., Lajnef, N., Alavi, A.H., A comprehensive review of self-powered sensors in civil infrastructure: State-of-the-art and future research trends. *Eng. Struct.*, 234, 111963, 2021 May 1.
128. Amutha, J., Sharma, S., Nagar, J., WSN strategies based on sensors, deployment, sensing models, coverage and energy efficiency: Review, approaches and open issues. *Wireless Pers. Commun.*, 111, 2, 1089–115, 2020 Mar.
129. Dong, C., Li, S., Han, R., He, Q., Li, X., Xu, D., Self-powered wireless sensor network using event-triggered energy harvesters for monitoring and identifying intrusion activities. *IET Power Electron.*, 12, 8, 2079–85, 2019 May 7.
130. Sultana, S., Akand, T., Dey, M., Design and implementation of home security system using PZT, in: *2016 9th International Conference on Electrical and Computer Engineering (ICECE)*, 2016 Dec 20, IEEE, pp. 554–557.
131. Tran, T.V., Dang, N.T., Chung, W.Y., Battery-free smart-sensor system for real-time indoor air quality monitoring. *Sens. Actuators B: Chem.*, 248, 930–9, 2017 Sep 1.
132. Yue, X., Kauer, M., Bellanger, M., Beard, O., Brownlow, M., Gibson, D., Clark, C., MacGregor, C., Song, S., Development of an indoor photovoltaic energy harvesting module for autonomous sensors in building air quality applications. *IEEE Internet Things J.*, 4, 6, 2092–103, 2017 Sep 20.
133. Jiang, Y., Liang, P., Huang, X., Ren, Z.J., A novel microbial fuel cell sensor with a gas diffusion biocathode sensing element for water and air quality monitoring. *Chemosphere*, 203, 21–5, 2018 Jul 1.
134. Ma, J., Jie, Y., Bian, J., Li, T., Cao, X., Wang, N., From triboelectric nanogenerator to self-powered smart floor: A minimalist design. *Nano Energy*, 39, 192–9, 2017 Sep 1.
135. Khoshnoud, F., Esat, I.I., de Silva, C.W., Rhodes, J.D., Kiessling, A.A., Quadrelli, M.B., Self-powered solar aerial vehicles: Towards infinite endurance UAVs. *Unmanned Syst.*, 8, 02, 95–117, 2020 Apr 2.

136. Shearwood, J., Williams, S., Aldabashi, N., Cross, P., Freitas, B.M., Zhang, C., Palego, C., Localization and tracking bees using a battery-less transmitter and an autonomous unmanned aerial vehicle, in: *2020 IEEE/MTT-S International Microwave Symposium (IMS)*, 2020 Aug 4, IEEE, pp. 1263–1266.
137. Elahi, H., The investigation on structural health monitoring of aerospace structures via piezoelectric aeroelastic energy harvesting. *Microsyst. Technol.*, 3, 1–9, 2020 Sep.
138. Somov, A., Chew, Z.J., Ruan, T., Li, Q., Zhu, M., Piezoelectric energy harvesting powered WSN for aircraft structural health monitoring, in: *2016 15th ACM/IEEE International Conference on Information Processing in Sensor Networks (IPSN)*, 2016 Apr 11, IEEE, pp. 1–2.
139. Zelenika, S., Hadas, Z., Bader, S., Becker, T., Gljušćić, P., Hlinka, J., Janak, L., Kamenar, E., Ksica, F., Kyratsi, T., Louca, L., Energy harvesting technologies for structural health monitoring of airplane components—A review. *Sensors*, 20, 22, 6685, 2020 Jan.
140. Le, M.Q., Capsal, J.F., Lallart, M., Hebrard, Y., Van Der Ham, A., Reffe, N., Geynet, L., Cottinet, P.J., Review on energy harvesting for structural health monitoring in aeronautical applications. *Prog. Aerosp. Sci.*, 79, 147–57, 2015 Nov 1.
141. Hussain, M.A. and Kyung Sup, K., WSN research activities for military application, in: *2009 11th International Conference on Advanced Communication Technology*, 2009 Feb 15, vol. 1, IEEE, pp. 271–274.
142. Đurišić, M.P., Tafa, Z., Dimić, G., Milutinović, V., A survey of military applications of wireless sensor networks, in: *2012 Mediterranean Conference on Embedded Computing (MECO)*, 2012 Jun 19, IEEE, pp. 196–199.
143. Dong, Y., Wang, Y., Li, S., Cui, M., Wu, H., Demand-based charging strategy for wireless rechargeable sensor networks. *ETRI J.*, 41, 3, 326–36, 2019 Jun.
144. Alhawari, A.R., Majeed, S.F., Saeidi, T., Mumtaz, S., Alghamdi, H., Hindi, A.T., Almawgani, A.H., Imran, M.A., Abbasi, Q.H., Compact elliptical UWB antenna for underwater wireless communications. *Micromachines*, 12, 4, 411, 2021 Apr.
145. Saeed, N., Celik, A., Al-Naffouri, T.Y., Alouini, M.S., Energy harvesting hybrid acoustic-optical underwater wireless sensor networks localization. *Sensors*, 18, 1, 51, 2018 Jan.
146. Zhang, D., Liu, M., Zhang, S., Zhang, Q., Non-myopic energy allocation for target tracking in energy harvesting UWSNs. *IEEE Sens. J.*, 19, 10, 3772–83, 2018 Dec 28.
147. Chen, Q., Liu, Y., Liu, G., Yang, Q., Shi, X., Gao, H., Su, L., Li, Q., Harvest energy from the water: A self-sustained wireless water quality sensing system. *ACM Trans. Embedded Comput. Syst. (TECS)*, 17, 1, 1–24, 2017 Sep 20.
148. Guida, R., Demirors, E., Dave, N., Melodia, T., Underwater ultrasonic wireless power transfer: A battery-less platform for the internet of underwater things. *IEEE Trans. Mob. Comput.*, 21, 5, 1861–1873, 2020 Oct 8.

149. Carminati, M., Turolla, A., Mezzera, L., Di Mauro, M., Tizzoni, M., Pani, G., Zanetto, F., Foschi, J., Antonelli, M., A self-powered wireless water quality sensing network enabling smart monitoring of biological and chemical stability in supply systems. *Sensors*, 20, 4, 1125, 2020 Jan.
150. Adu-Manu, K.S., Tapparello, C., Heinzelman, W., Katsriku, F.A., Abdulai, J.D., Water quality monitoring using wireless sensor networks: Current trends and future research directions. *ACM Trans. Sens. Netw. (TOSN)*, 13, 1, 1–41, 2017 Jan 28.
151. Li, X.J. and Chong, P.H., Design and implementation of a self-powered smart water meter. *Sensors*, 19, 19, 4177, 2019 Jan.
152. Tasic, V., Staake, T., Stiefmeier, T., Tiefenbeck, V., Fleisch, E., Tröster, G., Self-powered water meter for direct feedback, in: *2012 3rd IEEE International Conference on the Internet of Things*, 2012 Oct 24, IEEE, pp. 24–30.
153. Pimenta, N. and Chaves, P., Study and design of a retrofitted smart water meter solution with energy harvesting integration. *Discover Internet Things*, 1, 1, 1–5, 2021 Dec.
154. Chen, B.D., Tang, W., He, C., Deng, C.R., Yang, L.J., Zhu, L.P., Chen, J., Shao, J.J., Liu, L., Wang, Z.L., Water wave energy harvesting and self-powered liquid-surface fluctuation sensing based on bionic-jellyfish triboelectric nanogenerator. *Mater. Today*, 21, 1, 88–97, 2018 Jan 1.
155. Ordoñez, V., Arcos, R., Romeu, J., Josefsson, A., Investigation on electromagnetic vibration energy harvesting in water distribution control valves. *IEEE Sens. J.*, 21, 3, 2734–41, 2020 Oct 1.
156. Gould, C.A., Thermoelectric water meter energy harvesting. *J. Phys.: Conf. Ser.*, 1534, 1, 012010, 2020 Apr 1, IOP Publishing.
157. Mohamed, M.I., Wu, W.Y., Moniri, M., Power harvesting for smart sensor networks in monitoring water distribution system, in: *2011 International Conference on Networking, Sensing and Control*, 2011 Apr 11, IEEE, pp. 393–398.
158. Zhong, Y., Zhao, H., Guo, Y., Rui, P., Shi, S., Zhang, W., Liao, Y., Wang, P., Wang, Z.L., An easily assembled electromagnetic-triboelectric hybrid nanogenerator driven by magnetic coupling for fluid energy harvesting and self-powered flow monitoring in a smart home/city. *Adv. Mater. Technol.*, 4, 12, 1900741, 2019 Dec.
159. Bhatti, N.A., Alizai, M.H., Syed, A.A., Mottola, L., Energy harvesting and wireless transfer in sensor network applications: Concepts and experiences. *ACM Trans. Sens. Netw. (TOSN)*, 12, 3, 1–40, 2016 Aug 29.
160. Raza, U. and Salam, A., On-site and external energy harvesting in underground wireless. *Electronics*, 9, 4, 681, 2020 Apr.
161. Tan, A., Wang, S., Xin, N., Shi, Y., Peng, Y., A multi-channel transmission scheme in green Internet of Things for underground mining safety warning. *IEEE Access*, 8, 775–88, 2019 Dec 13.

162. Molaei, F., Rahimi, E., Siavoshi, H., Afrouz, S.G., Tenorio, V., A comprehensive review on internet of things (IoT) and its implications in the mining industry. *Am. J. Eng. Appl. Sci.*, 13, 3, 499–515, 2020 Sep 3.
163. Singh, U.K., Piezo-electric power scavenging for mining applications, University of Newcastle, 2007 Feb.
164. He, H., Dong, C., Fu, Y., Han, W., Zhao, T., Xing, L., Xue, X., Self-powered smelling electronic-skin based on the piezo-gas-sensor matrix for real-time monitoring the mining environment. *Sens. Actuators B: Chem.*, 267, 392–402, 2018 Aug 15.
165. Cammarano, A., Spenza, D., Petrioli, C., Energy-harvesting WSNs for structural health monitoring of underground train tunnels, in: *2013 IEEE Conference on Computer Communications Workshops (INFOCOM WKSHPS)*, 2013 Apr 14, IEEE, pp. 75–76.
166. Kahrobaee, S. and Vuran, M.C., Vibration energy harvesting for wireless underground sensor networks, in: *2013 IEEE International Conference on Communications (ICC)*, 2013 Jun 9, IEEE, pp. 1543–1548.
167. Wischke, M., Masur, M., Kröner, M., Woias, P., Vibration harvesting in traffic tunnels to power wireless sensor nodes. *Smart Mater. Struct.*, 20, 8, 085014, 2011 Jul 8.
168. Ge, R., Pan, H., Lin, Z., Gong, N., Wang, J., Chen, X., RF-powered batteryless wireless sensor network in structural monitoring, in: *2016 IEEE International Conference on Electro Information Technology (EIT)*, 2016 May 19, IEEE, pp. 0547–0552.
169. Shaikh, F.K. and Zeadally, S., Energy harvesting in wireless sensor networks: A comprehensive review. *Renewable Sustain. Energy Rev.*, 55, 1041–54, 2016 Mar 1.
170. Bello, H., Xiaoping, Z., Nordin, R., Xin, J., Advances and opportunities in passive wake-up radios with wireless energy harvesting for the internet of things applications. *Sensors*, 19, 14, 3078, 2019 Jan.

2

Self-Powered Wireless Sensor Networks in Cyber Physical System

Srividya P.

Department of Electronics and Communication Engineering, RVCE, Bangalore, Karnataka, India

Abstract
With the advancements in wireless communication, cyber physical system (CPS) has originated as a leading technology to support interactions among humans and objects by integrating abstract computations and physical processes. CPS has raised the demand for interconnected devices with physical input and output. Sensors, actuators, and embedded devices are the key components required for sensing, monitoring, and controlling the physical world. If wireless sensors are not deployed in CPS development, real-time decision making becomes difficult, and this leads to unreliable and unpredictable CPS design.

The sensors are usually power driven. But usage of a wired power supply is not preferable as it increases the complexity and leads to resource wastage. The power supply lines will also cause possible hazards and limit the growth of the wireless sensor network. As an alternative, battery driven sensors can be used. But it demands maintenance and recharge issues. The energy storage unit comes with the constraint of having a set lifetime; the sustainable operation of the sensors becomes the major challenge faced by the CPS systems. The wide spread distribution of the sensors, their mobility, and wireless operations also add to the drain of the storage unit. This necessitates for a self-powered sensor system that is either capable of harvesting ambient energy in order to sustain continuous operation without relying on external power sources or by using a power source that is self-generated. Therefore, a self-sustaining, maintenance-free power source needs to be integrated inside an individual sensor node for powering it by harvesting energy from its operational environment.

There is an abundance of energy available in nature in different forms like thermal energy, mechanical energy, solar energy, chemical energy, and many more. The conversion and harvesting of such energies can be used to build self-powered

Email: srividyap@rvce.edu.in

Rathishchandra R. Gatti, Chandra Singh, Rajeev Agrawal and Felcy Jyothi Serrao (eds.) Self-Powered Cyber Physical Systems, (41–56) © 2023 Scrivener Publishing LLC

devices and also to develop clean renewable energy sources. The widely available form of energy at any place and time is mechanical energy that includes mechanical vibration, air flow, raindrops, and hydraulic pressure. It is available specifically in the household, industrialized, and transport sectors where there is huge scope for the application of sensor networks. Hence, the focus will be on harvesting of mechanical energy. The two popular energy-harvesting devices that transform mechanical energy into electrical energy are piezoelectric nanogenerators and triboelectric nanogenerators. Utilizing these devices, self-powered sensors can be built. This chapter gives an overview of how utilization of nanogenerators can lead to development of self-powered sensors and its applications.

***Keywords*:** Self-powered, wireless sensor networks, piezoelectric generators and triboelectric nanogenerator, thermoelectric nanogenerator, pyroelectric naogenerator, hybridized nanogenerator

2.1 Introduction

The advent of new technologies like automation, smart cities, smart vehicles, smart grids, and others have led to a constant growth in the ecology. These technologies have aided in making lives simpler and convenient by providing access to wide range of services from anywhere and at any time. Achieving these targets poses several research challenges in integrating various digital capabilities like network connectivity and computational capability with the available physical devices and systems, providing higher level of security, energy consumption, higher data transfer rate, and many more.

Advances in networking, embedded system technology and wireless communication systems have been a driving force in developing CPS for diverse purposes. CPS supports interactions among humans, object and humans and between objects itself both physically and virtually. CPS is an assimilation of computations, networking and physical processes. The system involves sensing, monitoring, and controlling of physical systems using sensors, actuators and embedded controllers. CPS includes a broad network of devices that have varying computing, networking and storage capabilities. The networked objects interact with one another and exchange various sensor and multimedia data. The CPS makes a real-time decision on the inputs received from the sensor nodes and generates control signals to the actuators. In order to build an efficient CPS, the following should be satisfied:

1. The appliances should be connected together to form a single network using various communication protocols like Wi-Fi, zigbee, bluetooth, etc.
2. Changing network topology.

3. Remote access to internet.
4. Wireless sensor network (WSN) technology.

CPS finds wide range of applications in health care sectors, agricultural sectors, environment monitoring, transportation etc. For example if it is possible to predict disasters that might happen in near future, suitable precautions can be taken to reduce the damage caused by it. Similarly, if it is possible to predict the internal health conditions of geriatric people or patients, precautionary measures can be adopted. If soil fertility can be checked and reported from time to time, the quality of cultivation can be improved [1].

2.2 Wireless Sensor Networks in CPS

WSN forms a basic component of all CPS designs. CPS includes a number of spatially dispersed sensors that interacts with each other to solve complex tasks further beyond the capabilities of the individual sensor. WSN aids CPS in decision making by acquiring real time data. Without WSN, CPS designs become unpredictable and unreliable. WSNs help in monitoring physical and environmental parameters like sound, pressure, temperature, humidity, and many more and convey the information to a central site for processing. A WSN has vast static sensor nodes that have limited power and processing abilities. The sensor nodes are deployed in random fashion and can communicate over short-range radio links. Essential requirements of WSN are they must be self-sufficient, require less maintenance, and must work all the time. Although WSNs are long-lasting, supplying energy to the nodes is a great challenge.

The quiescent static power drawn by WSNs is very low. But most of the power is consumed in powering up the circuit to take measurements and in communicating the data. The energy requirements of the sensor nodes are met by using the batteries. The batteries have a limited life span and are sometimes shorter than the life of sensors. This demands a replacement after a few years. Battery replacement limits the network self-sustainability and increases the cost. Another drawback of batteries lies in the fact that they are bulky and generally dominate the size of the electronics. This hinders the device miniaturization. The existing drawbacks of battery driven sensor nodes have drawn significant research and development in energy harvesting technique from the ambient environment surrounding the sensor nodes.

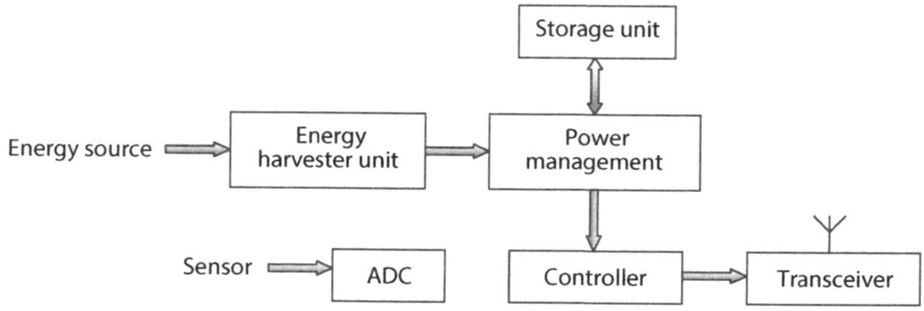

Figure 2.1 Architecture of WSN with energy harvester [3].

2.3 Architecture of WSNs with Energy Harvesting

The main units involved in incorporating energy harvesting in WSNs include (i) sensors; (ii) analog to digital converters to convert the analog data received by the sensors into digital form; (iii) microcontroller for processing the digital data; (iv) energy harvester unit that is capable of converting ambient or artificially generated energy into electricity; (v) power management unit to collect the electrical energy from the energy harvester and either supply it to other units for immediate usage or store it for future usage; (vi) transceiver unit to transmit or receive the information; (vii) memory to store the code and other data [2] as shown in Figure 2.1.

2.4 Energy Harvesting for WSN

Energy harvesting is the most prominent solution in developing self-sustained WSNs and in eliminating the limitations of battery life time. When energy harvesting is applied to WSNs, it reduces the installation charges and maintenance cost. It removes the dependencies of CPS devices on batteries. The reduction of dependence on batteries has increased the flexibility, scalability and life span of CPS. Thus, energy-harvesting technology has become the key to Industry 4.0.

Energy harvesting can be done either using naturally available sources or artificially by human activities. Natural sources include various ambient environmental energy like light, heat, airflow, vibrations, and some lesser-developed biochemical domain like biodegradable materials, water droplets, tree trunks, acoustic noise, sweat of human beings, bio-fluids, like glucose, and others, whereas human activities include radio frequency (RF) signals as shown in Figure 2.2. These sources are converted to

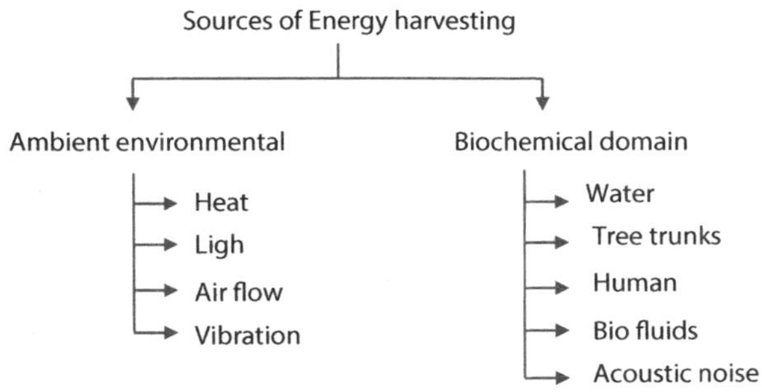

Figure 2.2 Sources of energy harvesting.

electrical energy by energy harvesting techniques. The challenges involved in designing an energy harvesting system lie in the construction of a compact, low cost system that provides an uninterrupted and stable power supply to the load even when there is uneven or unpredictable availability of energy sources. The type of energy harvesting technique depends on the location of the node, the feasibility, and suitability of the technique with the node. As the demand for low-power, self-sustained electronics is on the rise; there is a drastic increase in demand for converting ambient energy into electrical energy.

2.5 Energy Harvesting Due to Mechanical Vibrations

Energy harvesting by vibrations is the concept of converting the ambient variations like sound pressure waves, stress or strain into electrical energy. The various mechanisms that can be used in the conversion are electromagnetic, electrostatic, and piezoelectric.

1. Electromagnetic transduction or conversion is based on the relative motion of the magnet and the coil that helps in electric current induction. This type of conversion is suitable for applications that demand low power sources. Although the method can work at low frequencies and can be applied to microscales and macroscales, they have low output voltage. Integrating electromagnetic generators with WSNs is a difficult task because of interference of magnetic field with the electronic device operation.

2. Electrostatic conversion or transduction uses a variable capacitor in which the relative movement of two plates results in capacitance change and generates electricity. Although this type of conversion is simple, has high output voltage and can be integrated easily with WSNs, it demands for dual power sources to begin the energy harvesting process and the output power is low of the order of nanowatts to microwatts. The research to design these circuits to bypass the need for external power supply is still under developmental phase. This makes the technology less suitable to power WSNs.
3. High electromechanical coupling factor of piezoelectric transduction has made it the most popular mechanical energy harvesting technique compared to electromechanical or electrostatic transductions. When piezoelectric material is subjected to mechanical stress or strain, voltage difference is developed. The produced electric field is proportional to the applied stress. Development in nanoscale materials has further boosted the fabrication of miniaturized piezoelectric generators with properties like high piezoelectric coefficient, electromechanical coupling factor, flexibility, and integrating ability. At present, research on bioinspired, inorganic, organic, and composite piezoelectric materials are being carried out.

2.6 Piezoelectric Generators

Piezoelectric generators (PEG) are fabricated by placing the piezoelectric material between flexible substances that protect the piezoelectric material from ambient variations. Quartz, topaz, Rochelle salt, barium titanate is some crystalline materials that exhibit piezoelectricity. When these materials deform due to external mechanical pressure they generate electricity. This is called positive piezoelectric effect. Similarly, on applying electrical energy if the material deforms, it is called negative piezoelectric effect. The electrical displacement obtained by applying mechanical pressure is given by expression (2.1).

$$D = dT + \varepsilon E \tag{2.1}$$

where d represents piezoelectric coefficient, T stress, ε material permittivity and E the electric field.

2.7 Piezoelectric Materials

Active materials like ferroelectrics when subjected to mechanical strain undergo polarization change resulting in piezoelectricity. The performance of piezoelectric energy harvester is mainly dependent on the type of material selected [4]. This has drawn attention to investigating broad varieties of materials that include organic, inorganic and composite materials.

Although piezoceramics, like lead zirconate titanate, barium titanate, potassium titanate, etc., have larger dielectric, piezoelectric coefficients, electromechanical coupling capacity, and higher energy conversion rates, they are brittle. This makes it unsuitable to build piezoelectric generators as it cannot endure huge strains without being damaged. Piezopolymers, like polyvinylidene fluoride (PVDF), on the other hand, exhibit low electromechanical coupling but are extremely flexible. This has led to the development of piezoelectric composites that combines the benefits of both piezoceramics and piezopolymers. Piezoelectric materials under various categories are shown in Table 2.1. The materials used for piezoelectric energy harvesting exhibit wurtzite or perovskite crystal structure.

Table 2.1 Piezoelectric materials under various categories [6].

Category	Piezoelectric materials
Natural crystal	Quartaz, Rochelle salt, topaz, sucrose
Natural materials	Silk, wood, enamel, viral proteins, DNA, bone
Synthetic crystals	Gallium orthophosphate, diisopropylammonium bromide
Synthetic ceramics	Barium titanate, lead titanate, potassium niobate, lithium tantalite, and others
Polymers	Polyvinylidene fluoride
Organic nanostructures	Self-assembled diphenylalanine peptide nanotubes
Lead-free piezoceramics	Bismuth ferrite

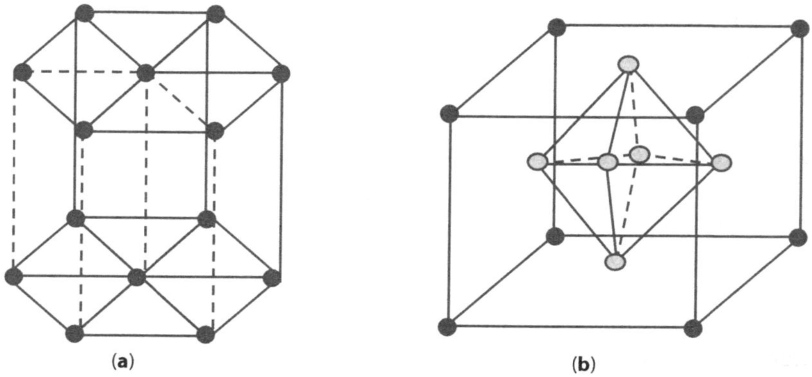

Figure 2.3 Piezoelectric material structures (a) Wurtzite structure and (b) Perovskite structure.

The atoms in wurtzite structure are arranged in tetrahedral symmetry as shown in Figure 2.3(a) and the perovskite structure has a cubic unit cell as shown in Figure 2.3(b). Compared to materials with wurtzite structure, the materials with perovskite structure shows better piezoelectric property. However, perovskite structures can be forced to generate piezoelectricity only by a process called polling, which is to be carried out at temperatures below Curie point where the structures causes an electric dipole to be created.

2.8 Types of Piezoelectric Structures

The most popular structures used for piezoelectric generators are:

1. Cantilever beam with one or two layers of piezoelectric material is called as unimorph and bimorph respectively and shown in Figure 2.4. The shaded region represents the piezoelectric layer and the un-shaded region represents the elastic layer. It is the most extensively used device configuration for piezoelectric generation [3]. Piezoelectric cantilever beam is

Figure 2.4 Unimorph and Bimorph structures of the cantilever.

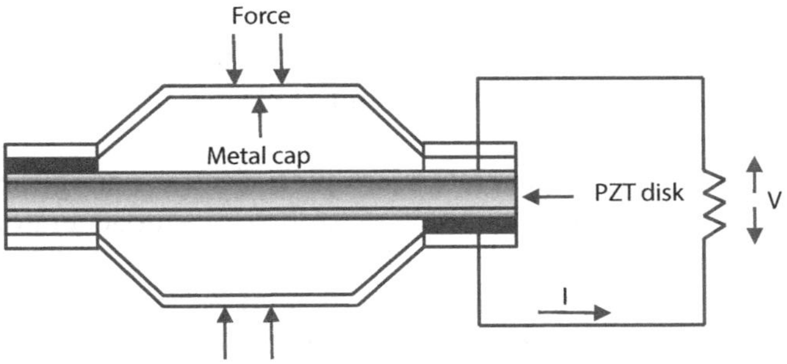

Figure 2.5 Cymbal transducer.

highly flexible, can be easily encapsulated, and is quite stable under low frequency conditions.

When the beam is subjected to vibrations, dynamic strain is induced and alternating voltage generates through the electrodes that covers the piezoelectric material layers.

2. Cymbal transducer is suitably designed for piezoelectric energy harvesting under high stress environments [5]. It is capable of generating larger output power compared to cantilever structure. It consists of a piezoelectric disk placed between two cymbal shaped metal endcaps as shown in Figure 2.5. The disk consists of electrodes on both top and bottom surfaces. The air-filled cavity on the inner surface of the endcaps supports mechanical transformation by converting the axial stress into tangential stress of opposite sign. When external force is applied on the endcap, piezoelectric disk produces electric field owing to piezoelectric effect. The generated electricity can be improved by increasing the diameter of the transducer and the base of the cavity. The endcaps provide support to the piezoelectric disk when subjected to higher loads.

2.8.1 Nanogenerators

Although many piezoelectric materials are being investigated since the discovery of piezoelectricity, over the past few years the investigation on one-dimensional nanostructures (1D nanostructures) is being carried out. Nanofibers, nanorods, naotubes, and nanowires are some of the 1D nanostructure that have recently been developed as a potential structures in fabricating ultracompact and more efficient nanogenerators. This has led to miniaturization of piezoelectric generators. Nanofiber is a common term that refers to nanosized fibers. A rigid nanofiber is called a nanorod, a hollow nanofiber is called a nanotube and an electrically conducting nanofiber is called a nanowire.

Various types of nanogenerators are piezoelectric nanogenerator, triboelectric nanogenerator (TENG), pyroelectric nanogenerator (PNG), and thermoelectric nanogenerator (TEG). Piezoelectric and triboelectric nanogenerators convert mechanical energy into electrical energy, whereas a pyroelectric nanogenerator and thermoelectric nanogenerator convert the external thermal energy into electricity.

2.8.2 Piezoelectric Nanogenerators

Zinc oxide was used as an impending material for building nanogenerators owing to its formation at relatively lower temperature. But recent advancements shows that lead zirconate titanate and barium titanate have higher energy conversion efficiency and output power due to their higher piezoelectric coefficients.

2.8.3 Triboelectric Nanogenerators

Triboelectric nanogenerator (TENG)-based sensors are low cost, highly stable and flexible. This makes it a potential candidate for energy harvesting. TENGs provide an infrastructure to collect information though large number of self-powered sensors [4]. TENG fundamentally is based on the principle of triboelectric effect and electrostatic induction effect.

Triboelectric effect is experienced when an electrically charged material comes in contact with another material by friction. When there is contact between the materials, few parts of the surfaces form a chemical bond and charges are exchanged between the two to equalize the electrochemical potential called electrostatic induction. The charges can be ions, electrons or molecules. When the materials are separated, few of the bonded atoms hold the electrons and some give them away. This introduces triboelectric charges of the surface.

TENG shows higher power performance, higher efficiency under multi frequency operation as compared to pyroelectric generators. TENGs also exhibit incredible potential in harvesting energy at lower frequencies (< 5–10 Hz).

TENG involves four working modes- (a) Vertical contact separation mode (b) Linear sliding mode (c) Single electrode mode (d) Freestanding mode. These approaches demand two dissimilar triboelectric materials having appropriate electrode connection and good insulation in between the layers. Any combination of dielectric-dielectric or metal-dielectric can be used. The principle behind the modes of operation is that whenever a displacement occurs in any of the triboelectric layers, the movement of electrostatic charge breaks the electrostatic status and results in the potential difference across the electrodes. The recurring mechanical movement in forward and reverse directions in the layers generates forward and reverse voltage between the electrodes. The modes of TENG are shown in Figure 2.6.

a) **Vertical contact separation mode:** The construction includes two unlike dielectric films facing each other with electrodes deposited on the top and bottom surfaces. Oppositely charged surfaces are created between the dielectric films when the dielectric films are in contact. Potential drop is then created on separation by a small gap by lifting due to external force. By connecting the electrodes to the load, free electrons then flow from one electrode to the other creating opposite potential in order to balance the electrostatic field. Upon closing the gap, the potential disappears and electrons flow back.

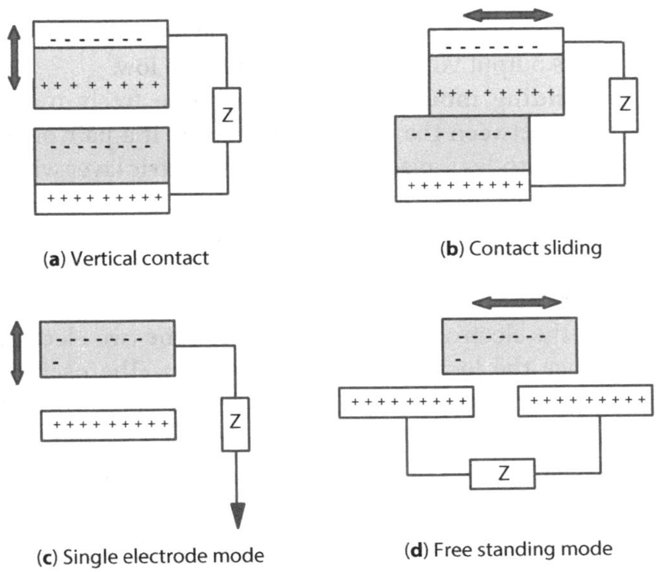

Figure 2.6 TENG modes of working.

This model is easy to fabricate, has simple design and of lower cost. Enhanced performance can be obtained by using multi-unit stacking. Due to the little surface abrasion, this mode has the advantage of long working life easily up to 1,000,000 cycles and is commonly used in wearable electronics, mechanical energy harvesting, blue energy, implantable medicine, and self-powered sensor.

b) **Linear sliding mode:** In this mode, charge is generated due to the sliding in-between the layers of TENG. The assembly is similar to vertical contact separation mode with the electrodes affixed behind the triboelectric layers. The movement is in sideward direction. This helps in efficient charge generation as a result of higher contact area.

c) **Single electrode mode:** In vertical contact separation and linear sliding modes, load is connected between the two electrodes. They can move freely and can work for mobile applications. When object like humans in motion on the floor cannot be connected electrically to the load, a single electrode TENG must be used. In this mode, the bottom electrode is grounded. If the top object is detached from bottom one, electric field distribution changes and electron exchange takes place between bottom electrode and ground so as to maintain the electrode potential change. Although this is the simplest mode of TENG, due to small transfer of charge its output voltage and current is too low.

d) **Free standing mode:** This mode has one freely movable electrode between the triboelectric layers. If a pair of symmetric electrodes is placed beneath a dielectric layer with the size of the electrodes and the gap between the two remaining in the same order as the size of the moving entity, the entity's approaching toward and/or departing from the electrodes creates an asymmetric charge distribution in media, causing the electrons to flow in-between the two electrodes to balance the local potential distribution. The oscillating electrons between the pair of electrodes produce power. The moving entity does not have to directly touch the top dielectric layer of the electrodes so that a free rotation is possible without direct mechanical contact in rotation mode, which reduces the wearing of the surfaces drastically. This approach is good for extending the durability of the TENGs.

This mode exhibits high output efficiency and can be integrated easily into real time applications.

2.8.4 Pyroelectric Nanogenerators

Pyroelectric nanogenerators (PYG) convert the external heat in the ambience into electrical energy as a result of impulsive polarization within the pyroelectric materials. The time dependent temperature variations cause oscillations of electric dipoles, which further result in polarization.

Temporary voltage is generated by pyroelectric materials when they are either heated or cooled. This is called pyroelectricity. Pyroelectric materials exhibit polar symmetry with spontaneous polarization even when the electric field is not applied across it. The variations in temperature results in the alteration of position of atoms within the crystal structure causing a change in the polarization of the material. This polarization change causes a potential difference across the crystal. If the temperature remains stable at its changed value, the pyroelectric voltage that was developed gradually decreases due to the current leakage. The movement of electrons through the crystal plus leakage through the voltmeter connected across the crystal is the main cause of the current leakage.

The complete charging and discharging process can be described as follows. At initial state when temperature is almost zero ($dT/dt = 0$), the electric diploes are at equilibrium, and there is no flow of electrons. As the temperature increases gradually ($dT/dt> 0$), impulsive polarization occurs and results in electron migration. When the temperature increases beyond a certain value, the dipoles cannot generate further electricity. Now as the temperature starts decreasing ($dT/dt< 0$), the reverse flow of electrons occurs and at temperature equal to zero, electric dipoles return to their initial state. The current generated and the temperature gradient (dT/dt) are related by expression (2.2).

$$i_p = A.p.\frac{dT}{dt} \tag{2.2}$$

where A represents surface area of the material and p represents pyroelectric coefficient given by expression (2.3).

$$p =.\frac{dPs}{dt} \tag{2.3}$$

where *Ps* represent spontaneous polarization.

Owing to its high conversion efficiency, pyroelectric materials do not require immense heat sink to generate temperature gradient. However, temperature oscillations are required to generate fluctuations.

2.8.5 Thermoelectric Nanogenerator

Thermoelectric nanogenerators (TEG) also convert the external heat energy into electricity like pyroelectric nanogenerators. While PYG generates electricity due to temperature fluctuations, the TEG generates electricity due to temperature gradients. TEG works on the theory of thermoelectric effect also called seeback effect. According to the theory, when a temperature gradient occurs between two electrodes of thermoelectric materials, there is an electron flow from one end to another resulting in a potential difference.

The π type model of thermoelectric effect is illustrated in Figure 2.7. When one end of P or N type material is heated and the other is cooled, the majority charge carriers (holes in P type and electrons in N type) drift from hot side to cold side and the current direction in both P and N type materials remains the same. Working circuit is formed when an external circuit is connected.

Manganese, calcium and lead telluride are the thermoelectric materials used in making TEGs as they have the capacity to resist at very high temperature. Zinc oxide is reported to have excellent thermoelectric and piezoelectric properties. Zinc oxide doped with antimony or aluminium proves to have higher comprehensive property and hence improved performance. Comparison between pyroelectric and thermoelectric nanogenerators is shown in Table 2.2.

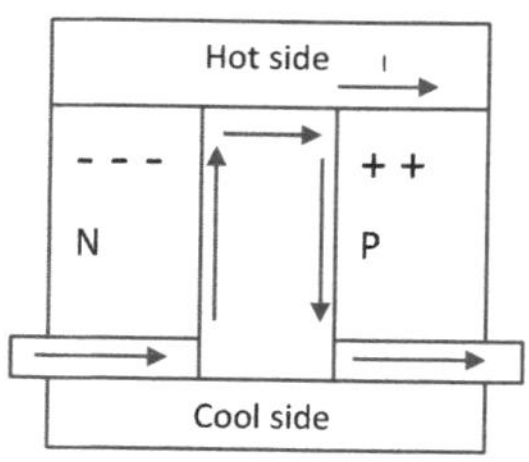

Figure 2.7 π type TE device.

Table 2.2 Comparison of pyrolectric and thermoelectric nanogenerator.

Pyroelectric nanogenerator	Thermoelectric nanogenerator
Heat energy is transformed to electricity due to impulsive polarization within the pyroelectric materials.	Heat energy is transformed to electricity due to seeback effect.
The time dependent temperature variation of the ambience will lead to overall temperature variation and the temperature remains same throughout the device.	Temperature gradient exists across TEGs and the temperature varies along the device.
Structure is simple.	Compared to pyroelectric nanogenerator, structure is complex.

2.9 Hybridized Nanogenerators for Energy Harvesting

The surrounding environment has various types of energies. To grab the advantages of these energies effectively in energy harvesting the research is now focused on developing various forms of hybridized generators. Hybridized nanogenerators aim at utilizing various available energy sources individually or simultaneously by combining many nanogenerators into one unit. This enables the use of available ambient energy at any time, boosts the output power and increases the efficiency of space utilized. The system also assists in satisfying the energy demands of electronic devices and to adapt to various working environments.

The piezoelectric nanogenerator (PENG) has the capability to offer higher conversion efficiency and provide stable output power even under low and weak mechanical stimuli. The active layers of both TENG and PENG are intrinsic thin film in nature. This allows easy integration of the layers into any desired structure.

Different kinds of hybridized nanogenerators developed are:

1. Hybridized piezoelectric-electromagnetic nanogenerators.
2. Hybridized piezoelectric-triboelectricnanogenerators.

3. Hybridized triboelectric and electromagnetic nanogenerators
4. Hybridized triboelectric-piezoelectric nanogenerators.
5. Hybridized triboelectric-piezoelectric-electromagnetic nanogenerators.

2.10 Conclusion

The sudden rise in demand for portable electronics, which mostly run on batteries, has increased the demand for innovation in energy storage technologies. Although the devices consume less power, the total number of such devices being used is huge. At present, there are billions of people using mobile phones that require enormous number of sensor networks to be installed across the globe. Powering such networks consisting of huge number of sensors using batteries would be practically impossible due to maintenance issues. Energy harvesting from the ambience is the only feasible solution. The type of energy harvesting depends on the location of the device. This chapter highlighted the various methods of energy harvesting techniques due to mechanical vibrations. Further development on the discussed devices can be achieved by researching on the materials, structural designs, performance improvement, and developing more device applications.

References

1. Lin, C.-Y., Zeadally, S., Chen, T.-S., Chang, C.-Y., Enabling cyber physical systems with wireless sensor networking technologies. *Int. J. Distrib. Sens. Netw.*, 2012, Article ID 489794, 21, 2012.
2. Williams, A.J., Torquato, M.F., Cameron, I.M., Fahmy, A.A., Sienz, J., Survey of energy harvesting technologies for wireless sensor networks. *IEEE Access*, 9, 77493–77510, 2021.
3. Wang, Z.L., Chen, J., Zi, Y., Niu, S., Lin, L., *Triboelectric nanogenerators*, Friction, Springer, 17, 481–506, April 2020.
4. Sezer, N. and Koç, M., A comprehensive review on the state-of-the-art of piezoelectric energy harvesting. *Nano Energy*, 80, 1–25, 105567, 2021.
5. Chua, H.G., Kok, B.C., Goh, H.H., Modelling and design analyses of a piezoelectric cymbal transducer (PCT) structure for energy harvesting application. *WIT Trans. Ecol. Environ.*, 186, 103–114, 2014.
6. Wang, Z., Pan, X., He, Y., Hu, Y., Gu, H., Wang, Y., Piezoelectric nanowires in energy harvesting applications. *Adv. Mater. Sci. Eng.*, 2015, 103–114, 2015.

3

The Emergence of Cyber-Physical System in the Context of Self-Powered Soft Robotics

Darwin S.[1] **and Fantin Irudaya Raj E.**[2*]

[1]*Department of Electronics and Communication Engineering, Dr. Sivanthi Aditanar College of Engineering, Tiruchendur, Tamil Nadu, India*
[2]*Department of Electrical and Electronics Engineering, Dr. Sivanthi Aditanar College of Engineering, Tiruchendur, Tamil Nadu, India*

Abstract

Biological systems that use soft materials are operated with the intervention of activated electrical materials and become frequently used as an inspiration for soft robots. Soft robots have various benefits over traditional robots, including compatibility with wearable devices, secure interaction between machine–human beings, locomotion, and gripping. The soft actuator is an important part of a soft robot since it gives a malleable body and enables it to perceive the world to accomplish a specific actuation behavior in terms of locomotion. The soft robotics-based applications are based on various actuation methodologies such as magnetic, electric fields, and heat which becomes the basic incitement. This review article takes into consideration of various conventional actuator methodologies with the recent non-contact sensing equipment suitable for soft robots due to the low modulus factor. The inadequate nature of control over the soft robots in terms of conventional methods is overcome with the present changes in terms of material fabrication, design structure is taken into account to make flexible and reliable soft robots. The power consumption is essential in the case of sensors or battery-operated devices, so proper self-powered sensors are needed for precise control.

***Keywords*:** Soft robots, self-powered sensors, actuators, material structure, self-powered robots

**Corresponding author*: fantinraj@gmail.com

Rathishchandra R. Gatti, Chandra Singh, Rajeev Agrawal and Felcy Jyothi Serrao (eds.) Self-Powered Cyber Physical Systems, (57–90) © 2023 Scrivener Publishing LLC

3.1 Introduction

The field of amenable, continuous, and adaptive robotics has received a lot of attention in recent years, because of significant breakthroughs in artificial intelligence technology [1]. It has the potential to facilitate a wide range of useful applications, from the manufacturing point of view to producing safety and health protection [2]. The mechanism in the biological process along with the locomotion stimulates scientists to research multifunctional systems [3]. Soft robotics [4, 5] is a new discipline of robotics that has emerged as a result of such innovative and unique research. The discriminate features related to soft robots play a vital role in the replacement of the traditional structural design of robots. Materials with high rigidity like aluminum and steel have been utilized which affects the flexibility of the robots. Soft robots are of particular importance because it is different from the traditional rigid kind of robots, they can easily deform, adapt to the external environment without harming persons, and, eventually, permit minimal manufacturing cost [6]. The amalgamation of actuators along with sensors makes a robotic system communicate with the external surrounding environment. The sensing capability of soft robots and the actuator functionality intermixing with the robotic body become one of the challenges in designing soft robots. The minimization in dimension of soft robots makes the functionality of the above-mentioned parameters even much more difficult [7, 8]. The materials with soft nature play a vital role in the formation of soft robots and these materials include fluids, alloys with shape memory, polymer-based electroactive nature, and 2-dimensional materials [9–15]. The communication between human and soft robots are very safe compared to the traditional robots by minimizing the danger of injury by itself and reducing the collision with the surrounding environment [16]. The efficient control algorithm and flexibility in the material structure enhance the opportunities for soft robots in various application platforms. The control algorithm required for stability in the case of traditional robots is much more complex compared to soft robots. Because soft robots work under the platform of shock absorption nature of the materials used, they do not need any complex algorithms. For this reason, soft robots are performed well in difficult and unstructured environments.

In this review, the robotic actuation types with the influence of thermal, pressure, photo responsivity, and explosive nature of actuators were discussed in section 3.2. The electrode utilization in soft robots was analyzed in section 3.3. In sections 3.4 and 3.5 the soft robot structures and control models were categorized. The wide application areas of soft robotics are discussed in section 3.6. The outlook on the future scope of the soft robot is discussed in section 3.7 and final conclusion with section 3.8.

3.2 Actuators and Its Types

3.2.1 Nature of Actuation

In conventional robots, due to their rigidity, the freedom to move around is somewhat difficult when compared to soft robots. The soft robots have flexible body structures making them free to move around the environment without any disturbance. The soft segments that generate a constant force to distort the segment in a preferred manner are achieved with variable length tendons and SMA amalgamated with soft segments. Figure 3.1 shows the various actuation process in soft robots and their advantages. The technology development in the field of integrated circuits paves the way for soft actuators/sensors to combine along with the microelectromechanical

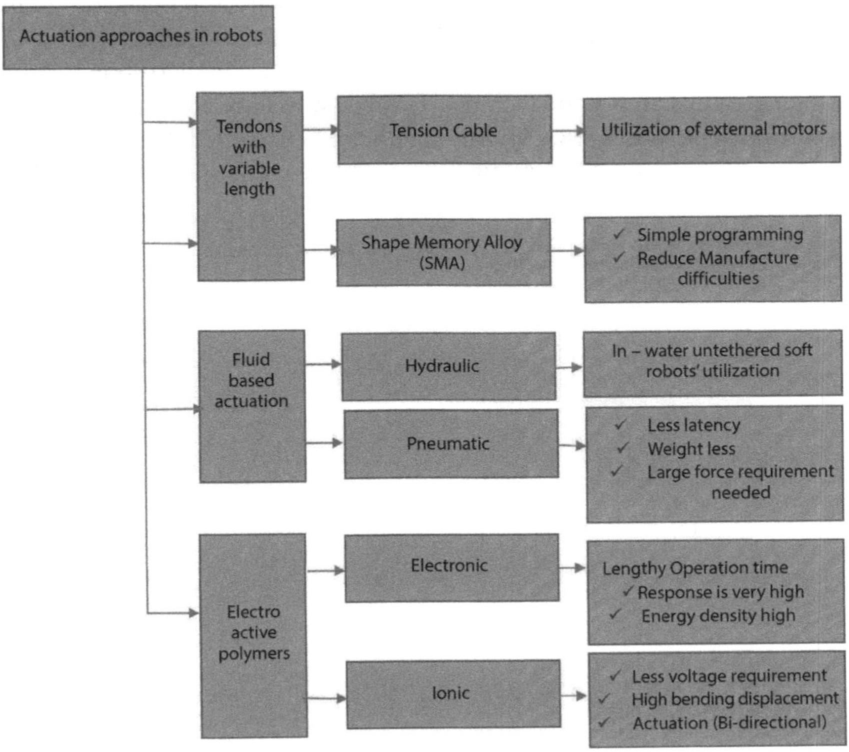

Figure 3.1 Various actuation processes and its merits.

systems (MEMS). The semiconductor industry mainly relies on silicon material which becomes the reliable one for the fabrication of MEMS with miniaturization in the device dimensions compact for robots [17].

Figure 3.2 depicts the functioning of the silicone arm structure of the octopus stimulating robot and the curvature movement is increased to recreate a muscular-based design. Due to the coiling the arms bend and enhance the curvature from the base point to the tip. The wrapping motion of the arm around the target was increased due to the similar spiral structure.

3.2.1.1 *Actuators Based on Thermal Materials*

Some of the actuators give a response to thermal stimulation. This kind of actuator provides better safety for applications related to biomedical. On the other hand, the thermal-based actuators are slower and less precise. The efficiency and responsiveness could be improved by using finer sheets, more power, and materials that can absorb more heat. The shape memory alloy (SMA) is one of the thermal actuator materials which has reduced programming concepts and a simple fabrication process. The reduced cost and readily available SMA is CuZnAl, FeMnSi, and CuAlNi. Furthermore, because of their greater stability and technical feasibility, NiTi-related SMAs are chosen for most purposes [18, 19]. Any shape can be obtained due to the deformation of the material by exerting external force. The material transitions

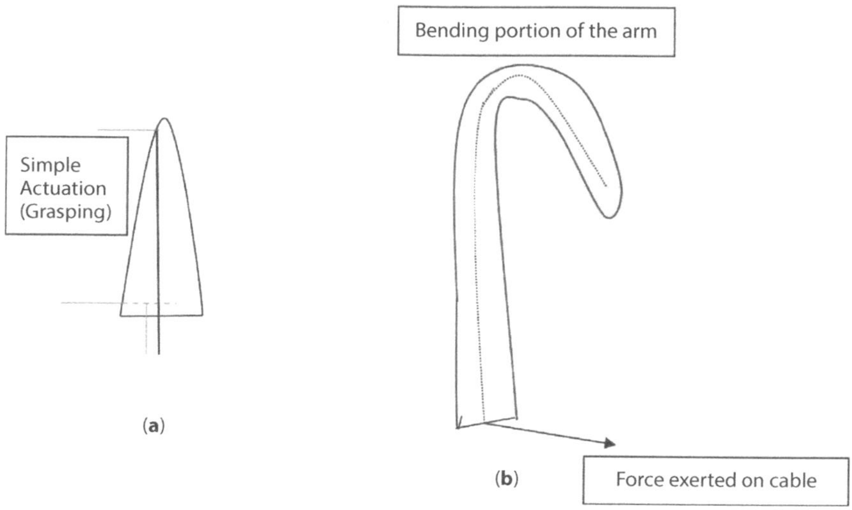

Figure 3.2 Octopus arm. (a) Grasping action. (b) Base to tip curvature during arm bending.

to the austenite crystal state when the temperature goes up beyond a particular threshold and restores to its previous natural shape before it could be deformed. Furthermore, the SMA will employ steady stress in order to reclaim its former shape [20]. GoQBot is a smooth-bodied rolling robot modelled by caterpillars that also uses SMA coils to provide more strain than the linear straight shape [21]. Another module consists of SMA wires which are placed longitudinally in a flexible polymer to operate as a muscle, causing the inchworm to contract when it moves [22]. Many variables affect the material's bending qualities, and one is the eccentric property of the SMA wires. Figures 3.3 (a) & (b) shows the inchworm schematic in the pull and push moment of anchor due to the reduced number of legs in the body area. A passive anchoring method based on friction pads is utilized for better movement. SMA-based actuation has some demerits in terms of minimum strain produced, which is overcome with device structure modification. SMA wires with nickel-titanium combined in a fiber-reinforced polymer with restrained nature [23].

The octopus inspired mechanical behavior-based soft robot with the combination of spiral SMA and tension cable [24]. The actuators are based on a soft hinge along with the amalgamation of SMA wires with point-to-point soft joints which is amenable. Magnets are also included in the framework, acting as interconnections, allowing the 3D-printed sections to be quickly assembled and disassembled to construct sophisticated soft deployable structures. This structure requires a reduced amount of storage

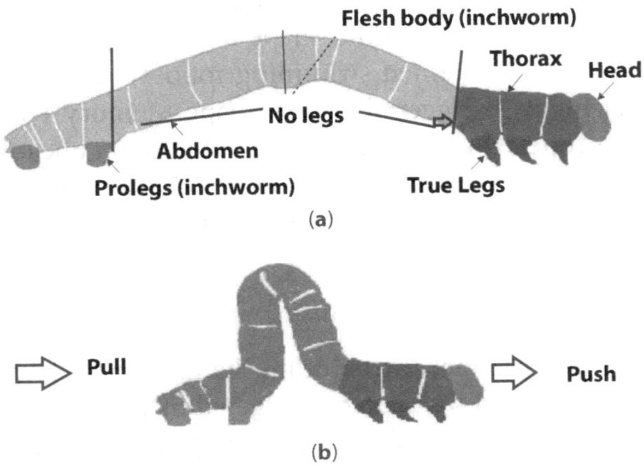

Figure 3.3 (a) Pull moment of anchor of inchworm. (b) Inchworm anchor push and pull directions.

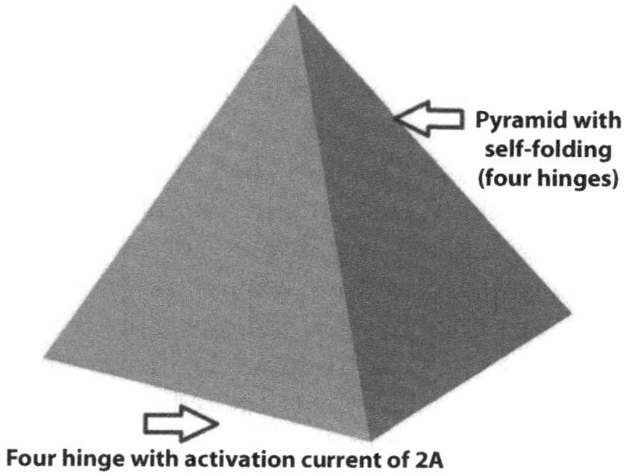

Figure 3.4 Pyramid with self-folding SMP.

area due to its aligned and dismantled facility [25]. Elastomer layers-based thermal SMA is introduced to introduce fast transfer of heat which is the merit compare to other methods. SMA wire with a U-like shape model makes better flexibility due to the layers one with the capability of able to stretch other one is fixed which provides a muscle with artificial nature used in soft robotics for speedy locomotion [26].

PU and thermoplastic polyurethane are the most prevalent types of SMPs produced for biomedical purposes [27]. SMPs depend on the self-folding technique used to meet out folding with localized address procedure, the actuators are the combination of paper, SMPs, and circuits with resistive nature. When the temperature is increased, the SMP shrinks by about half a percent. A pyramid shape folding type SMP is shown in Figure 3.4 [28]. The unmanned swimming robot's architectural concept. The bistable element is triggered by the Shape Memory Polymer (SMP) muscle, which propels the fins backward and the robot ahead. The hollow floaters keep the robot from sinking. In Figure 3.5, the schematic of floaters and fin structure of the swimming robot is shown.

3.2.1.2 Actuators Based on Pressure

To enable efficient actuation, amenable mechanisms must layout the rigidity qualities spatially, and they can be triggered by external forces or pressure to produce the required deformations. Despite the existence of a broad range of compatible or amenable mechanisms, research on machines with high-precision and MEMS has been thoroughly analyzed and reviewed [29, 30]. The main target is to achieve a better optimization for the structures based on a continuum [31–35]. Despite though pressure-related actuators are less prevalent than thermo and electrotriggered NEMS and MEMS systems, due to the miniature hydraulically pressure-driven flexible capability it ensures a prominent method in soft robotics [36]. Even at small dimensions, soft actuators have demonstrated the potential to create high force outputs. Researchers are working to ease the workflow of miniature inflatable chambers, achieve more precise motility, and generate larger forces in this field. Furthermore, while pressure-driven actuators have the ability to create enormous forces while remaining lightweight, they typically require a big, stiff external pump.

Pneumatic actuators are also known as air cylinders or pneumatic cylinders are extremely effective and secure motion controllers that use pressurized gas or air to convert energy into linear motion. They are especially well suited to frequent valves that facilitate fluid movement. The metal backbones are being utilized to create origami robots [37] with built-in wireless connectivity and tension sensing capabilities. The created Pt origami stability is enhanced by a thin elastomer that can be reconfigured and serves as the backbone network of the origami robot.

Hydraulic actuation is accomplished by injecting and propelling fluids into specially engineered chambers that can be utilized to induce the desired outcome, such as bending or twisting. Hydraulically driven soft robots, like pneumatic actuators, are challenging to miniaturize because they require rigorous control and power systems to generate appropriate force applied for mobility. A robot that looks like a robot with an elastomer tail along with a peduncle in the posterior, fin, and the front face of the fish. There is a fixed non-stretchable layer at the middle part of the fish and the ribs are available on both sides of the actuators. Figure 3.6 shows soft tail fish actuator based on hydraulic elastomer.

The inflate and the deflate of the rib action fall on the hydraulic pressure. A microcontroller is used to monitor and control the movement of the fluid entering and leaving the ribs with the external force through the pump [38]. The actuators in some soft robots were powered not only by the ubiquitous air, and even by other chemicals. Its actuation force was self-regulated via the chemical decomposition of h_2o_2 into oxygen gas in a sealed space [39].

3.2.1.3 Actuators Based on Photo Responsivity

Soft actuators that are activated by light are interesting since they are wirelessly connected and can be controlled even when they are small [40]. Photochromic molecules receive optical impulses and translate them into

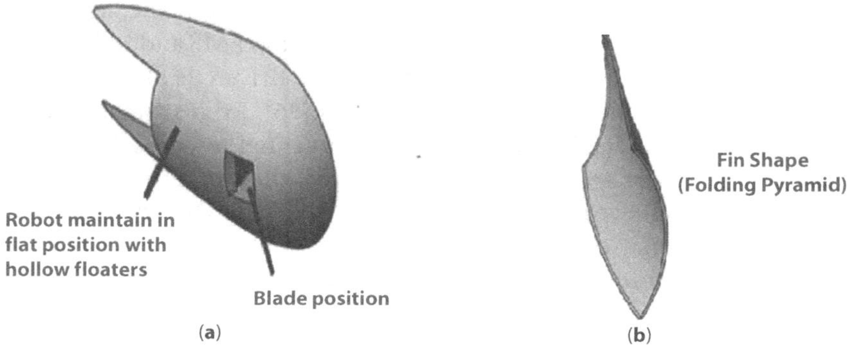

Figure 3.5 (a) Floater to assure that the robot is in an afloat state. (b) Fin shape to support the robot in a dynamic position.

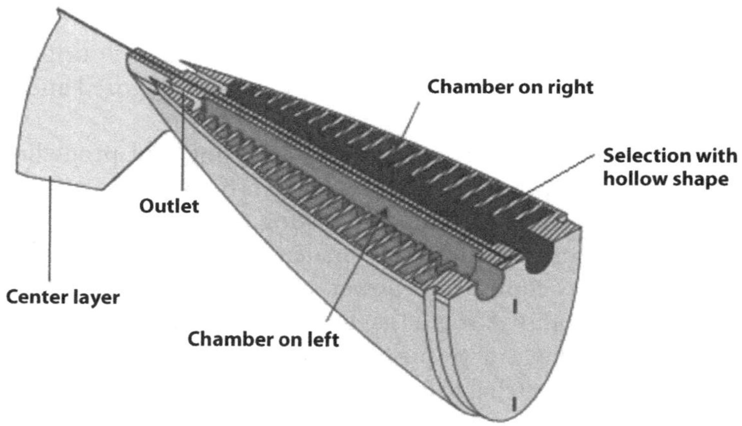

Figure 3.6 Soft tail fish actuator based on hydraulic elastomer.

various property improvements in photo responsive substances. The molecules are incorporated into fluids, gels, and polymers [41]. The ability to use sunlight to trigger actuators leads to the emergence of soft robots that can perform in the present world without consuming too much energy [42, 43]. The liquid crystal polymer network is the widely used actuator in photo responsive method [44]. Other than this liquid crystal type actuator, materials based on carbon utilization in soft robotics is high [45, 46]. Light-based actuators often require a maximum intensity of light with little change or deformation. If uninterrupted action of the photo responsive actuator is required for a particular application, a sequential loading of light is typically required so that the actuator can deform while illuminated and then back to its initial condition when the illumination is turned off.

This may limit the uses of direct sunshine because turning the illumination on and off at regular intervals would be challenging. This limitation can tame with frequent flapping-wing actuation beneath sunlight utilizing a nanocrystal line-based metal-polymer bilayer to address this obstacle. Without the use of any additional technologies, the structure converts sunlight into mechanical energy. As a result, deformation is obtained beneath the sunlight and then relaxed due to a self-shadowing mechanism which is shown in Figure 3.7.

The highest frequency of 4.49 Hz was recorded while using the approach, with light responsiveness of 0.39 s. The observed frequency of 2.7 Hz is sufficient to generate a butterfly-like flying movement [47]. Laser-induced graphene (LIG) structure was introduced to manage the actuator shape based on the polymer and the layers of LIG.

Biological cells with materials imply by synthetic or artificial polymers have been used in the design of photo-responsive soft actuators. In truth, nature presents a plethora of biological cells with well-defined movement patterns. As a result, biohybrid systems were developed by combining these cells with artificial materials. Motile cells that can swim (e.g., bacteria), contractile cells that can generate tension in an appropriate direction (e.g., muscular system), and immune cells with innate chemotaxis capabilities are among the biological cells employed (e.g., blood cells with white color) [48, 49]. Light-responsive soft actuators have received a lot of recognition in recent times, owing to the potential of longer wavelength NIR light to permeate biomaterials with little loss. Due to this NIR has evolved as one of the most suitable methods of biomedical-based wireless device actuation [50, 51].

Low actuation rates and deteriorated mechanical features arising from inadequate fabrication methods limited the capability of NIR-stimulated

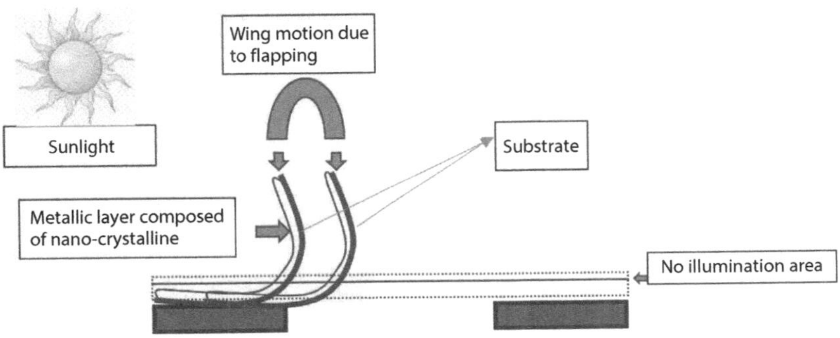

Figure 3.7 Wing motion based on flapping schematic underneath the sunlight.

actuators [52–55]. The reagents with photothermal nature such as carbon nanotubes and graphene are added as impurities with polymers to develop soft actuators stimulated by NIR [56, 57]. The energy transformation from photon to thermal is achieved with the agents of photothermal materials which receive NIR light. When the number of photothermal reagents is reduced, the actuation speed falls this is the major drawback of this method. The solubility of the reagent with polymers.

3.2.1.4 *Actuators Based on Explosive Function*

PneuNet-based actuation depends on the chemical combination which creates an explosion that leads to generating maximum temperature gas. The motion of the actuator is done with the help of an explosion. The 3-D multi-material printing technology is also utilized in soft robots [58]. The reactions indulge with chemicals with the spark generated electrically react with hydrocarbon create explosion. One drawback of this method was the difficulty to alter the jump's position or orientation [59]. The deformation of the combustion chamber is shown in Figure 3.8. However, if the alleged robot landed on its back, it will be unable to place itself for the next jump. controlled jumping in 3D-printed soft robots driven by butane and oxygen combustion. The robot's guided jumping (up to 1.12 m height over 100 cycles) was made possible by pneumatic legs [60] that helped the robot bend before jumping. However, actuation relying on explosive materials has a short life span since chemicals must be replaced; they are really not scalable, have restricted uses, and necessitate extremely strong mechanical resiliency to withstand the explosive impact event.

3.2.1.5 *Electric Actuation Methods*

The functioning of di-electric-based elastomer actuators is dependent on Coulombic two flexible electrodes on opposing edges of an abridgment

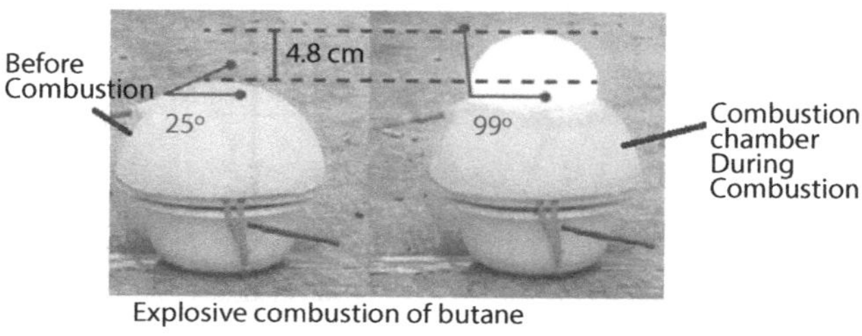

Figure 3.8 Deformation of combustion chamber.

membrane attract each other with a potential difference. The advantages of dielectric actuators are highly flexible, have a maximum energy density, can withstand high stresses, have soul-sensing properties, and precise simulation models represent better biological muscles for soft robots [61]. However, the drawbacks of dielectric actuators are the huge voltages necessary to run this actuator in the kV region and current leakage, both of these factors make it easier to attain breakdown [62]. Depending on the applications the dielectric elastomer actuator can be expanded or bent. Soft grippers based on electrocohesive nature and versatile actuators are introduced. This dielectric actuator has the potential for dissimilar deformation of the thin sheet dielectric elastomer. Furthermore, when the DEA is combined with a buckling frame, substantial and often faster deformations are possible [63–65].

DEAs are used in a multilayer construction approach to creating a crawling soft robot. The actuators related to DEA were created without the need for pre-stretching, reducing, or eliminating the need for rigid parts. The development of bilayer actuators is composed of a series of procedures that include spin-coating the precursor, UV curing the elastomer, and eventually stamping the electrodes. The urethane/polybutadiene acrylic copolymer is the major component in dielectric actuators. The implementation of this type of actuator is in an inchworm robot. Soft and deformable systems with very few stiff components are described. This was accomplished by the layering of elastomers, which produced a quick turnaround time, a large value of energy density, and reduced actuation voltage. In the case of the fabrication aspect, advances in roll-to-roll processing have been used to create rolled dielectric elastomer actuator, which is beneficial in terms of design improvement, high electroelastomer to weight ratio, customizability, and compactness all of which are required to accomplish multiple actuation modes.

To get rid of this type of actuator, liquid-based elastomers with soul healing capabilities which is not available in polymeric solids. Hydraulically Amplified Self-Healing Electrostatic (HASEL) actuators use liquid elastomers to create an enhanced transducer with soft nature that mimics muscle. HASEL actuators are built using an electrohydraulic approach that combines the flexibility of fluid-related actuators with the muscular efficiency of dielectric actuators, as well as self-sensing capabilities. On the other hand, HASEL is not like soft fluidic actuators, which lose actuation efficiency due to fluid flow in the channels. The schematic of the HASEL actuator is shown in Figure 3.9.

Liquid dielectrics are integrated into a flexible structure, resulting in hydraulic pressure when electrostatic force is exerted. The liquid dielectrics

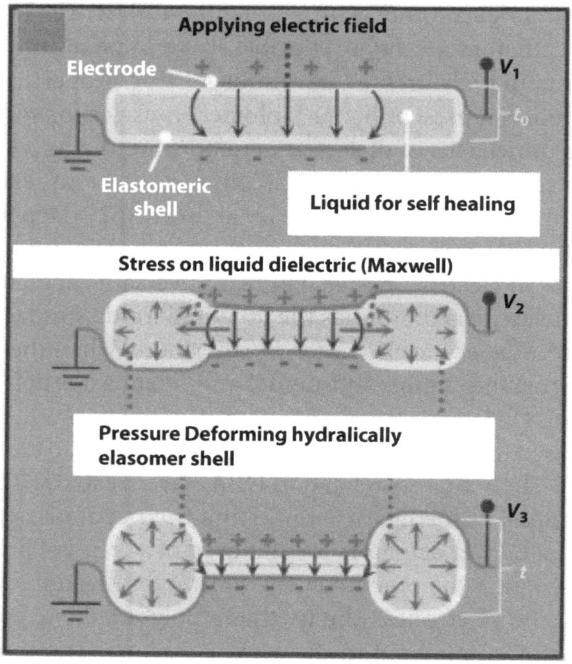

Figure 3.9 Schematic of HASEL actuator.

utilization in HASEL allows for repeated self-healing after several malfunctions. The basis of functioning of a HASEL actuator consists of an elastomeric shell partially wrapped with two electrodes and packed with a dielectric liquid. An electric field is formed through the dielectric liquid when a potential drop is placed betwixt the electrodes.

The Maxwell electrostatic pressure on the dielectric (liquid) leads to the movement of the dielectric liquid of the specified volume between the two electrodes. If there is an increase in strain due to the rise in voltage. The applied bias crosses the threshold value the force outweighs the mechanical force (restoring) creating an attraction between the electrodes and pulling them towards each other. The different actuation modes can be achieved with the alteration of HASEL design based on the needs of fluidic actuators. The drawback of HASEL is due to the large electrical voltages to achieve actuation.

The material with Piezoelectric property is used in actuators when a potential difference occurs due to the action of vibrational or mechanical forces and the deformation is reached with the help of exerted electric field [66]. Large voltages are required for piezoelectric materials in soft robotics, which limits their uses. The piezoelectric effect has been employed to enable a polyvinylidene fluoride (PVDF) layer to prolong and shrink under an AC voltage, resulting in a structural modification in the soft robot.

3.3 Soft Actuator Electrodes

The actuation mechanism of the actuator response to stimulation determines the best electrode materials and architectures to combine with soft actuators. Soft actuators that may be electrically or thermally stimulated, in general, necessitate electrodes that are very flexible and elastic and can maintain a consistent and accurate connection with other elements of the actuator [67–69]. For the creation of flexible and elastic electrodes, many manufacturing approaches have been investigated, including dimensional design of electrodes relying on fractal patterns or electrode deposition on compacted thin films. Furthermore, a variety of electrode materials have been investigated. metallic sheets and diaphragms, and carbon-based elements Nanoparticle, liquid metals, carbon powders, graphene, as well as many composite materials [70–72].

Carbon-based elements are the most widely employed because of their conductivity however at high electrostatic stress/strain levels, with CNTs are the most prevalent of all carbon-based materials. CNT-based multilayer-perforated electrodes for DEAs allow the DEA to have a reduced capacitance as the number of holes' increases [73]. Because of their excellent electrical and thermal conductivity at ambient temperature, liquid crystal materials are widely investigated in the creation of electrodes. Furthermore, the electrodes should be made of a material that is efficacious and dependable in case of enhanced conductivity, insolubleness, and no corrosion in humid settings. Even if the components seemed to fabricate the electrodes are well-established and have outstanding results, economic viability, and energy consumption, electrode stabilization, viability, high polarizability, and handling remain challenges that must be overcome in enhancing the actuators' reliability of the system. Improvements in flexible and elastic conforming electronics allowed electronics to be placed on curvilinear and uneven surfaces, revolutionizing healthcare and the wearable technology-related industries. The interface between real human flesh and artificial sensor systems, known as e-skin, necessitates high conformability, which is also the driving force for the choice of materials for soft robots. Furthermore, the electrodes must be flexible, conformable, and stretchy, as the actuators would bend and extend when exposed to a stimulus.

The electrodes can even be translucent and morphological, depending on the application's needs.

Finally, encapsulation is commonly utilized to extend the life and dependability of soft actuator electrodes. The electrodes have been encapsulated with a variety of solid and liquid substances, including Eco flex, hydrogels, silicone, and polyimide all of which have varied flexing and stretching capacities as well as diverse characteristics, making them appropriate for a variety of applications [74]. The merits and demerits of the various actuators are summarized in Table 3.1. Table 3.2 summarizes the behavior of various soft robots based on the speed of operation responsiveness, stimulation method, driving source, and actuation method.

Table 3.1 Actuator types and their merits and challenges.

Actuators	Stimulation method	Merits	Challenges
Pressure	Hydraulic method	Mechanical Movement is not required Stability is high Improved velocity Accept large loads Force of actuation is high	Degrade performance due to leakage Control cost is high The additional energy source is required for large scale maintenance
	Pneumatic method	Ease of use Insensate to drift in temperature Very quick operating period Manageable cost	Air compressibility is required Not recommended for heavy loads Segregation among energy and source is limited

(*Continued*)

Table 3.1 Actuator types and their merits and challenges. (*Continued*)

Actuators	Stimulation method	Merits	Challenges
Explosive		Actuation speed is high The maximum jump is valid Widely used in military applications Rigid and uncompressed	Need explosive materials to perform actuation Preparation and installation are not very safe. Lifetime is very short
Photo responsivity		Controlling and altering the response rate is simple Quick turnaround period and outstanding resolution Complete remote-control capability	Complex materials with a highly ordered and polarized pattern of molecules are required. The intensity of light requirement is high
Thermal	SMA (Shape Memory Alloy)	Extremely accurate output Rapid response time Ability to break through many materials Nature friendly The frequency response is high	Efficiency is low Power consumption is high Hysteresis The greater annealing temperature is used to the pre-program motion

(*Continued*)

Table 3.1 Actuator types and their merits and challenges. (*Continued*)

Actuators	Stimulation method	Merits	Challenges
	SMP (Shape Memory Polymer)	Density is low Cost is negligible Bio-degradable Deformable elasticity is high Withstand for maximum drift in temperature	In most cases, stiff components are required to restore the original form Organized synthesizing and handling procedures are required
Electrical	Elastomer-dielectric	Energy density is high Power to weight ratio is high Scalable	Response time is poor Capacity payload is low
	Piezoelectric	Perfect for situations requiring considerable force High bandwidth operation	Require displacement is high The drive voltage is very high
	Motor	Conversion of power is efficient Reduce pollution Easy maintenance Cost is low Suitable for industrial applications	It's tough to create a large actuating force Huge size Flexibility is limited

3.4 Sensors

Sensors enable the robotic gear to be proprioceptive, however, the deformable nature of soft robots limits the utilization of several traditional sensors like strain gauges and measurement depends on inertia. Other sensing technologies are so preferable to traditional sensors. Such as contactless

Table 3.2 Nature of soft robots.

Reference	Actuators	Stimulation method	Actuation mode	Speed in [mm s^{-1}]	Driving source
[80]	Pressure	Hydraulic method	Method of gripping	50	Fluid-41.4 kPa
[37]		Pneumatic method	Crawling and Twisting	2	Air-11.0 kPa
[81]	Explosive		Jumping action	9	Combustion using Butane
[47]	Photo responsivity		Wing Flapping motion	Response time – 0.39	Sunlight
[82]	Thermal	SMA (Shape Memory Alloy)	Swimming, Crawling	140	Voltage heating – 30V
[28]		SMP (Shape Memory Polymer)	Folding (self)	Response time – 75sec	Current 2 Ampere
[7]		Liquid metal method	Crawling	1.2	200 Degree Celsius
[83]	Electrical	Elastomer – Dielectric	Forwarding	30	450V
		Piezoelectric method	Forward/backward and climbing	200	200Hz
		Motor	Crawling	200	6V

sensors. For proprioception of curvilinear structures, sensors using low modulus liquid-phase substances and elastomers are acceptable, and these sensors provide a modest impedance shift to soft robot gear. Soft lithography is used to create tiny elastomer layers containing microfluidic channels. Liquid conductors, including eutectic gallium-indium, are used to feed these microfluidic channels [75, 76]. The shape of the channels and their electrical resistance vary when the structure deforms, and strain may be determined by detecting the resistance changes. Sensors of this sort may be modified by changing channel shapes and can record a variety of stresses. Alternative techniques to sensor manufacturing have recently been developed, including the deposition (mask) and printing (3D) of conductive material [77]. In terms of sensing techniques, inductive and capacitive variations as well as resistive variations due to sensor hardware deformation might be employed [78].

The capacitive technique is utilized to monitor the motion of McKibben actuators printed with microfluidic channels on elastomer surfaces. The inductance varies as the dimensions of fibers around the McKibben actuator alter as it shrinks or extends. When semi-movement is imparted to the tip of IPMC, it produces a linear output voltage. IPMC also reacts to dynamic stimuli like impact, resulting in a damped electrical fluctuation [79].

Despite the fact that both the IPMC and the piezo sensor can recognize mechanical stimuli, the IPMC is more sensitive. IPMC can also detect any type of deformation, including bends, strain, compression, torsional activities, and shear. Despite soft sensors are not widely used in soft robotics, it is the major component for achieving precise one for controlling soft robots. The sensors with low impedance, are flexible and stretchable and are suitable for electronic healthcare, and wearable devices.

3.5 Soft Robotic Structures and Control Methods

Soft robots have different architectures than rigid body robots. Soft robots are often made of soft materials, making them unsuitable for use with ordinary electric motors. In this regard, researchers have devised a number of unique architectures for soft robotic devices, including those constructed of soft materials, namely silicone rubber and SMA, as well as soft materials encased in a medium. Soft robotic technologies have also been designed, with some media encapsulated within the silicone substance. Manta with actuation mode of swimming model is implemented using silicone substance [84]. Silicone rubber with two chambers (internal) was employed to achieve freedom with two degrees to achieve motion of bending. A soft snake robot uses the same construction [85]. They conceived and created a snake-like robot system based on a fluidic system for this study. The suggested system is made up of four fluidic actuator pieces with passive wheels between each pair. Based on this framework, a multigait soft robot is also being created [86].

Jamming skin-based locomotion which is the combination of five jammable serial components. Ground coffee and six axes robotic arm grabber was also developed [87–89]. Other soft robotic systems include ones that employ soft material simply for their actuation portions rather than the entire system. Soft robotic fish is an example of a construction that replicated the body of a fish. Excluding the area from the dorsal to the anal segment, the system may be seen as an inflexible robot system in its entire process. This is due to the tail's job of providing pressure to traverse the entire body, just like a real fish performs. A smart bird is a kind of robot developed. The title of the system implies that it is modelled after birds in order to construct a flying system. Like soft robotic fish, the majority of its system may be classified as rigid-body. Nevertheless, in order to achieve greater functionality, it uses soft material on the system's wings that mimic the material properties of a genuine bird wing.

Soft robot control methods are divided into two areas: open/close control. The most common idea of soft robots is to make use of material qualities, particularly softness. It's critical to think about compliance, which allows the robots' bodies and activities to be tailored to their surroundings. It is hard to deploy appropriate sensors or get a trustworthy dynamic model under which to construct the controller. As a result, various efforts were made to construct soft robots without employing specific sensors or feedback control, a technique known as open-loop control. Unlike open-loop control which uses compensatory input to decide output, loop control (closed) allows for a more precise and reliable actuation of the developed system. Closed-loop control mirrors a system's outputs to produce control inputs by using proprioceptive sensors. Adaptive controllers with the linear model are developed [90]. Despite the fact that various examples of soft motion control were recorded, control remains the most difficult task in soft robotics since most control systems are not yet well defined relative to traditional robots. Moreover, development in soft robot navigation for autonomous behavior should be made so that the robot may be employed in a variety of situations, including high-level activities and communications.

Power sources, actuators, and sensors have interconnected with the help of electronics and control algorithms. Soft robots have just recently begun to include a soft electronic system for all of their soft robotic elements, but the usage of stretchy or flexible electronics is critical for improved performance and greater adaptation to the environment, particularly in biological applications. Rigid electronics are now either built into the soft robot body and therefore become a part of it, or they are placed externally in a physically separate location from the soft robot body [91–93].

The advancement of soft material fabrication techniques, such as shape deposition manufacturing (SDM) and 3D printing technology, is closely linked to the rise of soft robot research efforts. This is due to the fact that the fabrication process for rigid robot systems is ineffective for soft material production, and hence soft material fabrication technique is in great demand. Conventional robotic systems need sophisticated calculations and continual system analysis, and the environment restricts activities to repetitive and non-versatile duties that are only dependable under specific conditions. More adaptable and dependable robotic systems that can adapt to unpredictable and/or diverse environments and accomplish essential tasks are in high demand. Continuous robotics using inverse kinetic algorithms and neural network algorithms are two forms of soft robot control systems. Piecewise-constant-curvature approximation is involved in the development of continuum robots [94–96].

3.6 Soft Robot Applications

Soft robotics is a developing subject, with new manufacturing processes and control approaches being developed. Researchers are drawing inspiration from nature's solutions to create flexible and responsive joints, insect/fish-like robots, and other imaginative systems. Human-made robots are often rigid in order to perform accurate, quick, and frequent jobs while maintaining mechanical integrity. In contrast, most creatures, including insects, have soft bodies while having hard exoskeletons. The human body, for example, has 11 percent skeleton, with muscles accounting for 42 percent of total body mass. Understanding how animals utilize their soft bodies to complete complicated movements and jobs might provide valuable insights for human aid, healthcare, and emergency management in uncertain circumstances. As a consequence, the robots may perform activities that are beyond the capabilities of typical rigid robots. Soft robots influenced by earthworms would be effective at rescuing confined humans following earthquakes, for example. Soft robotics aims to create simple mechanisms utilizing intelligence based on the mechanical aspect of creative pairings of soft materials, whereas animal movement is dependent on a relationship between neurological and mechanical controllers. In SMA-based actuators, the deformation is achieved with the help of heating. With simple activities, it provides better displacement when compared to other actuators. The scientist chose SMA based spring starfish robot because of the huge size and low force property of elastic materials which is shown in Figure 3.10. A ring oscillator is a device with an odd number of oscillators. NOT gates are used in a loop to produce a variable output. binary signal 1 and 0 or true and false on two distinct levels. The inverters, also known as NOT gates, are placed in series. The last inverter's output is sent back into the first one.

The creation of a soft ring oscillator that uses a single source of steady pressure to induce periodical motion in soft actuators [97]. The pneumatic type inverters with elastomeric valves consist of two chambers in a cylindrical shape and a tube to pass through the chambers. However, rigid robots have navigation difficulties in underwater and terrain. The locomotion capability of soft robots is high and efficient compared to the rigid type. Several soft robots have been influenced by underwater animals to move effectively and in a managed way in the water. The soft robotic fish includes a stiff head with traditional electronics inside and a soft-bodied tail that allows it to dive, turn, and swim like a fish. The octopus' soft robot has eight legs to simulate the behavior of an octopus and can go via narrow apertures and irregular surfaces and also conduct grasping with its legs [98]. Soft robots have the benefit of being naturally compatible with human and live animal tissues. These are some of the experimental

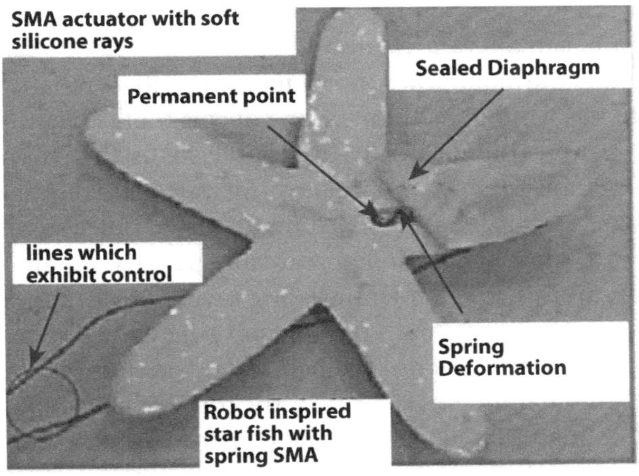

Figure 3.10 SMA spring deformation starfish robot.

areas with a high potential for soft robotics adoption in minimally invasive surgery (MIS). In endoscopy and laparoscopic surgery, a controlled stiffness in conjunction with a granular jamming mechanism can be used. For cardiac ablation in MIS, a soft endoscopic approach influenced by an elephant's trunk has been designed, making it accessible via a tight space and manipulation [99].

The key benefits of soft robots above rigid robotic devices are highly biocompatible and bio integrity, which are crucial factors for wearable and human helper applications. Because soft robots are made of elastic and soft materials, they have mechanical qualities that are identical to those of living beings, making them suitable for application in wearable devices that support human and maybe animal activities. Soft, elastomeric materials absorb mechanical strain and reduce injuries to both the robot and the operator. Soft actuators are included in soft robotic gloves to be used in hand rehabilitation for patients with grasp disorders. A soft robotic glove for thumb rehabilitation is also available. Another soft robotic glove for hand rehabilitation that conducts specific training tasks with human–machine interaction (with a clinician) [100].

There are numerous other studies on soft exoskeleton systems for hand rehabilitation [101]. Soft pneumatic actuators are used in connection with stiff structures in the gait rehabilitative soft robot for spinalized rodents, and their intrinsic softness makes them suited for dealing with soft tissues. The wearable rehabilitation system for ankle-foot uses pneumatic artificial muscle actuators to simulate muscle-tendon ligament shape and function. Because of their minimal degrees of freedom, complexity, and challenge in calculating grasping actions, rigid robots are limited in the types of manipulation they can execute. Soft robotic manipulators are more flexible than rigid robots and can operate fragile and unidentified things using a simple control method. The biological concept involved with synthetic to establish the soft materials will be the future scope of soft robots. Synthetic biology allows the creation of biologically inspired systems that mimic the performance of various biological structures of various sizes. Soft materials are able to generate versatile, compact, multifaceted full systems with great mechanical resistance for emulating real biological tissues and organisms, and these discoveries complement existing inventions in the soft robotics field [102–106]. The oscillators based on soft ring nature were introduced for motion-driven applications. The ball could travel forward by inflating and deflating the chambers in sequence. The robot of hexagonal shape with such a flexible foam-based border was constructed utilizing a soft ring oscillator in some other moving application. Figure 3.11 shows the rotational type ring oscillator soft robot.

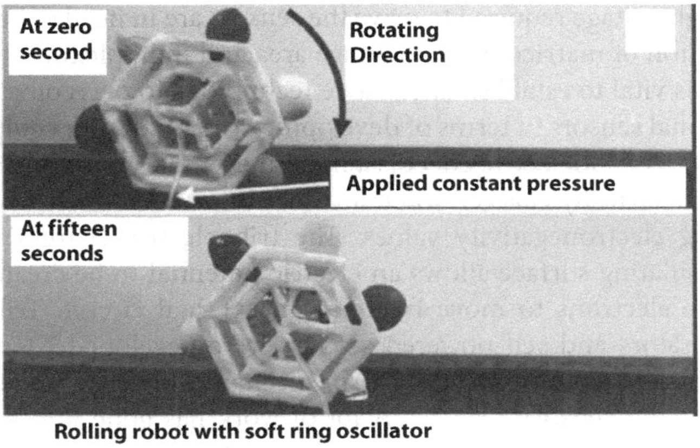

Figure 3.11 Rolling robot with ring-type oscillator.

3.7 Future Scope

The utilization of soft robotics in biomedical devices for surgical operations, treatment, and the precise processing of soft natural materials is currently the major driving force behind the sector. As a result, research into biocompatible and biodegradable soft actuators is gaining popularity. Soft actuators with self-sensing capabilities could contribute to low soft robots that communicate with their surroundings via perception and actuation. Nonetheless, more work is necessary to miniaturize a few of the electronic component's requisite for on-board application domains, such as softer and smaller pumps that can provide extreme pressures for pneumatic and hydraulic actuators, as well as smaller devices ability to produce high voltages requisite for some elastomers which are activated electrically.

The stretchable 3D-printed platform addresses the difficulty of rigid components size miniaturization. Furthermore, the incorporation of adjustable electrical components is required to build fully compatible and autonomous soft robots. Several CMOS-based logic and memory devices, as well as energy sources, have been described in this regard. Nevertheless, studies have continued to construct soft robotic systems using whole new techniques and materials. The soft robotic system needs fewer computations and so is more flexible than rigid robots when it comes to jobs like packing various commodities in factories. Soft robots are well adapted for interaction between people because of their soft and responsive character, as well as their ability to accomplish tasks in uncertain environments.

The high voltage required to drive the sensors are in need of huge power consumption of matrices with the huge area and multiplexing their application. It is vital to establish alternative technology to overcome the limits of traditional sensors in terms of developing critical sensing equipment for soft robots. A triboelectric effect is a natural event in which one substance becomes electrically charged when it comes into contact with another due to varying electronegativity values. The triboelectric charges' existence on the insulating surface allows an electric potential to be created, which allows the electrons to move freely in an external circuit. Triboelectric nanogenerators and self-powered sensors were developed based on this idea as efficient solutions to transform mechanical energy into electricity for energy and sensory reasons. A lot of work has gone into developing versatile triboelectric pressure sensors. For triboelectric pressure sensors, though, there is still an exchange between mechanical qualities, such as flexibility, increased stretching ability, and system performances, such as the sensitivity of pressure and least detection range.

Self-generating electricity has been used to demonstrate that various types of soft robots may proactively interpret and react to external stimulation and internal movements. They're made possible by self-powered, extremely stretchy, and very responsive triboelectric skins that use the triboelectrification effect to constantly detect proximity and pressure. The hundred percent excellent strain leads to perfect stretchability and enhances the sensitivity is up to 0.29 kPa in tribo-based skins and the surface made of micro prism in a triangular shape. The tribo-skins can be monolithically integrated into pneumatic actuators and permit soft robots to conduct numerous proactively detecting and responsive activities due to their qualities in all aspects. A conscious gripper can be actively knowledgeable about various operations in transporting an object, such as approaching, gripping, uplifting, descending, and even accidentally falling off the objects. If a baby wets his trousers, a perceptible robot finger can recognize it. A mindful robotic crawler can also sense very tiny human physiological signs and understand its muscle movements during oscillating gaits. For constantly multiplexing-sensing applications, such robots with the massive area covered by tribo-skins have been demonstrated. Furthermore, the actively reacting signals can operate optoelectronic components directly for transparent communication, and they could be further analyzed for more advanced uses like response with sound, light, phrases, and so on. The self-powered, extremely responsive, highly stretchy, and feasible tribo-skins can be used in a variety of applications that require soft communication interfaces.

The working principle tribo skin self-powered mechanism is shown in Figure 3.12. When compared to passive sensors, the tribo-driving skin's energy derives from the triboelectrification phenomenon, which is a combination of contact electrification and induction based on an electrostatic mechanism. The rubber with the largest electron affinity is a great negative triboelectric substance. Because of the electronegativity difference, electrons move from the surface of an object to the rubber's surface when it gets in contact with the tribo-skin. The positively and negatively charged interfaces split as the object moves away, creating a potential difference. The exposed negative charges on the rubber surface transport unpaired electrons to the ground from the embedded silver flakes, causing the electrode to accumulate positive charges and generate a current and voltage output.

Electrostatic equilibrium is achieved when the item is kept some distance away, and electrons stop motion. The spaced range between both the tribo-skin and the item, and also their potential difference, decreases as the object comes again. This process causes a retrograde transfer of electrons from the ground to the embedded electrode, resulting in an inverse voltage. Varied interface regions and deformation depths are produced on the micro prisms as a consequence of different pressures, generating pressure-induced changes in the flakes and pressure-dependence outputs to external circuits. The outputs in the low-pressure region dropped as the strain grew, whereas the outputs in the high-pressure region were essentially identical between strains. Since the device was stretched, partial triangular-micro prism structures were shifted out of the specified test range. When induced pressures were minimal, the triangular micro prisms created lesser contact surfaces, resulting in lesser outputs. Thus, in the

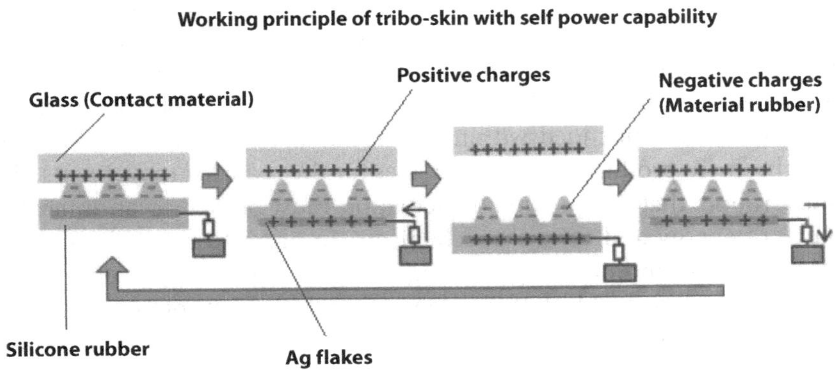

Figure 3.12 Self-powered tribo skin working principle.

reduced zone, the basic sensory structure predominates the results. The pressure-sensitive triangular micro prisms were slammed down whenever the pressures have been raised to a high-pressure zone, and the object fully contacted the rubber. As a result, the constant object region dictated outputs in the high-pressure zone. Because of its intrinsic stretchability and compatibility, tribo-skins can be integrated into soft robots as interactive robotic skins which can sense and respond to information using self-generated energy. The actuators with double-balloon shapes put on either front inflate and deflate sequentially, and the robot rolls from one side of the material to the other. The ring oscillator is used to govern chronological alterations in time—the distance in addition to motion. Soft robots can execute a variety of actively monitoring and interacting jobs due to the excellent combination of tribo-skins and actuators with soft nature. The optoelectronic systems can be operated with the support of self-powered signals which are utilized in various applications like visual-based communications due to their sophisticated functions.

3.8 Conclusion

A suitable relationship between the environment and a body is required for robotic actuation. To apply forces and govern the motion of the device, robots normally use nondeformable parts. Soft robots are a fast-emerging field in robotics, as traditional robot limitations, such as human–machine interaction and flexibility are resolved. We described some of the actuation techniques utilized in soft robots in this article, which vary from micrometer to centimeter scale and greater. The most relevant or important findings were presented, the fundamental concept behind the motions obtained in soft robots. A broad range of materials that can be triggered electrically, pneumatically, magnetically, thermally, and hydraulically have been examined. Other materials that responded to chemical and pressure differences are also reviewed. Currently, a large percentage of soft robots use pneumatic networks embedded within polymeric materials for actuation and propulsion. To produce sufficient stresses for the required movements, the softest robots that use pneumatic actuation must be tied to an inflexible source of energy and control mechanisms. In order to bring the sensors into operating condition, huge driving voltage is needed for smooth monitoring and response. Most of the elements with the soft robots are of sensory type, so a better way to consume the power and generate the power itself by means of triboelectric effect is also discussed.

References

1. Mirfakhrai, T., Madden, J.D., Baughman, R.H., Polymer artificial muscles. *Mater. Today*, *10*, 4, 30–38, 2007.
2. Morales, D., Palleau, E., Dickey, M.D., Velev, O.D., Electro-actuated hydrogel walkers with dual responsive legs. *Soft Matter*, *10*, 9, 1337–1348, 2014.
3. Hülse, M., Wischmann, S., Pasemann, F., The role of non-linearity for evolved multifunctional robot behavior, in: *International Conference on Evolvable Systems*, 2005, September, Springer, Berlin, Heidelberg, pp. 108–118.
4. Seok, S., Onal, C.D., Cho, K.J., Wood, R.J., Rus, D., Kim, S., Meshworm: A peristaltic soft robot with antagonistic nickel titanium coil actuators. *IEEE/ASME Trans. Mechatron.*, *18*, 5, 1485–1497, 2012.
5. Tolley, M.T., Shepherd, R.F., Mosadegh, B., Galloway, K.C., Wehner, M., Karpelson, M., Whitesides, G.M., A resilient, untethered soft robot. *Soft Robot.*, 1, 213–223, 2014.
6. Rus, D. and Tolley, M.T., Design, fabrication and control of soft robots. *Nature*, *521*, 7553, 467–475, 2015.
7. Fusco, S., Huang, H.W., Peyer, K.E., Peters, C., Häberli, M., Ulbers, A., Pané, S., Shape-switching microrobots for medical applications: The influence of shape in drug delivery and locomotion. *ACS Appl. Mater. Interfaces*, *7*, 12, 6803–6811, 2015.
8. Amjadi, M., Turan, M., Clementson, C.P., Sitti, M., Parallel microcracks-based ultrasensitive and highly stretchable strain sensors. *ACS Appl. Mater. Interfaces*, *8*, 8, 5618–5626, 2016.
9. Dickey, M.D., Stretchable and soft electronics using liquid metals. *Adv. Mater.*, *29*, 27, 1606425, 2017.
10. Cianchetti, M., Laschi, C., Menciassi, A., Dario, P., Biomedical applications of soft robotics. *Nat. Rev. Mater.*, *3*, 6, 143–153, 2018.
11. Shintake, J., Cacucciolo, V., Floreano, D., Shea, H., Soft robotic grippers. *Adv. Mater.*, *30*, 29, 1707035, 2018.
12. Xu, W. and Gracias, D.H., Soft three-dimensional robots with hard two-dimensional materials. *ACS Nano*, *13*, 5, 4883–4892, 2019.
13. Hines, L., Petersen, K., Lum, G.Z., Sitti, M., Soft actuators for small-scale robotics. *Adv. Mater.*, *29*, 13, 1603483, 2017.
14. Ghosh, A., Yoon, C., Ongaro, F., Scheggi, S., Selaru, F.M., Misra, S., Gracias, D.H., Stimuli-responsive soft untethered grippers for drug delivery and robotic surgery. *Front. Mech. Eng.*, *3*, 7, 2017.
15. Fantin Irudaya Raj, E., Implementation of machine learning techniques in unmanned aerial vehicle control and its various applications, in: *Computational Intelligence for Unmanned Aerial Vehicles Communication Networks*, pp. 17–33, Springer, Cham, 2022.
16. Gnanasekar, A.K., Deivakani, M., Bathala, N., Raj, E., Ramakrishna, V., Novel low-noise CMOS bioamplifier for the characterization of neurodegenerative diseases, in: *GeNeDis 2020*, pp. 221–226, Springer, Cham, 2021.

17. Petersen, K.E., Silicon as a mechanical material. *Proc. IEEE*, *70*, 5, 420–457, 1982.
18. Appadurai, M. and Fantin Irudaya Raj, E., Finite element analysis of lightweight robot fingers actuated by pneumatic pressure, in: *Recent Advances in Manufacturing, Automation, Design and Energy Technologies*, pp. 379–385, Springer, Singapore, 2022.
19. Rodrigue, H., Wang, W., Han, M.W., Kim, T.J., Ahn, S.H., An overview of shape memory alloy-coupled actuators and robots. *Soft Robot.*, *4*, 1, 3–15, 2017.
20. Lin, H.T., Leisk, G.G., Trimmer, B., GoQBot: A caterpillar-inspired soft-bodied rolling robot. *Bioinspiration Biomimetics*, *6*, 2, 026007, 2011.
21. Wang, W., Lee, J.Y., Rodrigue, H., Song, S.H., Chu, W.S., Ahn, S.H., Locomotion of inchworm-inspired robot made of smart soft composite (SSC). *Bioinspiration Biomimetics*, *9*, 4, 046006, 2014.
22. Kim, M.S., Chu, W.S., Lee, J.H., Kim, Y.M., Ahn, S.H., Manufacturing of inchworm robot using shape memory alloy (SMA) embedded composite structure. *Int. J. Precis. Eng. Manuf.*, *12*, 3, 565–568, 2011.
23. Huang, X., Kumar, K., Jawed, M.K., Mohammadi Nasab, A., Ye, Z., Shan, W., Majidi, C., Highly dynamic shape memory alloy actuator for fast moving soft robots. *Adv. Mater. Technol.*, *4*, 4, 1800540, 2019.
24. Laschi, C., Cianchetti, M., Mazzolai, B., Margheri, L., Follador, M., Dario, P., Soft robot arm inspired by the octopus. *Adv. Rob.*, *26*, 7, 709–727, 2012.
25. Wang, W., Kim, N.G., Rodrigue, H., Ahn, S.H., Modular assembly of soft deployable structures and robots. *Mater. Horiz.*, *4*, 3, 367–376, 2017.
26. Huang, X., Kumar, K., Jawed, M.K., Mohammadi Nasab, A., Ye, Z., Shan, W., Majidi, C., Highly dynamic shape memory alloy actuator for fast moving soft robots. *Adv. Mater. Technol.*, *4*, 4, 1800540, 2019.
27. Rani, E. F. I., Pushparaj, T. L., Raj, E. F. I., Escalating the resolution of an urban aerial image via novel shadow amputation algorithm. *Earth Sci. Inf.*, 15, 2, 905–913, 2022.
28. Felton, S.M., Tolley, M.T., Shin, B., Onal, C.D., Demaine, E.D., Rus, D., Wood, R.J., Self-folding with shape memory composites. *Soft Matter*, *9*, 32, 7688–7694, 2013.
29. Ananthasuresh, G.K. and Frecker, M.I., Chapter 9 in Compliant Mechanisms, L.L. Howell (Ed.), John Wiley & Sons Inc., New York, 2001.
30. Blanding, D.L., *Exact constraint: Machine design using kinematic principles*, Amer Society of Mechanical, New York, 1999.
31. Wang, M.Y., Wang, X., Guo, D., A level set method for structural topology optimization. *Comput. Methods Appl. Mech. Eng.*, *192*, 1-2, 227–246, 2003.
32. Kim, C.J., Moon, Y.M., Kota, S., A building block approach to the conceptual synthesis of compliant mechanisms utilizing compliance and stiffness ellipsoids. *J. Mech. Des.*, 130, 2, 022308–022319, 2008.

33. Jain, S., Chinnaiah, B., Rajakumar, M.P., Fantin, E., Raj, I., Jotiram, G.A., Improved IoT-based control system combined with an advanced control management server-based system. *J. Green Eng.*, *10*, 10, 8488–8496, 2020.
34. Pham, M.T., Teo, T.J., Yeo, S.H., Synthesis of multiple degrees-of-freedom spatial-motion compliant parallel mechanisms with desired stiffness and dynamics characteristics. *Precis. Eng.*, *47*, 131–139, 2017.
35. Lum, G.Z., Teo, T.J., Yeo, S.H., Yang, G., Sitti, M., Structural optimization for flexure-based parallel mechanisms–towards achieving optimal dynamic and stiffness properties. *Precis. Eng.*, *42*, 195–207, 2015.
36. Tsai, N.C. and Sue, C.Y., Review of MEMS-based drug delivery and dosing systems. *Sens. Actuators A: Phys.*, *134*, 2, 555–564, 2007.
37. Appadurai, M. and Raj, E.F.I., Finite element analysis of composite wind turbine blades, in: *2021 7th International Conference on Electrical Energy Systems (ICEES)*, 2021, February, IEEE, pp. 585–589.
38. Appadurai, M. and Raj, E., Epoxy/silicon carbide (sic) nanocomposites based small scale wind turbines for urban applications. *Int. J. Energy Environ. Eng.*, *13*, 1, 191–206, 2022.
39. Onal, C.D., Chen, X., Whitesides, G.M., Rus, D., Soft mobile robots with on-board chemical pressure generation, in: *Robotics Research*, pp. 525–540, Springer, Cham, 2017.
40. Fantin Irudaya Raj, E. and Balaji, M., Analysis and classification of faults in switched reluctance motors using deep learning neural networks. *Arabian J. Sci. Eng.*, *46*, 2, 1313–1332, 2021.
41. Jiang, H.Y., Kelch, S., Lendlein, A., Polymers move in response to light. *Adv. Mater.*, *18*, 11, 1471–1475, 2006.
42. Jenish, I., Felix Sahayaraj, A., Appadurai, M., Fantin Irudaya Raj, E., Suresh, P., Raja, T., Manikandan, V., Fabrication and experimental analysis of treated snake grass fiber reinforced with polyester composite. *Adv. Mater. Sci. Eng.*, *2021*, 1–13, 2021.
43. Deng, J., Li, J., Chen, P., Fang, X., Sun, X., Jiang, Y., Peng, H., Tunable photothermal actuators based on a pre-programmed aligned nanostructure. *J. Am. Chem. Soc.*, *138*, 1, 225–230, 2016.
44. Hu, Y., Liu, J., Chang, L., Yang, L., Xu, A., Qi, K., Wu, Y., Electrically and sunlight-driven actuator with versatile biomimetic motions based on rolled carbon nanotube bilayer composite. *Adv. Funct. Mater.*, *27*, 44, 1704388, 2017.
45. Neelakandan, S., Beulah, J. R., Prathiba, L., Murthy, G. L. N., Irudaya Raj, E. F., Arulkumar, N., Blockchain with deep learning-enabled secure healthcare data transmission and diagnostic model. *Int. J. Model. Simul. Sci. Comput.*, 13, 04, 2241006, 2022.
46. Gelebart, A.H., Jan Mulder, D., Varga, M., Konya, A., Vantomme, G., Meijer, E.W., Broer, D.J., Making waves in a photoactive polymer film. *Nature*, *546*, 7660, 632–636, 2017.

47. Dong, X., Xu, J., Xu, X., Dai, S., Zhou, X., Ma, C., Ding, J., Sunlight-driven continuous flapping-wing motion. *ACS Appl. Mater. Interfaces*, *12*, 5, 6460–6470, 2020.
48. Yao, S., Zhang, X., Zhou, W., Gao, R., Xu, W., Ye, Y., Ma, D., Atomic-layered Au clusters on α-MoC as catalysts for the low-temperature water-gas shift reaction. *Science*, *357*, 6349, 389–393, 2017.
49. Raj, E.F.I., Available transfer capability (ATC) under deregulated environment. *J. Power Electron. Power Syst.*, *6*, 2, 85–88, 2016.
50. Randall, C.L., Gultepe, E., Gracias, D.H., Self-folding devices and materials for biomedical applications. *Trends Biotechnol.*, *30*, 3, 138–146, 2012.
51. Drury, J.L. and Mooney, D.J., Hydrogels for tissue engineering: Scaffold design variables and applications. *Biomaterials*, *24*, 24, 4337–4351, 2003.
52. Zhao, Y., Song, L., Zhang, Z., Qu, L., Stimulus-responsive graphene systems towards actuator applications. *Energy Environ. Sci.*, *6*, 12, 3520–3536, 2013.
53. Li, C., Liu, Y., Huang, X., Jiang, H., Direct sun-driven artificial heliotropism for solar energy harvesting based on a photo-thermomechanical liquid-crystal elastomer nanocomposite. *Adv. Funct. Mater.*, *22*, 24, 5166–5174, 2012.
54. Ko, H. and Javey, A., Smart actuators and adhesives for reconfigurable matter. *Acc. Chem. Res.*, *50*, 4, 691–702, 2017.
55. Fantin Irudaya Raj, E. and Appadurai, M., Static 2D-finite element analysis of eccentricity fault in induction motor, in: *Smart Technologies for Energy, Environment and Sustainable Development*, vol. 1, pp. 409–422, Springer, Singapore, 2022.
56. Liu, X., Wang, X., Liu, T., Keller, P., Gold nanoparticles incorporated nematic gel micropillars capable of laser actuation at room temperature. *Macromolecules*, *49*, 21, 8322–8331, 2016.
57. Liu, X., Wei, R., Hoang, P.T., Wang, X., Liu, T., Keller, P., Reversible and rapid laser actuation of liquid crystalline elastomer micropillars with inclusion of gold nanoparticles. *Adv. Funct. Mater.*, *25*, 20, 3022–3032, 2015.
58. Jin, M. and Zhang, X., A new topology optimization method for planar compliant parallel mechanisms. *Mech. Mach. Theory*, *95*, 42–58, 2016.
59. Shepherd, R.F., Stokes, A.A., Freake, J., Barber, J., Snyder, P.W., Mazzeo, A.D., Whitesides, G.M., Using explosions to power a soft robot. *Angew. Chem. Int. Ed.*, *52*, 10, 2892–2896, 2013.
60. Bartlett, N.W., Tolley, M.T., Overvelde, J.T., Weaver, J.C., Mosadegh, B., Bertoldi, K., Wood, R.J., A 3D-printed, functionally graded soft robot powered by combustion. *Science*, *349*, 6244, 161–165, 2015.
61. Kwon, G.H., Park, J.Y., Kim, J.Y., Frisk, M.L., Beebe, D.J., Lee, S.H., Biomimetic soft multifunctional miniature aquabots. *Small*, *4*, 12, 2148–2153, 2008.
62. Shintake, J., Rosset, S., Schubert, B., Floreano, D., Shea, H., Versatile soft grippers with intrinsic electroadhesion based on multifunctional polymer actuators. *Adv. Mater.*, *28*, 2, 231–238, 2016.

63. Priyadarsini, K., Raj, E.F.I., Begum, A.Y., Shanmugasundaram, V., Comparing DevOps procedures from the context of a systems engineer. *Mater. Today: Proc.*, 2020.
64. Hedenström, A., Norevik, G., Warfvinge, K., Andersson, A., Bäckman, J., Åkesson, S., Annual 10-month aerial life phase in the common swift Apus apus. *Curr. Biol.*, *26*, 22, 3066–3070, 2016.
65. Appadurai, M., Raj, E., Jenish, I., Application of aluminium oxide–water nanofluids to augment the performance of shallow pond: A numerical study. *Process Integr. Optim. Sustainability*, *6*, 1, 211–222, 2022.
66. Sohn, J.W. and Choi, S.B., Various robots made from piezoelectric materials and electroactive polymers: A review. *Int. J. Mech. Syst. Eng.*, *3*, 122, 2, 2017.
67. Rogers, J.A., A clear advance in soft actuators. *Science*, *341*, 6149, 968–969, 2013.
68. Rogers, J.A., Someya, T., Huang, Y., Materials and mechanics for stretchable electronics. *Science*, *327*, 5973, 1603–1607, 2010.
69. Rosset, S. and Shea, H.R., Flexible and stretchable electrodes for dielectric elastomer actuators. *Appl. Phys. A*, *110*, 2, 281–307, 2013.
70. Saidaminov, M.I., Abdelhady, A.L., Murali, B., Alarousu, E., Burlakov, V.M., Peng, W., Bakr, O.M., High-quality bulk hybrid perovskite single crystals within minutes by inverse temperature crystallization. *Nat. Commun.*, *6*, 1, 1–6, 2015.
71. Kim, O., Kim, H., Choi, U.H., Park, M.J., One-volt-driven superfast polymer actuators based on single-ion conductors. *Nat. Commun.*, *7*, 1, 1–8, 2016.
72. Dickey, M.D., Stretchable and soft electronics using liquid metals. *Adv. Mater.*, *29*, 27, 1606425, 2017.
73. Jenish, I., Sahayaraj, A.F., Suresh, V., Appadurai, M., Irudaya Raj, E.F., Nasif, O., Kumaravel, A.K., Analysis of the hybrid of mudar/snake grass fiber-reinforced epoxy with nano-silica filler composite for structural application. *Adv. Mater. Sci. Eng.*, *2022*, 1–10, 2022.
74. Li, K., Cheng, X., Zhu, F., Li, L., Xie, Z., Luan, H., Zhang, Y., A generic soft encapsulation strategy for stretchable electronics. *Adv. Funct. Mater.*, *29*, 8, 1806630, 2019.
75. Vogt, D.M., Park, Y.L., Wood, R.J., Design and characterization of a soft multi-axis force sensor using embedded microfluidic channels. *IEEE Sens. J.*, *13*, 10, 4056–4064, 2013.
76. Kramer, R.K., Majidi, C., Sahai, R., Wood, R.J., Soft curvature sensors for joint angle proprioception, in: *2011 IEEE/RSJ International Conference on Intelligent Robots and Systems*, 2011, September, IEEE, pp. 1919–1926.
77. Muth, J.T., Vogt, D.M., Truby, R.L., Mengüç, Y., Kolesky, D.B., Wood, R.J., Lewis, J.A., Embedded 3D printing of strain sensors within highly stretchable elastomers. *Adv. Mater.*, *26*, 36, 6307–6312, 2014.
78. Felt, W., Chin, K.Y., Remy, C.D., Contraction sensing with smart braid McKibben muscles. *IEEE/ASME Trans. Mechatron.*, *21*, 3, 1201–1209, 2015.

79. Mojarrad, M. and Shahinpoor, M., Ion-exchange-metal composite sensor films, in: *Smart Structures and Materials 1997: Smart Sensing, Processing, and Instrumentation*, 1997, June, vol. 3042, pp. 52–60, International Society for Optics and Photonics, United States, 1997.
80. Mazzeo, A., Aguzzi, J., Calisti, M., Canese, S., Vecchi, F., Stefanni, S., Controzzi, M., Marine robotics for deep-sea specimen collection: A systematic review of underwater grippers. *Sensors*, *22*, 2, 648, 2022.
81. Loepfe, M., Schumacher, C.M., Lustenberger, U.B., Stark, W.J., An untethered, jumping roly-poly soft robot driven by combustion. *Soft Rob.*, *2*, 1, 33–41, 2015.
82. Jin, H., Dong, E., Xu, M., Liu, C., Alici, G., Jie, Y., Soft and smart modular structures actuated by shape memory alloy (SMA) wires as tentacles of soft robots. *Smart Mater. Struct.*, *25*, 8, 085026, 2016.
83. Chen, S., Tan, M.W.M., Gong, X., Lee, P.S., Low-voltage soft actuators for interactive human–machine interfaces. *Adv. Intell. Syst.*, *4*, 2, 2100075, 2022.
84. Raj, E. F. I., Appadurai, M., Rani, E. F. I., Jenish, I., Finite-element design and analysis of switched reluctance motor for automobile applications. *Multiscale Multidiscip. Modeling, Experiments Design*, 5, 3, 269–277, 2022.
85. Onal, C.D. and Rus, D., Autonomous undulatory serpentine locomotion utilizing body dynamics of a fluidic soft robot. *Bioinspiration Biomimetics*, *8*, 2, 026003, 2013.
86. Shepherd, R.F., Ilievski, F., Choi, W., Morin, S.A., Stokes, A.A., Mazzeo, A.D., Whitesides, G.M., Multigait soft robot. *Proc. Natl. Acad. Sci.*, *108*, 51, 20400–20403, 2011.
87. Amend, J.R., Brown, E., Rodenberg, N., Jaeger, H.M., Lipson, H., A positive pressure universal gripper based on the jamming of granular material. *IEEE Trans. Robot.*, *28*, 2, 341–350, 2012.
88. Brown, E., Rodenberg, N., Amend, J., Mozeika, A., Steltz, E., Zakin, M.R., Jaeger, H.M., Universal robotic gripper based on the jamming of granular material. *Proc. Natl. Acad. Sci.*, *107*, 44, 18809–18814, 2010.
89. Marchese, A.D., Onal, C.D., Rus, D., Autonomous soft robotic fish capable of escape maneuvers using fluidic elastomer actuators. *Soft Rob.*, *1*, 1, 75–87, 2014.
90. Caldwell, D.G., Medrano-Cerda, G.A., Goodwin, M., Characteristics and adaptive control of pneumatic muscle actuators for a robotic elbow, in: *Proceedings of the 1994 IEEE International Conference on Robotics and Automation*, 1994, May, IEEE, pp. 3558–3563.
91. Cianchetti, M., Ranzani, T., Gerboni, G., De Falco, I., Laschi, C., Menciassi, A., STIFF-FLOP surgical manipulator: Mechanical design and experimental characterization of the single module, in: *2013 IEEE/RSJ International Conference on Intelligent Robots and Systems*, 2013, November, IEEE, pp. 3576–3581.
92. Stokes, A.A., Shepherd, R.F., Morin, S.A., Ilievski, F., Whitesides, G.M., A hybrid combining hard and soft robots. *Soft Rob.*, *1*, 1, 70–74, 2014.

93. Chen, Z., Um, T.I., Zhu, J., Bart-Smith, H., Bio-inspired robotic cownose ray propelled by electroactive polymer pectoral fin, in: *ASME International Mechanical Engineering Congress and Exposition*, 2011, January, vol. 54884, pp. 817–824.
94. Chen, F.J., Dirven, S., Xu, W.L., Li, X.N., Soft actuator mimicking human esophageal peristalsis for a swallowing robot. *IEEE/ASME Trans. Mechatron.*, *19*, 4, 1300–1308, 2013.
95. Marchese, A.D., Komorowski, K., Onal, C.D., Rus, D., Design and control of a soft and continuously deformable 2d robotic manipulation system, in: *2014 IEEE International Conference on Robotics and automation (ICRA)*, 2014, May, IEEE, pp. 2189–2196.
96. Palagi, S. and Fischer, P., Bioinspired microrobots. *Nat. Rev. Mater.*, *3*, 6, 113–124, 2018.
97. Appadurai, M., Fantin Irudaya Raj, E., LurthuPushparaj, T., Sisal fiber-reinforced polymer composite-based small horizontal axis wind turbine suited for urban applications—A numerical study. *Emergent Mater.*, 5, 2, 565–578, 2022.
98. Case, J.C., Yuen, M.C., Jacobs, J., Kramer-Bottiglio, R., Robotic skins that learn to control passive structures. *IEEE Rob. Autom. Lett.*, *4*, 3, 2485–2492, 2019.
99. Deng, T., Wang, H., Chen, W., Wang, X., Pfeifer, R., Development of a new cable-driven soft robot for cardiac ablation, in: *2013 IEEE International Conference on Robotics and Biomimetics (ROBIO)*, 2013, December, IEEE, pp. 728–733.
100. Polygerinos, P., Galloway, K.C., Savage, E., Herman, M., O'Donnell, K., Walsh, C.J., Soft robotic glove for hand rehabilitation and task specific training, in: *2015 IEEE International Conference on Robotics and Automation (ICRA)*, 2015, May, IEEE, pp. 2913–2919.
101. Yap, H.K., Lim, J.H., Nasrallah, F., Goh, J.C., Yeow, R.C., A soft exoskeleton for hand assistive and rehabilitation application using pneumatic actuators with variable stiffness, in: *2015 IEEE International Conference on Robotics and Automation (ICRA)*, 2015, May, IEEE, pp. 4967–4972.
102. El-Atab, N., Qaiser, N., Bahabry, R., Hussain, M.M., Corrugation enabled asymmetrically ultrastretchable (95%) monocrystalline silicon solar cells with high efficiency (19%). *Adv. Energy Mater.*, *9*, 45, 1902883, 2019.
103. Bahabry, R.R., Kutbee, A.T., Khan, S.M., Sepulveda, A.C., Wicaksono, I., Nour, M., Hussain, M.M., Corrugation architecture enabled ultraflexible wafer-scale high-efficiency monocrystalline silicon solar cell. *Adv. Energy Mater.*, *8*, 12, 1702221, 2018.
104. Ghoneim, M.T., Zidan, M.A., Alnassar, M.Y., Hanna, A.N., Kosel, J., Salama, K.N., Hussain, M.M., Thin PZT-based ferroelectric capacitors on flexible silicon for nonvolatile memory applications. *Adv. Electron. Mater.*, *1*, 6, 1500045, 2015.

105. Rojas, J.P., Torres Sevilla, G.A., Hussain, M.M., Can we build a truly high performance computer which is flexible and transparent? *Sci. Rep.*, *3*, 1, 1–8, 2013.
106. Torres Sevilla, G.A., Ghoneim, M.T., Fahad, H., Rojas, J.P., Hussain, A.M., Hussain, M.M., Flexible nanoscale high-performance FinFETs. *ACS Nano*, *8*, 10, 9850–9856, 2014.

4

Dynamic Butterfly Optimization Algorithm-Based Task Scheduling for Minimizing Energy Consumption in Distributed Green Data Centers

Sengathir Janakiraman[1] and Deva Priya M.[2]*

[1]*Department of Information Technology, CVR College of Engineering, Mangalpally, Vastunagar, Hyderabad, Telangana, India*
[2]*Department of Computer Science & Engineering, Sri Eshwar College of Engineering, Coimbatore, Tamil Nadu, India*

Abstract
Data Center (DC) industry is the chief energy consumer in the world as it utilizes approximately 300 billion kWh per year for facilitating diversified categories of heterogeneous services to its users. Distributed Green Data Center (DGDC) providers are being paid by the users around the globe for their service that provision expected and actual degree of Quality of Service (QoS) required for tasks. DGDCs involve Internet Service Providers (ISPs) with diverse bandwidth prices and potentialities for handling user tasks. Moreover, the cost of solar energy, wind and power grid related to GDCs completely depend on their geographical locations. Hence, the process of task scheduling among DGDCs ensuring high level of QoS and profit is an uphill task. In this chapter, Improved Dynamic Butterfly Optimization Algorithm (IDBOA)-based Task Scheduling (IDBOATS) for DGDCs is proposed to attain potential task scheduling with high level of QoS and minimized cost overhead. The proposed IDBOATS scheme is a multi-objective optimization method which is designed for DGDCs to target on the reduction of likelihood determined between mean loss of 2 tasks. It also concentrates on improving the profit of DGDC providers by partitioning the tasks among different ISPs and rate of task services associated with each GDC. It adopts the merits of Mutation-based Local Search Algorithm (MLSA) for improving diversity of

**Corresponding author*: devapriya.m@sece.ac.in

Rathishchandra R. Gatti, Chandra Singh, Rajeev Agrawal and Felcy Jyothi Serrao (eds.) Self-Powered Cyber Physical Systems, (91–114) © 2023 Scrivener Publishing LLC

solutions and preventing issue of local optima by intelligently scheduling tasks of diversified applications and allocating available resources within the bounds of response time. The simulation experiments confirm the potential of IDBOATS in decreasing task loss and increasing profit on par with the competitive task scheduling schemes taken for comparative investigations.

Keywords: Task scheduling, Distributed Green Data Centers (DGDC), energy consumption, Mutation-based Local Search Algorithm (MLSA), Improved Dynamic Butterfly Optimization Algorithm (IDBOA)

4.1 Introduction

The fast spread of cloud computing endorses quick development of Data Centers (DCs). Cloud Service Providers (CSPs) use Distributed Green Data Centers (DGDCs) to deal with the growing cloud applications and minimizing carbon footprint. In DGDC, the rudimentary and pressing demands include inter-DC data transmission for intermittent data storage and recovery, software sharing and Virtual Machine (VM) cloning [1]. Though several efforts are made to construct DGDCs or offer optimal scheduling amid DCs, an optimal scheduling is expected. Distributed Data Centers (DDCs) include several physical servers and cooling services that involve several Mega Watts (MWs) of energy. In 2017, the energy spent by DCs in the United States was about 0.078 Trillion kWh which was about 2.9% of the whole energy consumption. This increased to 4% in 2020. More than 80% of the energy spent in 2017 was produced by burning fossil fuel that brings permanent pollution to global surroundings. Subsequently, to drop the carbon footprint, growing number of DDCs like Facebook and Google implement green energy devices to reduce the pollution of fossil fuel and move to DGDCs for lessening brown energy consumption [2, 3]. With the increase in the kinds and quantity of varied applications, the cost of energy involved by a DGDC provider dramatically increases. The implementation of tasks associated with every application usually offers consistent revenue to the provider in terms of items specified in the Service Level Agreement (SLA) assigned among provider and users [4, 5].

Increasing amount of services like search engine, computing, etc., simultaneously run and compliantly offer services. With increase in users' demands, it is essential to assure reduced latency and increased accessibility of applications [5]. To deal with this issue, the services are positioned and applied in DCs in several topographical sites. Currently, many organizations are deploying several varied applications in DGDCs to malleably offer low cost and improved QoS to users. The spatial and temporal differences in costs of Power Grid (PG) and accessibility of Renewable Energy (RE) make it demanding to reduce the energy consumed by the DGDC providers by perceptively scheduling tasks of varied applications amid DGDCs with reduced delay [6, 7]. Maximizing profit at GDCs, reducing energy consumption and optimally specifying task service rates demand much focus. DGDCs are gaining popularity as they are involved in hosting and operating increased number of cloud services from the recent past.

Several applications simultaneously run on DGDCs and demand more infrastructure resources that normally comprise of several servers and cooling services. To ensure energy efficiency, applications are typically positioned on DGDCs in different locations. Every DGDC consumes 10s of MWs of PG and REs for energizing and cooling while implementing diverse tasks. Furthermore, owing to the growing energy cost related to the DGDCs, it is significant to improve the number of servers. The DGDCs improve the utilization of RE to the maximum extent possible. Nevertheless, as the amount of available RE is restricted and do not meet the demands of the DGDCs, it is essential to deal with the purchase of power from PG to implement tasks in various applications [8]. As DGDCs involve enormous amount of power, the provider should take into account both reduction of energy cost and enhancement of revenue. These two goals are often in conflict, and hence an optimum trade-off among them is needed. Many researchers have dealt with reducing the energy price of DCs or improving the profit, but no previous studies have focused on the trade-off. Furthermore, factors like cost of PG, speed of wind, air density, solar radiation, overall power and quantity of accessible servers in diverse DGDCs show spatial disparities. They provide a better opportunity to improve the revenue and power cost of DGDC providers. Further, to assure well-timed task completion and cost reduction, every DC is associated with several Internet Service Providers (ISPs) that distribute data amid DDCs and several global users. It is seen that the energy and bandwidth prices contribute to much of the operative cost of a DDC provider.

The existing works focus on the revenue generated by the implementation of each task and the cost of energy spent by every DC provider determined based on queuing theory concepts. They fail to focus on the variety of bandwidth expenses and dimensions of ISPs, and hence may cause task loss and reduced profit. User tasks are to be conveyed through a Wide Area Network (WAN), including several ISPs before they reach the back-end DGDCs. DGDCs use paid services of ISPs to get user tasks and responses exchanged between them and their associated users. As they are distributed in different geographical positions, the costs of PG show spatial difference. The air based on green energy of diverse GDCs differ based on topographical sites. The revenue of a DGDC provider can be maximized with the price of ISP bandwidth, obtainability of green energy, cost of PG and revenue showing spatial variations.

In this chapter, Improved Dynamic Butterfly Optimization Algorithm (IDBOA)-based Task Scheduling (IDBOATS) for DGDCs is proposed to attain potential task scheduling among DGDCs with high level of QoS and minimized cost overhead. This proposed IDBOATS scheme is a multi-objective optimization method designed for DGDCs targeting on reducing the likelihood of mean task loss and maximizing the profit of DGDC providers by partitioning the tasks amid different ISPs and rates of task service associated with each GDC. It adopts the merits of Mutation-based Local Search Algorithm (MLSA) for improving the diversity of solutions and preventing the issue of local optima for intelligently scheduling tasks of diversified applications and allocating available resources within bounds of response time. The simulation experiments confirm the potential of IDBOATS in decreasing task loss and increasing profit on par with the competitive task scheduling schemes taken for investigation.

4.2 Related Work

In this section, researches related to Green DCs and energy-based scheduling of tasks are discussed in detail.

4.2.1 Green Data Centers

Yang *et al.* [9] have designed energy and computation proficient embedding schemes for robust green virtual DCs. The propounded schemes have considered diverse network topologies and scales by associating them with numerous current embedding schemes. Nevertheless, these studies consider a collection of servers positioned at a DC. Hasan *et al.* [10] have focussed on green energy management to make use of RE in a DC. Green energy virtualization deals with ambiguity in accessing green energy, and a green SLA based on threshold parameters is propounded. The green SLA algorithm based on RE is propounded without involving greater costs. Nevertheless, the spatial difference in the accessibility of green energy is not considered and hence fails to implement this feature for comprehending cost-effective energy management in DGDC. Cheng *et al.* [11] have propounded a dynamic power-based resource provisioning scheme for varied workloads in DCs driven by green energy. It focuses on maximizing system throughput and reducing the energy depletion based on RE supply.

It efficiently explores the optimum resource-allocation scheme based on Simulated Annealing (SA) and fuzzy modeling. Anastasopoulos *et al.* [12] have designed a service-provisioning mechanism using Stochastic Linear Programming (SLP). The authors have employed dimensionality-reduction schemes that include Lagrangian relaxation and average approximation to handle computational complexity. The authors claim that the scheme offers quick convergence and lessens CO_2 emissions around 60% for diverse tasks.

Wu *et al.* [13] have proposed two multi-objective optimization schemes to deal with the tradeoff amid cost and brown energy involved in DC placement. The authors have dealt with the optimal quantity of DCs and their positions based on USNET and NSFNET topologies. Nevertheless, the authors have not considered the revenue involved in the implementation of tasks in green DCs. Niu *et al.* [14] have propounded a green based and cost-efficient data handling framework to incorporate the benefits of RE and active pricing schemes for a MapReduce framework. The tasks are scheduled by considering diverse factors comprising of RE source, battery usage, energy efficacy of workloads and live pricing. Furthermore, an efficient energy-based model is designed to make scheduling choices. Khosravi *et al.* [15] have considered the total energy cost based on the amount of energy consumed and the overhead energy. Several effectual VM positioning schemes are propounded to assess the performance and find the parameters with the maximum influence on the amount of brown and green energy consumed, cost as well as carbon footprint.

Zhang *et al.* [16] have focused on green energy-based inter-DC VM relocation in a flexible framework and have framed it as an Integer Linear Program (ILP). The brown energy cost spent by the DCs is reduced. Numerous heuristic schemes are propounded for ensuring optimal solution in large-scale networks. Lu *et al.* [17] have considered the scalability of DDCs driven by RE for dropping the energy cost in a defensive environment. Increasing the use of RE and reducing the cost of energy in a grid for huge data transmissions among supportable and green DCs are challenging. A heuristic scheme is proposed to address this issue by explicitly focussing on maximized use of green energy and minimized use of grid energy cost in inter-GDC bulk data transmissions for green and supportable DCs in a multi-electricity market. Investigations are carried out based on topology, power consumption charges and wind power to improve RE use and save energy costs over the prevailing data transfer schemes. Yuan *et al.* [18] have propounded Spatio-Temporal Task Scheduling (STTS) to reduce the energy cost by efficiently scheduling tasks to handle delay restrictions. They have dealt with the spatial and temporal disparities in DGDCs.

In every time slot, power cost reduction is framed as a non-linear controlled optimization problem and is focussed on using genetic SA-based Particle Swarm Optimization (PSO). The authors claim improved throughput and reduced energy cost when compared to diverse task scheduling schemes while firmly focussing on the delay restrictions of tasks. The hybrid scheme adopts a hybrid metaheuristic algorithm to deal with the non-linear optimization problem.

4.2.2 Energy-Aware Task Scheduling

Recently, energy-based task scheduling at the DCs has gained growing amount of attention. Aroca & Gonçalves [19] have assessed the likelihood of constructing servers based on power computers with reduced power by performing an investigation of applications on x86 and ARM computer frameworks. The assessment performed on web as well as database servers comprise of load, temperature, power, request delays and amount of requests tackled by every system. Performance in terms of Watts along with latency to deal with the HTTP or SQL requests is the best indicator of QoS supported by the system. ARM-based SoCs offer better performance in servers and clusters especially in terms of Watt. Goiri *et al.* [20] have designed a scheduler called GreenSlot to deal with the parallel batch tasks in a DC driven by photovoltaic solar array and PG. Future availability of solar energy is predicted and tasks are arranged to improve green energy consumption while maintaining task deadlines. PG is used when its price is less so as to avoid deadline violation. Beloglazov & Buyya [21] have propounded an architecture called Open-Stack Neat for dynamic association of VMs in OpenStack clouds. It is organized to use VM consolidation schemes and is combined with accessible OpenStack placements deprived of altering the configuration. Kiani & Ansari [22] have focussed on optimization problem involving G/D/1 queue model and have dealt with workload dissemination to improve profit by assigning them to servers driven by brown and green energy sources. Nevertheless, it does not take the bandwidth cost and variances in bandwidth expenses of ISP and capacities into consideration. The users' workloads are categorized into green and brown depending on the amount of energy required by them during the process of execution. Authors have focused on avoiding decomposition of workloads and executing them on the conforming servers driven by thermal power, wind and solar energy. They have considered the SLAs amid users and DCs along with varying electricity markets in diverse regions.

Lei *et al.* [23] have dealt with a multi-objective problem for carrying out energy proficient scheduling of tasks in a GDC. The nodes involved in computing are Dynamic Voltage and Frequency Scaling (DVFS) supported and partly driven by green energy. To overcome this, an enhanced multi-objective co-evolutionary scheme based on Preference Inspired Co-Evolutionary Algorithm with goal (PICEA-g) is proposed. General opposition-based learning is used to find the optimum supply voltage and clock frequency for seamless execution of tasks. Tripathi *et al.* [24] have dealt with the assorted ILP problem to diminish the Total Cost of Ownership (TCO) by involving appropriate capacity for highly accessible DGDCs. Availability is ensured by using a capacity provisioning scheme. Power consumption and server placement costs are reduced, provided the quantity of green energy consumed is diminished. The cost involved in the deployment of servers and the amount of power consumed are considered along with renewable green energy.

Katsaros *et al.* [25] have developed a method to evaluate and estimate the ecological efficacy of VMs positioned on private clouds. They have precisely focused on the information comprising of energy accessibility in an area along with the amount of energy and resource utilized and the ecological efficacy of VMs. An architecture is designed to evaluate and predict the ecological efficacy of a VM. Chen *et al.* [26] have propounded a task scheduling architecture for security dependent workflows. It provides theoretical investigation on how to replicate the predecessor tasks such that the start time of a task is not influenced by encryption as well as data transmission times. The current finish times of tasks are defined and a scheduling scheme with choosy task replication is propounded to lessen the economic costs and increase resource efficacy such that the workflow's makespan is not deferred. Chen *et al.* [27] have propounded a mobile cloudlet based service mode that compliantly realizes the cost-delay trade-off with the conventional distant cloud service mode. An energy reduction problem is framed and solved by negotiating amid mobile cloudlets, distant cloud and the propounded mode. Nevertheless, this study does not focus on the profit improvement of the provider.

Shah-Mansouri *et al.* [28] have considered delay, cost of cloud services and energy consumed and have framed a utility enhancement problem for mobile users. An optimal scheduling scheme is propounded for delay-tolerant as well as delay-sensitive applications. Non-convex profit intensification problem is framed for a Cloud Service Provider (CSP) and resolved to come out with an optimal pricing approach. The authors have not focussed on the accessibility of green energy in GDCs. Zhang & Zhou [29] have propounded a two-stage scheme to lessen needless task assignment and improve task scheduling performance on clouds. A job classifier is developed to categorise tasks involving historical data. Therefore, an optimum amount of VMs of diverse types are established. Dynamic scheduling of tasks is possible at a single DC. Cai *et al.* [30] have propounded a green energy-driven scheme that proficiently facilitates a DC to improve its computing performance using computational sprinting. Different sprinting schemes are proposed to deal with the time-varying and intermittent nature green energy source. Zhu *et al.* [31] have propounded an iterative heuristic framework for job scheduling in a hybrid cloud. Two job accumulating and scheduling schemes are proposed to enhance utilization, time of utilization and number of hired VMs. They have not focussed on the accessibility of green energy. Juarez *et al.* [32] have propounded a scheduling scheme to reduce the execution time and amount of energy consumed by proficiently implementing task-based applications on a distributed platform. A dynamic algorithm in polynomial time is proposed to focus on resource distribution and heuristic rules are applied to get sustaining solutions in reasonable time. Srichandan *et al.* [33] have propounded a hybrid bacteria foraging scheme that applies crossover and mutation of Genetic Algorithm (GA) to co-operatively reduce energy consumption and makespan. Then, an effectual scheduling scheme is propounded for varied clouds. Barzegar *et al.* [34] have propounded a dual-phase algorithm called Energy-Aware Task Scheduling with Duplication Clustering Dynamic (EATSDCD) voltage/frequency scheme. The scheme involves duplication and clustering to plan for the precedence-based task graph on processors using DVFS. The makespan and energy of processors are reduced while implementing Directed Acyclic Graph (DAG) mainly focusing on the throughput. The key idea is to reduce energy consumption. Once the critical path and dependent tasks along the non-critical paths are found, the slack time for tasks in these paths is shared among reliable tasks existing in the path. The DVFS' frequency is scaled down to implement these tasks along with idle and communication stages without prolonging the task execution time.

Yuan *et al.* [35] have designed a multi-objective optimization scheme for DGDCs to improve the income of providers and lessen the mean task loss likelihood of applications by cooperatively determining the splitting of tasks amid several ISPs as well as service rates of every GDC. A problem is framed and solved using SA-based Bi-objective Differential Evolution (SBDE) scheme to get an estimated Pareto-optimal set. Minimum Manhattan distance is approved to choose a solution that identifies the task service rates and splitting tasks amid ISPs in every time slot. The profit and task loss driven by RE and smart grid are improved by cooperatively splitting the tasks among ISPs. They miss to perceptively plan tasks of varied applications and assign resources based on response times.

4.3 Improved Dynamic Butterfly Optimization Algorithm (IDBOA)-Based Task Scheduling (IDBOATS)

In this section, the problem definition, delay constraint, Green energy model, Energy Consumption model, Constraints-imposed optimization problem and detailed view of Improved Dynamic Butterfly Optimization Algorithm (IDBOA)-based Task Scheduling (IDBOATS) used for solving optimization problem are presented. Efficient task scheduling schemes facilitate improved balancing of loads [36, 37].

4.3.1 Problem Definition

The problem considered for multi-objective objective optimization essential for establishing a DGDC is presented as follows.

Initially, users forward their requests or tasks through diverse kinds of electronic devices which are then introduced for classification. This process of classification is performed to segregate them depending on types. Further, separate First-In-First-Out (FIFO) queues are used for enqueuing tasks associated with the same application. An application is employed in different GDCs which are positioned at different locations to ensure the tasks' QoS and application availability. Then, specific number of ISPs is used for delivering the tasks to DGDCs. It is assumed that it determines energy depending on the kinds of sources that includes non-renewable power grid, wind turbines and solar panels. This problem concentrates on the process of task scheduler that executes profit and QoS-optimized task scheduling exploiting the architecture of DGDCs. It particularly specifies the service rates of Pareto optimal task, and tasks are allocated among ISPs depending on each DGDC in every time slot.

4.3.2 Delay Constraint

Let 'N' and 'C' represent the amount of applications and GDCs in the system environment. The proposed approach includes the merits of Google cluster trace 1 for evaluating the profit and QoS-optimized task scheduling process. The time incurred for servicing each incoming task service follows the properties of exponential distribution. It is also confirmed that the Google cluster trace meets the complete set of assumptions considered during the application of the proposed scheme. This inclusion of the Google cluster trace utilizes the M/M/1 queuing model for quantifying the performance of the proposed scheme with simulators of realistic server's clusters in the data centres. Thus, this proposed scheme includes a reliable queuing system to model the application server's performance.

In this context, $RT^{c,n}_{Avg(\tau)}$ representing the mean response time of application tasks executed in c^{th} GDC in the τth time slot is determined based on Equation (4.1).

$$RT^{c,n}_{Avg(\tau)} = \frac{\theta^{c,n}_{(\tau)}}{\mu^{c,n}_{(\tau)}\left(1 - W^{c,n,0}_{(\tau)}\right)} \leq \varepsilon_n \tag{4.1}$$

where,

$$\theta^{c,n}_{(\tau)} = \frac{\rho^{c,n}_{(\tau)}}{1 - \rho^{c,n}_{(\tau)}} - \frac{\left(\omega^{c,n}_{(\tau)}\right)\left(\rho^{c,n}_{(\tau)}\right)^{\omega^{c,n}_{(\tau)}+1}}{1 - \left(\rho^{c,n}_{(\tau)}\right)^{\omega^{c,n}_{(\tau)}+1}} \tag{4.2}$$

$$W^{c,n,0}_{(\tau)} = \frac{1 - \rho^{c,n}_{(\tau)}}{1 - \left(\rho^{c,n}_{(\tau)}\right)^{\omega^{c,n}_{(\tau)}+1}} \tag{4.3}$$

$$\rho^{c,n}_{(\tau)} = \frac{\pi^{c,n}_{(\tau)}}{\mu^{c,n}_{(\tau)}} \tag{4.4}$$

In specific, the value of $RT^{c,n}_{Avg(\tau)}$ must be less than ε_n as specified in Equation (4.5)

$$RT^{c,n}_{Avg(\tau)} \leq \varepsilon_n \tag{4.5}$$

At this juncture, let $\pi^{c,n}_{(\tau)}$ depicts the tasks' arrival rate computed for n^{th} application in the cth DGC delivered using the kth ISP in the time slot (τ). Then, the stability of the task queue is guaranteed based on the condition which specifies that $\pi_{(\tau)}^{c,n}$ must be less than $\mu_{(\tau)}^{c,n}$ as specified in Equation (4.6).

$$\pi_{(\tau)}^{c,n} = \sum_{k=1}^{K} \pi_{(\tau)}^{c,n,k} < \mu_{(\tau)}^{c,n} \tag{4.6}$$

Let $\pi_{(\tau)}^{n}$ depict the rate of task arrival with respect to n^{th} application in time slot (τ). Then, the cumulative sum of tasks' arrival rate of nth application delivered using the complete set of ISPs associated with the comprehensive number of GDCs need to be equal to $\pi_{(\tau)}^{n}$ as specified in Equation (4.7).

$$\pi_{(\tau)}^{n} = \sum_{c=1}^{C} \pi_{(\tau)}^{c,n} < \sum_{c=1}^{C} \sum_{k=1}^{K} \mu_{(\tau)}^{c,n,k} \tag{4.7}$$

4.3.3 Green Energy Model

In this green energy model, the models of wind and solar are introduced. Let $E_{N(\tau)}^{C,Wnd}$ represent the quantity of wind energy produced by the cth GDC in the τth time slot. In this context, the amount of wind energy generated using the wind energy model is determined based on Equation (4.8).

$$E_{N(\tau)}^{C,Wnd} = \frac{1}{2} \omega^{WE} \alpha^{D} \beta^{3} (S_{Wnd}^{C})^{3} L_{T-Slot} \tag{4.8}$$

where ω^{WE} and α^{D} refer to the wind-to-electricity transformation efficiency and density of air respectively.

Further, S_{Wnd}^{C} and β represent the speed of wind and area of wind rotor area in the time slot (τ). In addition, L_{T-Slot} represents the length of the time slot. On the other hand, if $E_{N(\tau)}^{C,Solar}$ represents the quantity of solar energy produced by the cth GDC in the τth time slot, then the amount of solar energy generated using the solar energy model is determined based on Equation (4.9).

$$E_{N(\tau)}^{C,Solar} = (AI_{Area}^{C})(\varphi_{Slot}^{C})(I_{Solar}^{C})L_{T-Slot} \tag{4.9}$$

Where, I_{Solar}^{C} and φ_{Slot}^{C} represent the solar irradiance and solar-to-electricity conversion efficiency with AI_{Area}^{C} as the area of active irradiation.

4.3.4 Energy Consumption Model

In this model of energy consumption, it is expected that the servers associated to similar kind of application are homogeneous in characteristics to uniformly spend their energy. Let, φ_n^c and μ_n^c represent the task count executed by each energized server in a minute through GDC. Then, $MSC_{n(\tau)}^c$ highlighting the maximum servers count related to n^{th} application in c^{th} DGC is computed based on Equation (4.10).

$$MSC_{n(\tau)}^c = \frac{\mu_n^c}{\varphi_n^c} \tag{4.10}$$

Moreover, the value of $MSC_{n(\tau)}^c$ should not exceed the value of φ_n^c as specified in Equation (4.11).

$$MSC_{n(\tau)}^c \leq \varphi_n^c \tag{4.11}$$

In this model, the total energy incurred by DGDCs is computed by adding up the energy incurred by the complete set of servers and their associated facilities, including the benefits of lighting and cooling. In specific, effective of power utilization is a predominant measure used for representing the energy efficiency related to the GDC with the value generally varying between 1.2 and 2.0 in the energy model.

4.3.5 Constraint-Imposed Optimization Problem

In the proposed IDBOATS scheme, the multi-objective optimization depends on the decision variable ranges related to $\pi_{(\tau)}^{c,n,k}$ and $\mu_{(\tau)}^{c,n}$ respectively. The total revenue gained by the tasks related to the complete set of all applications in the comprehensive GDCs in the time slot (τ) is determined based on Equation (4.12).

$$\varnothing_{(\tau)}^c = \sum_{n=1}^{N} \vartheta_{(\tau)}^c \tag{4.12}$$

where $\vartheta_{(\tau)}^c$ represents the total revenue earned by tasks of complete set of applications in the comprehensive collection of GDCs in time slot (τ).

The primary objective concentrates on maximizing $\varnothing^{c}_{(\tau)}$ such that the condition specified in Equation (4.13) is satisfied.

$$\underset{\pi^{c,n,k}_{(\tau)},\mu^{c,n}_{(\tau)}}{\text{Max}} \varnothing^{c}_{(\tau)} \tag{4.13}$$

The second objective focusses on minimizing $\Delta^{c}_{(\tau)}$ such that the condition presented in Equation (4.14) is satisfied.

$$\underset{\pi^{c,n,k}_{(\tau)},\mu^{c,n}_{(\tau)}}{\text{Min}} \Delta^{c}_{(\tau)} \tag{4.14}$$

Finally, the optimization problem associated with the proposed IDBOATS scheme is presented in Equation (4.15).

$$\begin{gathered}\underset{\pi^{c,n,k}_{(\tau)},\mu^{c,n}_{(\tau)}}{\text{Max}} \varnothing^{c}_{(\tau)} \\ \underset{\pi^{c,n,k}_{(\tau)},\mu^{c,n}_{(\tau)}}{\text{Min}} \Delta^{c}_{(\tau)} \\ \text{Subject to the constraint } \pi^{c,n,k}_{(\tau)} \geq 0 \text{ and } \mu^{c,n}_{(\tau)} \geq 0\end{gathered} \tag{4.15}$$

This above-mentioned optimization problem is solved through the merits of Dynamic Butterfly Optimization Algorithm (DBOA) as detailed below.

4.3.6 Primitives of Dynamic Butterfly Optimization Algorithm (DBOA)

In this section, initially the complete information about the classical Butterfly Optimization Algorithm (BOA) is presented which is then followed by the detailed view of the DBOA.

4.3.7 Classical Butterfly Optimization Algorithm

Classical BOA is proposed as a potential metaheuristic optimization algorithm inspired by food foraging and mating partner identification behavior of butterflies. To achieve food foraging activity, butterflies use their chemoreceptors that are present scattered throughout their body. These chemoreceptors aid in smelling or sensing food or flowers fragrance as well as aid in identifying the optimal partner for their mating. The butterflies determine their mating partner based on the intensity level of fragrance emitted by them during the change of locations. In specific, the search agents are guided by the intensity of fragrance during their movement in the search space. In BOA, the phase of exploitation (local search) is facilitated when the search agent is not able to sense the location of food or other mating partner in the search apace during the random movement. In contrast, exploration is introduced when the search agent is capable of sensing the food or mating partner location in the search space and possesses the possibility of moving towards other butterflies. In this BOA, vital fragrance function (P_{Frag}) is defined based on the intensity of stimulus as determined using Equation (4.16).

$$P_{Frag} = M_{Sense}\,(Fr_{Int})^{\alpha} \tag{4.16}$$

Where, Fr_{Int} and M_{Sense} refer to the intensity of fragrance and modality of sense with 'α' as the power exponent.

The value of M_{Sense} and 'α' are identified to vary between 0 and 1. In specific, 'α' is responsible for controlling the fragrance absorbed by the intensity of stimulus generated through search agents. Depending on the value of P_{Frag}, two core equations that aid in updating the location of search agents depending on the global and local search process are shown in Equations (4.17) and (4.18).

$$S_{(i)}^{t+1} = S_{(i)}^{t} + \left(R_{ND}^{2} \times S_{Best}^{*} - S_{(i)}^{t}\right) \times P_{Frag} \tag{4.17}$$

$$S_{(i)}^{t+1} = S_{(i)}^{t} + \left(R_{ND}^{2} \times S_{(j)}^{t} - S_{(k)}^{t}\right) \times P_{Frag} \tag{4.18}$$

Where, P_{Frag} and S_{Best}^{*} refer to the fragrance intensity related to i^{th} search agent and the best solution determined during the current iteration. Moreover, $S_{(j)}^{t}$ and $S_{(k)}^{t}$ represent the possible solutions determined by the jth and k^{th} search agents in the search space.

4.3.8 Transformation of BOA into DBOA using Mutation-Based Local Searching Strategy (MLSS)

In this section, a detailed view of DBOA is presented with the steps through which the traditional BOA is transformed using the merits of Mutation-Based Local Searching Strategy (MLSS). It operates towards the enhancement of current best solution or the improvement of solutions' quality. Initially, the local searching derives the current best solution (input task) as input at the end of each iteration. MLSS employs mutation over the determined current best solution (optimal task allocation attained so far). In this context, the global best solution determined so far is replaced by the newly generated solution generated through mutation. Further, MLSS selects an arbitrary solution from the existing solutions and performs the process of comparison and replacement. This method of comparison and replacement are repeated for maximum number iterations as the termination criterion. In specific, this MLSS applies mutation over the current best solution at the end of every iteration. It is the responsibility of MLSS to verify whether the fitness of new solution is better than the existing current solution. After this replacement, the new mutated solution substitutes the current best solution in the population space. Otherwise, the selection of solution is facilitated randomly from the other feasible solutions. MLSS is developed to offer the best solution in each iteration and prevent the problem of local optima and other solutions considered for ensuring the solution diversity by the process of replacing the worst solution by the best one.

The algorithm of DBOA utilized for solving the proposed IDBOATS-based optimization problem is presented as follows.

Algorithm: DBOA-based solution for optimal task assignment to GDC

Initialize the number of GDCs (flowers or mating butterfly), number of tasks (butterflies considered as search agents), size of the population and the maximum amount of iterations

Set the initial values of intensity of fragrance (Fr_{Int}), Sense of Modality (M_{Sense}), power exponent (α) and maximum number of iterations ($Iter_{Max}$)

While $\left(\underset{\pi_{(\tau)}^{c,n,k}, \mu_{(\tau)}^{c,n}}{\text{Min}} \Delta_{(\tau)}^{c} \right)$ do

For each task in the complete population search space do

Calculate the intensity of fragrance Fr_{Int} (optimization conditions) associated with each GDCs using Equation (4.16)

End For

Identify the optimal task

Allocate the optimal task to suitable and energy minimizing GDCs

For each task in the complete population search space do

Use interval [0,1] to generate a random value R_{nd}

If ($R_{nd} < P_{St}$) then

Update the task allocation to suitable GDC depending on Equation (4.17)

Else
Update the task allocation to suitable GDC depending on Equation (4.18)
End If
Estimate the fitness of new task
If the fitness value of new task is better than that of the task already allocated to GDC, update the allocation process by assigning the new incoming task to GDCs
End For
Update the value associated with the power exponent 'α'
Apply Mutation Operator-based local search over the current best solution to enhance the elements of the population or generate new potential tasks
Update the best global solution if determined during the current iteration
End While
Return the best optimal solution determined by DBOA (optimal allocation of tasks to suitable DGCs)

4.4 Results and Discussion

The experiments of the propounded IDBOATS-based optimization problem are conducted using the real-time data that includes the arrival rates of tasks pertaining to three applications which are sampled in the interval of 5 minutes in the Google cluster.

In the initial part of investigation, the proposed IDBOATS scheme and the benchmarked SBDE, EATSDCD, STTS and DVFS approaches are explored based on the convergence time and energy consumption with respect to increase in the scale factor. In this process of investigation, the convergence analysis is facilitated with respect to profit, and temporary knee distance with respect to IDBOATs in the time slot set to 100.

Figures 4.1 and 4.2 present the convergence time and energy consumption of the proposed IDBOATS and the baseline SBDE, EATSDCD, STTS and DVFS approaches for different scale factors used for investigation. From Figures 4.1 and 4.2, it is obvious that the propounded IDBOATS mechanism is proficient in attaining rapid convergence time and minimized energy consumption with respect to increase in the tasks' scaling factors. This potentiality of the proposed IDBOATS mechanism is mainly owing to the adoption of a bio-objective constrained optimization process that facilitates optimal task assignment in energy efficient DGCs considered in the systems model. Moreover, the inclusion of MTSS into DBOA

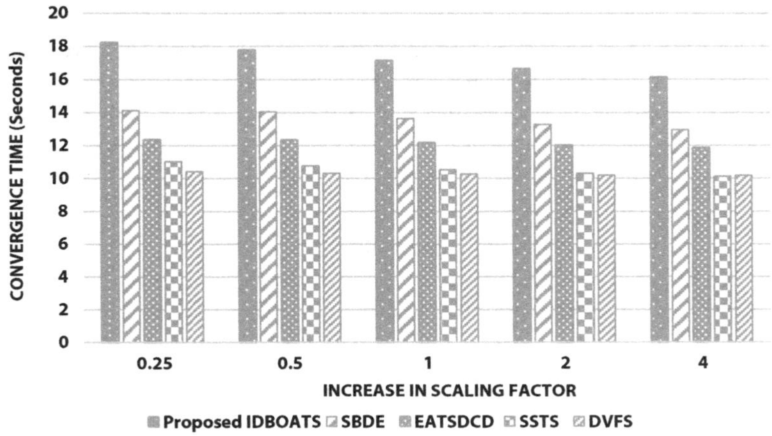

Figure 4.1 Convergence time of the proposed IDBOATS for increasing scaling factors.

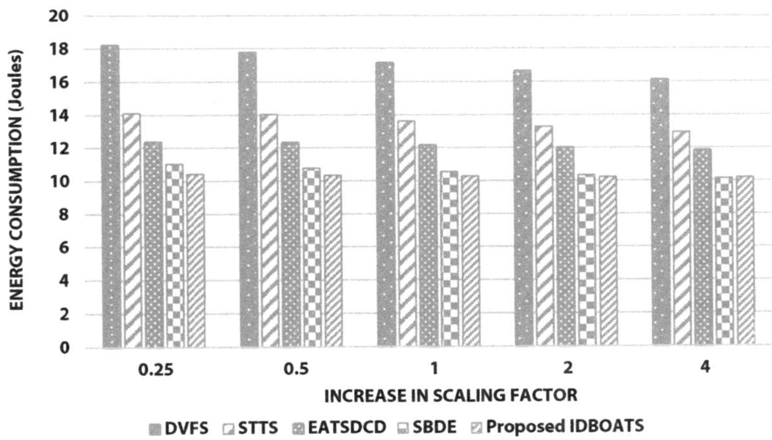

Figure 4.2 Energy consumption of the proposed IDBOATS for increasing scaling factor.

improves the convergence speed and minimizes the amount of energy consumed. Thus, the proposed IDBOATS scheme improves the convergence time by 11.28%, 15.78%, 18.92% and 21.98% when compared to the benchmarked SBDE, EATSDCD, STTS and DVFS approaches under different scale factors taken for investigation. Moreover, the energy consumed by the propounded IDBOATS scheme is minimized by 9.26%, 11.42%, 14.86% and 18.46% when compared to the standard SBDE, EATSDCD, STTS and DVFS schemes for different scale factors taken for investigation.

Figures 4.3 and 4.4 demonstrate the mean execution time and dollars of one day of the proposed IDBOATS and the baseline SBDE, EATSDCD,

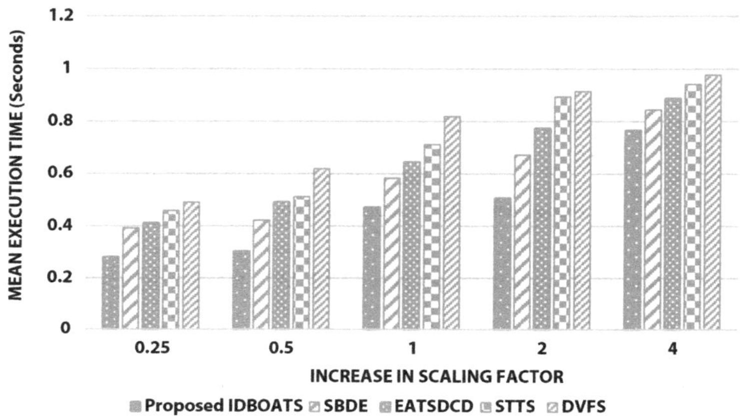

Figure 4.3 Mean execution time of the proposed IDBOATS for increasing scaling factor.

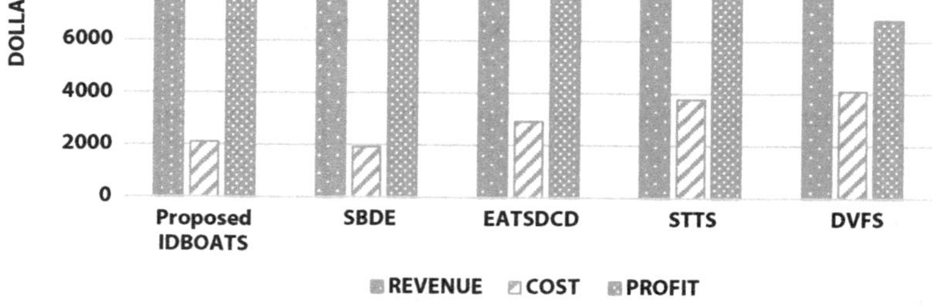

Figure 4.4 Dollars of one day of the proposed IDBOATS representing revenue, cost and profit for varying algorithms.

STTS, and DVFS approaches under different scale factors taken for investigation. It is identified that the proposed IDBOATS scheme minimizes the mean execution time independent of the increasing scaling factor which increases with respective increase in the amount of incoming tasks entering the environment. Minimization in mean execution time is mainly due to the derivation of contextual parameter inherited by the mutation operator improved DBOA to prevent unnecessary exploration and exploitation in the search space until it is essential.

On the other hand, dollars of one day of the proposed IDBOATS evaluated to determine revenue, cost and profit is also comparatively high on par with the compared SBDE, EATSDCD, STTS, and DVFS approaches. This revenue and profit with minimized cost is incurred by the proposed strategy due to the optimization problem that significantly alternates between maximization and minimization process during the phases of exploration and exploitation. Thus, the proposed IDBOATS scheme improves the mean execution time by 9.63%, 11.64%, 14.32%, and 16.94% when compared to the benchmarked SBDE, EATSDCD, STTS and DVFS approaches for varying scale factors taken for investigation. In addition, the cost incurred by the proposed IDBOATS scheme is minimized marginally by 10.32%, 12.86%, 15.74%, and 17.81% in contrast to the benchmarked approaches for varying scale factors considered for investigation.

Table 4.1 and Table 4.2 present the performance of the proposed IDBOATS scheme under Knee solution with ATLP and profit for varying amount of tasks and GDCs. The results confirm that the proposed IDBOATS scheme independent of the amount of tasks and GDCs handles the process of assigning the tasks to suitable GDCs without any delay and loss. This improved capability of IDBOATS scheme lay the foundation

Table 4.1 Knee solution with ATLP and profit of the proposed IDBOATS scheme for varying number of GDCs with tasks kept constant.

	Knee solution	
Compared algorithms	**ATLP**	**Profit ($)**
Proposed IDBOATS	0.0028	3741.9
SBDE	0.0032	3437.5
EATSDCD	0.0228	2842.9
STTS	0.0104	2716.4
DVFS	0.0047	2219.8

Table 4.2 Knee solution with ATLP and profit of the proposed IDBOATS scheme for varying number of tasks with GDCs constant.

Compared algorithms	Knee solution	
	ATLP	Profit ($)
Proposed IDBOATS	0.0022	4214.6
SBDE	0.0029	3421.8
EATSDCD	0.0212	3131.6
STTS	0.0198	2816.4
DVFS	0.0341	2519.2

for the minimization of ATLP that correspondingly increase the profit to expected level.

4.5 Conclusion

In this chapter, Improved Dynamic Butterfly Optimization Algorithm (IDBOA)-based Task Scheduling (IDBOATS) scheme is proposed for attaining potential task scheduling among DGDCs with high level of QoS and minimized cost overhead. It is formulated as a constrained multi-objective optimization method for optimal allocation of tasks to suitable DGDCs and minimizing the mean task loss which in turn maximizes the profit of DGDC providers by partitioning the tasks among different ISPs and rates of task service associated with each GDC. It includes the merits of MLSS and improves the solutions' diversity that prevents the problem of local optima and delayed convergence that are most common during the process of intelligently scheduling tasks associated with applications to assign the available resources within the bounds of response time. The simulation experiments of the propounded IDBOATS mechanism confirm better convergence time by 11.28%, 15.78%, 18.92% and 21.98% in contrast to the benchmarked SBDE, EATSDCD, STTS and DVFS approaches for varying scaling factors taken for investigation. Further, the energy consumptions incurred by the propounded IDBOATS scheme is minimized by 9.26%, 11.42%, 14.86% and 18.46% when compared to the benchmarked approaches for varying scaling factors taken for investigation. As a part of future scope, it is decided to formulate a Mayfly-based Optimization

Algorithm with more than two optimization factors and evaluate its importance over the proposed IDBOATS scheme.

References

1. Cao, J., Li, K., Stojmenovic, I., Optimal power allocation and load distribution for multiple heterogeneous multicore server processors across clouds and data centers. *IEEE Trans. Comput.*, 63, 1, 45–58, 2013.
2. Deng, X., Wu, D., Shen, J., He, J., Eco-aware online power management and load scheduling for green cloud datacenters. *IEEE Syst. J.*, 10, 1, 78–87, 2014.
3. Mei, J., Li, K., Ouyang, A., Li, K., A profit maximization scheme with guaranteed quality of service in cloud computing. *IEEE Trans. Comput.*, 64, 11, 3064–3078, 2015.
4. Chitra, P., Renuka, K.D., Senathipathi, K., Deepika, S., Geethamani, R., A survey on load balancing algorithms for cloud environment. *J. Comput. Theor. Nanosci*, 16, 2, 764–767, 2019.
5. Bi, J., Yuan, H., Tie, M., Tan, W., SLA-based optimisation of virtualised resource for multi-tier web applications in cloud data centres. *Enterp. Inf. Syst.*, 9, 7, 743–767, 2015.
6. Durai, K.N., Subha, R., Haldorai, A., Hybrid invasive weed improved grasshopper optimization algorithm for cloud load balancing. *Intell. Autom. Soft Comput.*, 34, 1, 467–483, 2022.
7. Chang, X., Wang, B., Muppala, J.K., Liu, J., Modeling active virtual machines on IaaS clouds using an M/G/m/m+ K queue. *IEEE Trans. Serv. Comput.*, 9, 3, 408–420, 2014.
8. Sathya Sofia, A. and Ganesh Kumar, P., Multi-objective task scheduling to minimize energy consumption and makespan of cloud computing using NSGA-II. *J. Network Syst. Manage.*, 26, 463–485, 2018.
9. Yang, Y., Chang, X., Liu, J., Li, L., Towards robust green virtual cloud data center provisioning. *IEEE Trans. Cloud Comput.*, 5, 2, 168–181, 2015.
10. Hasan, M.S., Kouki, Y., Ledoux, T., Pazat, J.L., Exploiting renewable sources: When green SLA becomes a possible reality in cloud computing. *IEEE Trans. Cloud Comput.*, 5, 2, 249–262, 2015.
11. Cheng, D., Rao, J., Jiang, C., Zhou, X., Elastic power-aware resource provisioning of heterogeneous workloads in self-sustainable datacenters. *IEEE Trans. Comput.*, 65, 2, 508–521, 2015.
12. Anastasopoulos, M., Tzanakaki, A., Simeonidou, D., Stochastic energy efficient cloud service provisioning deploying renewable energy sources. *IEEE J. Sel. Areas Commun.*, 34, 12, 3927–3940, 2016.
13. Wu, Y., Tornatore, M., Ferdousi, S., Mukherjee, B., Green data center placement in optical cloud networks. *IEEE Trans. Green Commun. Networking*, 1, 3, 347–357, 2017.

14. Niu, Z., He, B., Liu, F., JouleMR: Towards cost-effective and green-aware data processing frameworks. *IEEE Trans. Big Data*, 4, 2, 258–272, 2017.
15. Khosravi, A., Andrew, L.L., Buyya, R., Dynamic VM placement method for minimizing energy and carbon cost in geographically distributed cloud data centers. *IEEE Trans. Sustainable Comput.*, 2, 2, 183–196, 2017.
16. Zhang, L., Han, T., Ansari, N., Energy-aware virtual machine management in inter-data center networks over elastic optical infrastructure. *IEEE Trans. Green Commun. Networking*, 2, 1, 305–315, 2017.
17. Lu, X., Jiang, D., He, G., Yu, H., GreenBDT: Renewable-aware scheduling of bulk data transfers for geo-distributed sustainable datacenters. *Sustain. Comput.: Inform. Syst.*, 20, 120–129, 2018.
18. Yuan, H., Bi, J., Zhou, M., Spatiotemporal task scheduling for heterogeneous delay-tolerant applications in distributed green data centers. *IEEE Trans. Autom. Sci. Eng.*, 16, 4, 1686–1697, 2019.
19. Aroca, R.V. and Gonçalves, L.M.G., Towards green data centers: A comparison of x86 and ARM architectures power efficiency. *J. Parallel Distrib. Comput.*, 72, 12, 1770–1780, 2012.
20. Goiri, Í., Haque, M.E., Le, K., Beauchea, R., Nguyen, T.D., Guitart, J., Torres, J., Bianchini, R., Matching renewable energy supply and demand in green datacenters. *Ad Hoc Networks*, 25, 520–534, 2015.
21. Beloglazov, A. and Buyya, R., OpenStack Neat: A framework for dynamic and energy-efficient consolidation of virtual machines in OpenStack clouds. *Concurrency Comput.: Pract. Exper.*, 27, 5, 1310–1333, 2015.
22. Kiani, A. and Ansari, N., Profit maximization for geographically dispersed green data centers. *IEEE Trans. Smart Grid*, 9, 2, 703–711, 2016.
23. Lei, H., Wang, R., Zhang, T., Liu, Y., Zha, Y., A multi-objective co-evolutionary algorithm for energy-efficient scheduling on a green data center. *Comput. Oper. Res.*, 75, 103–117, 2016.
24. Tripathi, R., Vignesh, S., Tamarapalli, V., Optimizing green energy, cost, and availability in distributed data centers. *IEEE Commun. Lett.*, 21, 3, 500–503, 2016.
25. Katsaros, G., Stichler, P., Subirats, J., Guitart, J., Estimation and forecasting of ecological efficiency of virtual machines. *Future Gener. Comput. Syst.*, 55, 480–494, 2016.
26. Chen, M., Hao, Y., Lai, C.F., Wu, D., Li, Y., Hwang, K., Opportunistic task scheduling over co-located clouds in mobile environment. *IEEE Trans. Serv. Comput.*, 11, 3, 549–561, 2016.
27. Chen, H., Zhu, X., Qiu, D., Liu, L., Du, Z., Scheduling for workflows with security-sensitive intermediate data by selective tasks duplication in clouds. *IEEE Trans. Parallel Distrib. Syst.*, 28, 9, 2674–2688, 2017.
28. Shah-Mansouri, H., Wong, V.W., Schober, R., Joint optimal pricing and task scheduling in mobile cloud computing systems. *IEEE Trans. Wireless Commun.*, 16, 8, 5218–5232, 2017.

29. Zhang, P. and Zhou, M., Dynamic cloud task scheduling based on a two-stage strategy. *IEEE Trans. Autom. Sci. Eng.*, 15, 2, 772–783, 2017.
30. Cai, H., Zhou, X., Cao, Q., Jiang, H., Sheng, F., Qi, X., Yao, J., Xie, C., Xiao, L., Gu, L., Greensprint: Effective computational sprinting in green data centers, in: *2018 IEEE International Parallel and Distributed Processing Symposium (IPDPS)*, 2018, May, IEEE, pp. 690–699.
31. Zhu, J., Li, X., Ruiz, R., Xu, X., Scheduling stochastic multi-stage jobs to elastic hybrid cloud resources. *IEEE Trans. Parallel Distrib. Syst.*, 29, 6, 1401–1415, 2018.
32. Juarez, F., Ejarque, J., Badia, R.M., Rocha, S.N.G., Esquivel-Flores, O.A., Energy-aware scheduler for HPC parallel task base applications in cloud computing. *Int. J. Comb. Optim. Probl. Inf.*, 9, 1, 54, 2018.
33. Srichandan, S., Kumar, T.A., Bibhudatta, S., Task scheduling for cloud computing using multi-objective hybrid bacteria foraging algorithm. *Future Comput. Inf. J.*, 3, 2, 210–230, 2018.
34. Barzegar, B., Motameni, H., Movaghar, A., EATSDCD: A green energy-aware scheduling algorithm for parallel task-based application using clustering, duplication and DVFS technique in cloud datacenters. *J. Intell. Fuzzy Syst.*, 36, 6, 5135–5152, 2019.
35. Yuan, H., Bi, J., Zhou, M., Liu, Q., Ammari, A.C., Biobjective task scheduling for distributed green data centers. *IEEE Trans. Autom. Sci. Eng.*, 18, 2, 731–742, 2020.
36. Janakiraman, S. and Priya, M.D., Improved artificial bee colony using monarchy butterfly optimization algorithm for load balancing (IABC-MBOA-LB) in cloud environments. *J. Network Syst. Manage.*, 29, 4, 39, 2021.
37. Janakiraman, S. and Priya, M.D., Hybrid grey wolf and improved particle swarm optimization with adaptive intertial weight-based multi-dimensional learning strategy for load balancing in cloud environments. *Sustain. Comput.: Inform. Syst.*, 38, 100875, 2023.

5

Wireless Power Transfer for IoT Applications—A Review

Sasikala G. and Rajeev Ranjan*

School of Computer Science and Applications, REVA University, Bangalore, Karnataka, India

Abstract

For a very long time, classical wire networks were the primary method of transmitting electricity. Traditional cable (wire) transmission has issues, including ageing, wear, and electric spark production. In some environments, including the alpine and seafloor, it is challenging to utilize transmission lines since doing so compromises their usability's safety and dependability. The transmission of electricity can undergo a paradigm change thanks to wireless power transfer (WPT), which has radically revolutionized the process. WPT is a significant method for transmitting electrical energy over a long distance and across a vacuum or environment without the need for wires or other materials. A totally theoretical idea that is now a common feature in commercial items, particularly mobile phones and wearable smart gadgets, is wireless charging. 2014 has witnessed the release of the newest high-end smartphones from leading mobile providers including Huawei and Samsung. These gadgets may wirelessly recharge. Many businesses are now working on technologies that can transmit power via an air gap of 15 to 30 cm with tolerable efficiency, including Qualcomm, Evatran, and WiTricity. An overview of wireless power transmission techniques is given in this chapter. The chapter also focuses on current issues and potential growth in the field.

***Keywords*:** Wireless power transfer, wireless energy transfer, WiTricity, green internet of things, electromagnetic induction

**Corresponding author*: sasikala.g@reva.edu.in

Rathishchandra R. Gatti, Chandra Singh, Rajeev Agrawal and Felcy Jyothi Serrao (eds.) Self-Powered Cyber Physical Systems, (115–140) © 2023 Scrivener Publishing LLC

5.1 Introduction

Wireless electricity or WiTricity, as expressed in its short form, is an advanced technology for transmitting electrical energy in a wireless environment with roots in the past. This innovation could be attributed to Nikola Tesla, who invented the technology of transmitting electricity without wires in 1891 [1, 2]. This system works in tandem with the principle of wireless receivers, and the device must be within range of the transmitter. Such power transfer is classified as non-radioactive wireless power transfer. It is based on the magnetic close field coupling of the conductor loop [1]. There are subtypes defined as short-range and medium-range applications. The principle that resonant objects resonate energy efficiently while non-resonant objects remain unyielding forms the basic operating principle of such systems.

Applications for wireless transmission on AC machines have been widely reported by Nikola Tesla [3–5]. For example, energy transfer from the excited stator to the rotor cage (through the air gap) is done in the cage asynchronous machine [4]. Therefore, electromechanical machines are based on the principle of energy transfer through coupled windings. The primary and secondary coils form the basic system. The primary coil acts as a transmitter and the secondary coil are a receiver or load. The idea of an open secondary for Tesla's capacitive loads was highlighted in the mid-20th century [5]. More importantly, the favourable presence of the turn internal capacitance of the magnetic resonance receiving coil in Tesla's time would have helped him operate at frequencies of tens of megahertz. In recent years, there has been increasing interest in wireless technology.

Significant progress has been made, especially in the field of inductively coupled systems. The development of inductively coupled power transfer systems for a variety of applications has resulted in extremely high end-to-end system efficiencies of up to 80% [6]. Such examples include charging batteries for automotive systems and Wireless Sensor Networks. Transferring electricity wirelessly has emerged as a potential solution in recent years to address issues with energy and longevity in a sensor network.

5.2 Sensors

A device that generates an output signal with the intention of detecting some sort of physical occurrence is known as a sensor [8]. Sensors are utilised not just in commonplace objects such as tactile elevator buttons (touch sensors) and lights that dim or light up when the base is touched, but also in a wide variety of applications that the vast majority of people are completely unaware of. know. The use of sensors has expanded beyond the traditional areas of temperature, pressure, and flow measurement, for example to include (Magnetic, Angular Rate, and Gravity) (MARG) sensors [9, 10]. This expansion was made possible by advancements in micromechanics and platforms for microcontrollers that are simple to use.

Potentiometers and load sensors are two examples of analogue sensors that are still commonly used today [11, 12]. Their uses may be found in a wide variety of fields, including as manufacturing and machinery, aviation and aerospace, automotive, medical, and robotics, as well as many other facets of our everyday life. Other sensors, such as optical sensors [13] for measuring refractive index, vibration sensors [14] for measuring fluid viscosity, and electrochemical sensors [15] for monitoring the pH of the liquid, are also available for use in the process of determining the chemical and physical properties of materials. When something is being monitored, sensors are normally designed to have as little of an effect as possible; reducing the sensor smaller may frequently enhance this, and it may also have other advantages. It is necessary to employ an analogue signal to digital converter in order to transform an analogue sensor signal into a digital signal before it can be processed or used in a digital device.

The following is a list of the many kinds of sensors that are often used in a variety of applications [19]. All of these sensors can detect a wide variety of physical characteristics, including temperature, resistance, capacitance, conductivity, and heat transfer, amongst other things.

- Temperature Sensor
- Humidity Sensor
- Proximity Sensor
- Pressure Sensor
- Ultrasonic Sensor
- Light Sensor
- Touch Sensor
- Gas Sensor
- Colour Sensor
- Accelerometer

Recent advances in technology have made it feasible to manufacture an increasing number of micro-scale sensors, such as micro-sensors that use MEMS technology [16–18]. In the vast majority of situations, microsensors are capable of achieving substantially quicker measurement times and better sensitivity than macroscopic techniques. Disposable sensors are devices that are inexpensive and simple to operate; they are designed for short-term monitoring or one-time measurements. In today's world, there is a rising need for information that is rapid, economical, and dependable; as a result, their popularity has lately increased. With the help of this kind of sensor, anybody, everywhere, and at any time may acquire vital analytical information without the need for recalibration and without the concern that it will be contaminated.

A good sensor follows these rules:

- It has a high degree of sensitivity to the property that is being measured
- It is insensitive to any other attributes that could be encountered in the course of its application, and
- It does not have any effect on the attribute that was measured.

5.3 Actuators

A device that is capable of converting electrical impulses into mechanical action is called an actuator [20]. It is utilised to bring about motion or a change in the surrounding environment. Actuators are linked up to the output of the system. As input, it receives a signal from an electrical source, and as output, it generates a motion in a mechanical component. It receives input from a system or signal conditioning device (which is coupled to a sensor) and then transmits that information to the environment in the form of an output signal.

The encoder is responsible for providing the actuator with the data it needs [21]. The data is transmitted from the sensor to a signal conditioning device or system, which then processes the information and relays the results as orders to the actuators. A "temperature control system," in which the temperature is monitored by a temperature sensor, is an illustration of one of these kinds of systems. When the temperature reaches a certain threshold, the system issues a command to the fan (driver), instructing it to raise its rotational speed in order to bring the temperature down. One further illustration of this would be an automated door that swings open in response to the presence of a human.

Rotational or linear motion can be created by an actuator, depending on which one is selected [22]. Actuators [23] are used in the plants that make up industries in order to operate dampers, valves, fluid couplings, and other devices that are involved in the process of controlling industries. Actuators used in industry can be driven by air, hydraulic fluid, or electricity, depending on the kind of actuator. Actuators can be pneumatic, electrohydraulic, or electric, depending on their power source.

The main performance characteristics of the actuator are as follows [24–27]:

- Continuous working speed without limit number of starts per minute.
- Consistent performance and unaffected by load.
- The durable industrial design is able to function in severe situations without negatively affecting the performance of the product.
- Requires minimal routine maintenance.

5.4 Energy Requirement in Wireless Sensor Networks (WSNs)

Power consumption is the single most critical element in determining the lifespan of a sensor network [28–30]. This is due to the fact that sensor nodes are often powered by batteries. Energy optimization of sensor networks might be more difficult than it seems since it not only lowers the amount of energy that is consumed, but it also aims to make the network's lifespan as long as it possibly can [28]. Awareness of energy use across the whole design and operating processes is necessary for optimization to be realised. This guarantees that awareness of energy flows not only to individual nodes, but also to the group of sensor nodes that interact with one another and the complete network.

The design of a WSN should prioritise reducing the amount of power that is consumed by wireless sensor nodes in order to maximise the lifespan of the network [30]. The low power WSN regulates the busy active time or "wake time" of the device (such as a radio or microcontroller), and it also limits the current consumption while the device is "sleeping." This is done in order to lower the total power consumption of the system. Limit power. These networks accomplish this by modifying the power setting mode of the device, such as by switching it to an always-on, standby, or hibernation state, for example. The node and its components are capable of being utilised in a variety of power setting modes, which may be selected according to your preferences. As a result of the fact that energy policy varies depending on the application, there are some programmes that let you conserve power by turning off some of the nodes in the network, while others let all of the nodes operate at the same time. is needed. Assists in operating the WSN in a mode that conserves power.

A sensor node typically consists of four subsystems [31–33]:

1. **Computing Subsystem**: Composed of a microprocessor (also known as a microcontroller unit or MCU), which is responsible for operating sensors and putting communication protocols into effect. In order to better control their power consumption, MCUs often operate in a variety of modes. Because these modes of operation require power, you must take into account the power consumption values of the various modes when calculating the amount of time that each node will remain powered by its battery.
2. **Communication Subsystem**: Consists of radios with a limited range that allow for communication with neighbouring nodes as well as the outside world. Different modes of operation are available for the radio. When not sending or receiving, it is essential to power off the radio entirely rather than letting it sit in an inactive state so as to preserve the life of the battery.
3. **Sensing Subsystem**: Consists of a collection of sensors and actuators that serve to establish a connection between the node and the outside environment. Utilizing components with a low power consumption and conserving energy at the expense of power that is not needed are two ways to cut down on energy use.
4. **Power Subsystem**: Comprises a battery that serves as the source of power for the node. Check the amount of current that is being pulled from the battery to see how quickly it is being depleted even if the battery is designed to last for a long time. This is important to do even if the battery is designed to last for a long time. In most cases, the minimal power consumption is lower than the rated current capacity of the battery that is utilised for the sensor node. It is possible to lengthen the life of a battery by considerably lowering the current or by shutting it off more frequently.

To reduce the amount of energy that is used by sensor networks as a whole, several protocols and algorithms of varying sorts have been the subject of research on a global scale. One is able to considerably lengthen the lifespan of the sensor network by ensuring that the application layer, operating system, and network protocols are all designed with an awareness of energy consumption. These protocols and algorithms need to be able to identify the hardware and use the unique characteristics of microprocessors and transceivers in order to reduce the amount of power that sensor nodes use. This offers a bespoke answer for the many distinct sensor node designs that are out there. There are several varieties of sensor networks due to the numerous varieties of installed sensor nodes. In the realm of wireless sensor networks, this can also result in the development of a variety of collaborative algorithm types.

5.5 Wireless Sensor Network and Green IoT (G-IoT)

The proliferation of Internet of Things (IoT) technology and its incorporation into smart cities have altered the way in which we work and live, as well as contributed to the enrichment of our society [34]. The Internet of Things does bring certain issues, however, including an increase in energy consumption, the creation of hazardous pollutants, and the accumulation of e-waste in smart cities [35, 36]. Applications for smart cities are required to be safe for the environment, which necessitates a shift toward green internet of things. A more eco-friendly and sustainable environment may be achieved via the use of green internet of things. As a result, it is essential to work on developing technologies and strategies that can reduce the risks associated with pollution, waste generated by traffic, resource consumption, and energy consumption, as well as ensure public safety, quality of life, and environmental protection, and keep costs under control.

The green Internet of Things refers to a collection of technologies that include green RFID tags, green sensor networks, and green cloud computing networks [37]. A wireless sensor network (WSN) is a collection of numerous sensor nodes that share limited power and storage capacity. Achieving the Green Internet of Things (IoT) requires the successful implementation of an essential component known as the Wireless Sensor Network (WSN).

The advantages of the green WSN are:

- Renewable energy sources are charged and used.
- Optimizing energy consumption with an energy efficient strategy.
- To minimize the quantity of data and increase storage capacity, you should use algorithms that are data and context aware.

- To lower the amount of power needed for mobility, implement a routing strategy that is efficient with energy.

The green internet technology offered by the hardware solution for low power consumption devices needs to have both hardware and software aspects taken into consideration, and this needs to be done without sacrificing performance. On the other hand, software solutions provide designs that are both resource and energy efficient while simultaneously maximising utilisation of those resources. In addition to that, the method that you choose for the virtual machine needs to be one that is efficient with energy. The Internet of Things infrastructure includes smart cities, intelligent energy and grid systems, intelligent infrastructure, intelligent manufacturing, intelligent healthcare systems, and intelligent logistics systems.

5.6 Purpose of G-IoT

The Internet of Things (IoT) revolution is driving innovation in all fields of science and human life by making new information accessible anytime, anywhere, and by connecting every person, process, data, and thing, as well as locations, organisations, and facilities, in a way that has never been done before [34]. In spite of the many benefits that the Internet of Things (IoT) has to offer, the manufacturing, distribution, and utilisation of IoT devices and systems are extremely resource- and energy-intensive, which results in significant amounts of solid waste. as well as a growing threat.

It is critical to effectively address issues like rising energy consumption, waste generation, greenhouse gas emissions, and the use of natural resources and non-renewable raw materials in order to reduce the potential negative effects of technological advancement on people and the environment [38, 39]. Future energy will be clean. In order to achieve a sustainable smart society, it is imperative that technology, IoT, and economy be replaced with green alternatives. Green technology, green IoT, and a green economy have the potential to greatly enhance human health.

By utilising technologies, digital strategies, and solutions that simultaneously eliminate or minimise negative effects on human health and the environment, Green IoT has shown to have the potential to transform many fields and bring many benefits (including environmental protection, increased customer satisfaction, and increased profits) [40–42].

In the future, with the combination of sensors, IoT devices as well as 5G network connectivity and the strong support of AI [43], it will bring users green solutions [44]. This means that IoT devices will also be context-aware and able to perform certain functions, suggesting green forms of communication between people and things and with each other. Powering up in the design phase will soon make Green IoT ubiquitous. In fact, IoT devices must sleep when not in use and use routing algorithms while exchanging data.

Currently, the areas that Green IoT is focusing on include green RFID technology [45], green wireless sensor network technology [44], green cloud technology [46], green machine-to-machine technology [47], green data, green communication and network, green internet technology [48]. In the future, the potential of green IoT will increase if the following improvements are made [39, 49]:

- Radiofrequency energy harvesting should be considered to establish an energy balance to facilitate green connectivity [26], between IoT devices.
- In order to popularize green IoT devices, it is necessary to reduce the energy used by devices, while finding new and alternative resources.
- Finding the right technology techniques to improve QoS parameters [28], (bandwidth, throughput and latency) will help green IoT effectively and efficiently.
- M2M communication is crucial for reducing energy use. To enable automated processes, intelligent machines must become smarter [21]. Where rapid action is required, machine automation lag should be minimized.
- In the not too distant future, drones will take the place of a significant [17] number of Internet of Things devices, which will contribute to a reduction in both energy consumption and environmental pollution [20]. One such technology that might pave the way for an affordable, high-performance, and environmentally friendly Internet of Things is drone technology.
- Perceiving data and context. Collecting data without context unnecessarily consumes a lot of resources [32]. One of the toughest challenges going forward will be preserving data context so that future data processing makes more sense [33].

5.7 Motivation

In today's environment, when traditional wiring and distribution networks are struggling to keep up with demand, wireless power transfer is an absolute necessity [27, 50]. It eliminates all of the wires in order to make the power transmission and distribution system more straightforward and aesthetically pleasing. A system that efficiently transfers energy from one location to another through a vacuum or environment without the need for wires or other materials is referred to as wireless power transfer (WPT) [14]. This is the definition of WPT. Applications involving the transfer of energy provide a great number of challenges, in contrast to the fact that the transmission of information by radiation is an ideal method. Energy transmission is particularly inefficient when radiation is omnidirectional [23], and one-way radiation requires an unbroken line of sight as well as a sophisticated tracking device. Recent theoretical publications have provided a comprehensive investigation of the practicability of transferring medium-range energy utilising resonant objects linked through the edges of their non-radiative fields [18]. Two resonant objects that have the same resonant frequency have a natural tendency to exchange energy in an effective manner, whereas an external non-resonant item tends to use up a comparatively little amount of energy [20]. A "highly coupled" operating regime is typically seen in the behaviour of systems that feature coupled resonances. If a particular system can be run in this regime, it is anticipated that energy transfer will take place at a very high efficiency [21]. When carried out in this manner, mid-range power transmission is essentially omnidirectional, efficient, and unaffected by the contours of the surrounding space. Furthermore, there is less interference and loss caused by the objects in the surrounding area.

5.8 Contribution

The concept of wireless power transfer through various means is explored in this paper. Energy transferring schemes of different kinds are presented [33]. Wireless sensor networks are explored in detail, by reviewing their components, applications, and characteristics, along with their energy requirements for various subsystems and significant contribution to Green IoT. The need for this new system is reviewed against the drawbacks of traditional power transfer through cables [35]. Advantages of this system are studied, and challenges are presented with suggestions for future work and improvement.

5.9 Need of the Work

Thomas Edison is credited with being the inventor of electricity in 1882, when he constructed the first DC generators, radial line transmissions, and twisted copper wires that were used to power end consumers. Since that time, the same power grid technology that is based on copper has been used to distribute energy to large urban regions as well as industrial districts. Therefore [24, 25], the fact that something that has been in use for almost 141 years has a basic weakness may come as a surprise to some people. But there are many geographical obstacles and hurdles with evident subtleties that occur as a result of the fact that the grid employs enormous infrastructure regions to power both heavily inhabited and undeveloped areas [26]. This is the source of the problem. There is a prohibition on routing power lines near occupied or protected places. Being too near has a multitude of hazards, some of which might put one's life or health in jeopardy [28, 29]. In addition, it has been demonstrated that the production of substations and the maintenance of power lines both have a significant potential for the release of carbon emissions into the environment. The majority of people are aware that increasing levels of carbon dioxide in the atmosphere are one factor that contributes to the warming of the planet.

Although these power grids do provide electricity to the contemporary world, the seeming lack of accessibility and mobility [14], as well as the detrimental effects on the environment, need the development of new systems in order for society to be able to satisfy its expectations. Unfortunately [16, 18], the solutions that have been presented up until this point have fundamental limitations. Edison's approach to powering the globe was pitted against Nikola Tesla's, and the case for a new and improved electrical power distribution system has been closed and is finishing as it began when Edison's idea was challenged. The use of wireless power transfer, often known as WPT [23], as a method of power transfer is becoming increasingly used in a variety of systems. Despite the fact that its concepts are often thought to be conventional, they never fail to astound.

The WPT technology that has previously been developed often has a limited range. Despite this, the expansion of WPT will result in the development of ground-breaking technologies for transmission across medium and long distances [23–26]. Applications for "proximity" or "short-range" wireless power transmission include radio frequency identification, biomedical applications, and wireless charging, to name a few of the more common uses for this technology [16]. In point of fact, it is possible to charge certain common electronic devices, such as watches, ear buds, headphones, and tablets, wirelessly [25, 28]. Not just in the medical industry, but also in the military, where there is already a requirement for the creation of wireless networks that employ lasers to transport electrical energy, WPT for midrange applications is gaining more and more popularity. This is true for both of these industries.

25% of all carbon dioxide emissions come from the production of electricity and heat, but present distribution systems may be turned wireless by utilising the high-efficiency microwave technology that is now available and the whole spectrum of WPTs [12, 13]. increase. Future towers might receive electricity from neighbouring PV or industrial sites and output power at boost levels of up to 100 MW in these non-contact wireless power distribution systems [18]. With up to 90% efficiency, it incorporates short-range systems [14], from which medium-range systems send their energy to demonstration sites outfitted with digital buses for long-distance transmission [17]. WPT enables unlimited development of green energy power plants without the hassle of logistics and beyond all the requirements of conventional networks by merging the present advantages of wind, solar, and hydropower systems [18]. Additionally, replacing a current power grid with a wireless power system only entails swapping the cable for a wireless connection or a digital bus that supports a variety of use cases. Most crucially [19], WPT opens the door to the creation of a wireless local area network (WiGL) that may transmit electrical energy to end users in the same manner as WiFi transmits wired data.

5.10 Energy Transferring Schemes in WSAN

Wireless power transfer systems are basically the principle of wireless power transfer. The first principle was WiTricity [22], which stands for wireless power. Magnetic resonance is the foundation for his idea. In situations where a connecting line would be difficult, hazardous, or impossible, wireless communication makes sense [23]. High grade materials can boost the effectiveness of conventional power transmission systems, albeit at a substantial cost premium. Direct induction [7] is the most popular way to transmit wireless power, followed by resonant magnetic induction [26]. Other techniques include using electric wire with natural medium and electromagnetic radiation in the form of microwaves or lasers. Power consumption rises daily as a result of rising demand [34–37], which also causes power loss to rise. The transmission and distribution of power account for the majority of losses. This technique results in a loss of roughly 26% [35]. Figure 5.1 depicts various types of wireless energy transfer techniques.

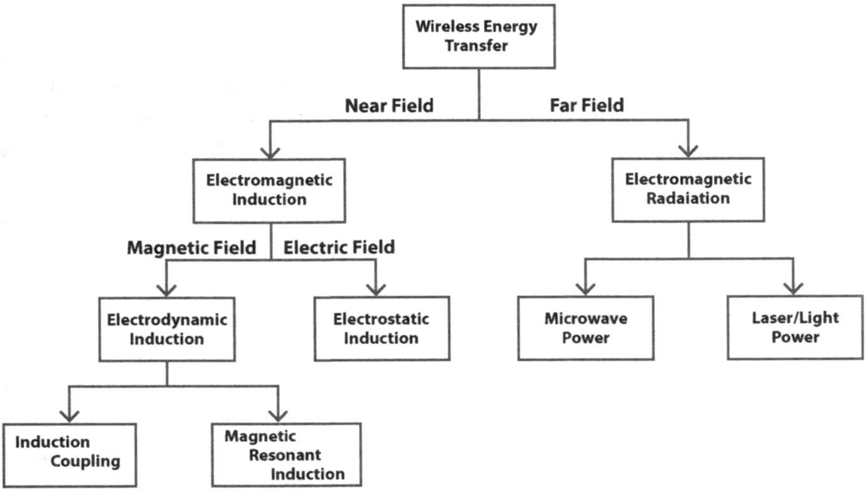

Figure 5.1 Classification of wireless transfer techniques [52, 53].

5.11 Electromagnetic Induction

There are several types of wireless power transfer (WPT), each with its own uses for sources, technologies, frequencies, and operating distances. Among them, the one that transfers electricity from a transmitting coil to a receiving coil utilising the principles of magnetic induction is becoming more and more significant [14, 15]. The transmitting coil is often subjected to a switching electric current, which creates a magnetic field at a predetermined frequency [13]. An electric current is produced in the receiving coil when it is positioned inside the magnetic field of the transmitting coil [16]. A battery can be charged or a DC circuit powered by rectifying this switching current into a DC voltage.

The spacing between the coils has a direct impact on how effectively electricity is transferred. As the distance widens [18, 19], the magnetic field gets progressively weaker. The magnetic resonance theory is used in the WPT approach [17], which is the most effective. According to this theory, the transmitting coil's total inductance and capacitance are used to compute the resonant frequency. When the receiving coil and the transmitting coil are both tuned to the same frequency, the coils will couple and be able to function across a larger distance [13–15]. These electromagnetic principles are of importance to both the medical professionals and their patients as a means of powering the interior medical equipment [17]. The following four categories of medical implanted microsystems (MIMs) will be discussed in this research article: implants, pumps, ultrasound imaging, and gastrointestinal (GI) endoscopy. The main drawback of the majority of AIMDs is their short battery life [19–21]. The battery may need to be replaced through invasive surgery. By using WPT techniques, surgery is avoided and external battery charging is made possible.

In the medical sector, mechanical pumps are used to transfer gases or liquids inside the body. The Ventricular Assist Device (VAD) [11–16] which takes the role of a damaged ventricle's blood pumping action, is a well-known medical pump [23]. These pumps may be driven by electromagnetic induction rather than an external battery as the VAD requires a steady stream of electricity [26]. A Doppler Flow metre watches the blood flow through a vascular graft using ultrasound imaging to look for possible failures. These clotting-related problems in the graft call for prompt replacement [41, 42]. An implanted Doppler flow metre system's single pacemaker battery has a 5-year lifespan. The battery must then be replaced during surgery. Invasive procedures can be avoided by utilising the basic principles of induction to power the circuit or charge the battery [38]. The gold standard for examining and identifying issues with the GI tract is an endoscopy.

At 300 kHz and 13.56 MHz, experiments were carried out in four distinct configurations. The tests used several coil turns and capacitor combinations to calculate each efficiency.

- In the first design, there are no organic tissues to divide the transmitting and receiving coils [26].
- The receiving coil is encased in living tissue in the second arrangement [25].
- The third design embeds a titanium pacemaker casing with a receiving coil into living tissues.
- The fourth option inserts the pacemaker case's receiving coil into the tissue before inserting a 1 mm ferrite shield nearby [25].

Figure 5.2 depicts a structure of a transcutaneous energy transfer (TET) system. In the process of electromagnetic induction, a conductor is placed in a certain location while the magnetic field changes, or the magnetic field is stationary while the conductor is moving. A voltage or electromotive force (EMF) [35–38] is created across the electrical conductor as a result. In 1830, Michael Faraday made the Law of Induction his own.

When electricity flows through a coil of metal wire, transient magnets called electromagnets are produced [42]. The magnetic field is created at the core or

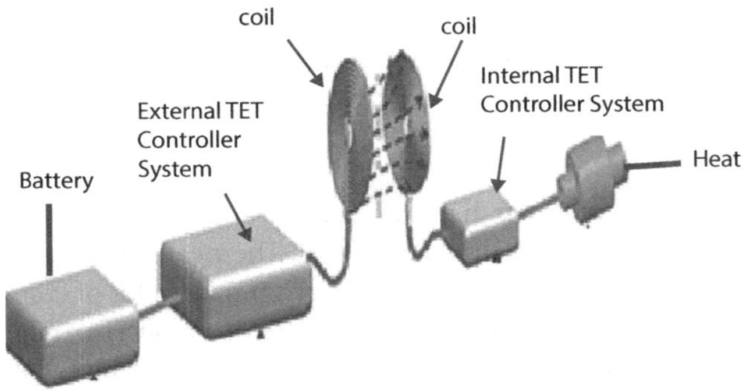

Figure 5.2 Basic structure of a TET system [55].

centre of the metal coil when electricity flows through it. By placing an iron rod in the coil's centre, this magnetic field may be strengthened. Iron metal naturally retains the magnetic field [45, 46], therefore inserting an iron rod at the centre makes it into a strong magnet, although it is only a transient magnet.

The iron rod with a strong magnetic core degrades into a regular iron rod when the current is switched off [42]. These magnets vary from permanent magnets in that the magnetic field is produced by electricity, whereas permanent magnets do not [44]. By altering the amount of electric current flowing through the coil, the magnetic field intensity produced by the current's passage may be changed [45].

Let us say you decide to purchase without cash and your parents pay using cards. Always swiped or scanned by the store clerk. The shopkeeper does not touch or photograph the card [32]. Why does he scan it or swipe it? And how does the money get taken off the card when it is swiped? Because of "electromagnetic induction," this occurs.

A. Applications of Electromagnetic Induction

1. Electromagnetic induction in AC generator
2. Electrical Transformers
3. Magnetic Flow Meter

Advantages:

Where the magnetic field must be interrupted and its poles changed, electromagnets are highly helpful [21–23]. This characteristic of electromagnets is effectively used in a magnetic separator. It draws iron scraps from a pile of scrap while current is flowing through it [25], and its operator transports the scrap to another location and cuts off the current supply to dump them. We acquire iron scrap by using magnetic separator.

Electromagnets also have the advantages of being affordable and simple to construct. They do not harm the test piece [30], which is a component of the electromagnet, and are lightweight.

Disadvantages:

The fact that electromagnets heat up quickly and waste a lot of electrical energy as a result is one of its drawbacks [18]. The steady magnetic field must be maintained for a continuous power supply. It takes a lot of coiling of copper wire, which takes up a lot of area [14], to create a powerful magnetic field. Electromagnets are therefore inappropriate for compact places. Since electromagnets' electric current produces the magnetic field [21], a short-circuit can seriously hurt the operator and the electromagnets themselves.

5.11.1 Electrodynamic and Electrostatic

The study of electric charges at rest is the focus of the physics subfield known as electrostatics. The terms "electro" and "static," which stand for stationary or resting position and electrical charges respectively [23], are combined to form the phrase "electrostatic." The phenomenon of a force of attraction or repulsion between two stationary charges gives birth to the name "electrostatic" [24].

Two other phrases that briefly describe it are created by the electrostatics [26]. Electrostatic force and electrostatic field are these concepts.

5.11.2 Electrostatic Field

The area around a static charge where the influence of the charge is perceived is known as the electrostatic field [32–36]. If another charge is placed near a stationary charge, whether positive or negative, that charge will either attract to the stationary charge or push it away.

Lines of force or field lines are used to represent the electrostatic field [33–35]. They depict the direction that an inside positive charge might take. A positive charge starts and a negative charge ends the electric field line produced by a stationary charge [28]. Therefore, the electric field lines of a stationary positive charge point outward, whereas the electric field lines of a stationary negative charge point inwards.

5.11.3 Electrostatic Force

Electrostatic force is the attraction or repulsion between two charges that are at rest [12]. A force of attraction exists between two differing charges as opposed to a force of repulsion between two identical charges [15].

Positive charges repel positive charges, while negative charges repel negative charges, to put it more properly [17]. Positive and negative charges are drawn to/attracted to one another.

5.11.4 Electromagnetic

The study of the electromagnetic field generated by charges in motion is the focus of the area of physics known as electromagnetism [19, 20].

The terms "electro" and "magnetic," which denote charges or electricity and magnets respectively, are combined to form the phrase electromagnetic [22]. Thus, the link between electricity and magnetism is described by electromagnetism.

5.11.5 Electromagnetic Field

Electric field (also known as electrostatic field) and magnetic field are combined to form the electromagnetic field [26]. In addition to its own electric field, a charge moving within a conductor also creates a magnetic field [27]. Therefore, the electrostatic field and the electromagnetic field can be said to be supersets of each other. The magnetic and electric fields coexist and are perpendicular to one another.

Engineering frequently makes use of electromagnetic fields, and machines mostly operate according to the electromagnetic induction theory [14]. As is common knowledge, magnets have their own magnetic fields, which are uncontrollably constant in strength [17]. The strength of the electromagnetic field may be readily controlled, whereas the electromagnetic field relies on the amount of charges.

5.12 Inductive Coupling

A device known as an inductive coupler transmits 24-volt DC electricity as well as 24-volt DC input and output signals over an air gap between its two halves [43]. The base and the remote are the parts of an inductive coupler. Not every base component works with every distant component. Before connecting any wires [45], you must ensure that the components of your inductive coupler are intended to function as a unit.

One technique that has been applied in wireless power transfer systems is inductive coupling [51]. This approach employs the mutual induction notion [40], exactly like electrical transformers, making it the most straightforward method out of all the others. This method employs two distinct coils, L1 as the first coil and L2 as the second coil [11–13]. The first coil, L1, will be attached to the source, acting as the transmitter, and the second coil, L2, will be attached to the load, acting as the receiver [12–14]. The electromagnetic wave produced by the first coil will be used to transport power from L1 to L2. This is because the magnetic flux will be created as the current passes through the conductor (coil). As a result, the magnetic flux generated by the first coil will migrate to the second coil, L2, and cut it. Electric current will induce when the magnetic flux cuts the conductor, according to the fundamentals of electromagnetic induction. Following that, an AC current will start to flow to the load. If the load needs a DC supply, the rectifier will change the AC current to a DC current [15].

5.13 Resonance Inductive Coupling

The electroinductive effect or resonant inductive coupling is superior than inductive coupling. This technique allows for longer induction times and more power transmission between the two coils, L1 and L2.

Utilizing resonance to communicate effectively across distance. Everything in the cosmos has its own inherent frequency, and resonance occurrences will happen when their natural frequencies coincide [16]. The amplitude will oscillate at its highest level during this phenomenon, which indicates that the item will get the greatest amount of energy transmission [13].

The electromagnetic wave will not be wasted or absorbed with this approach since it is not travelling through the air as it would otherwise [11]. Researchers at MIIT have created a resonant inductive coupling technique prototype that can transfer power over a 5-m distance without disrupting other electronic systems or harming living creatures [13], they claim. In other words, it is safe for the environment, can make many industries' jobs easier, and can save costs.

Two independent coils- the transmitter coil and the receiver coil - are adjusted to the same resonant frequency when using the resonant inductive coupling technique [15]. The notion of mutual inductance and the changing magnetic field created by the transmitter coil are what determine how much electricity is produced at the reception coil [16]. The receiver coil will be cut off by the fluctuating magnetic field that was created, and current will be generated there.

This system's Q factor has to be raised at the LC circuit in order to boost efficiency. Lowering the value of the resistance or capacitance would raise the Q factor [11–13]. By employing thicker wire or high-quality winding wire for the coil, the resistance value can be reduced.

5.14 Wireless Power Transmission Using Microwaves

Microwaves are electromagnetic waves with wavelengths between 0.1 mm and 10 cm, which are frequently utilised for communications (frequencies between 0.1 and 100 GHz) [12–15]. It is possible to transfer microwave-format electric power from space to the ground using antenna arrays made up of multiple antenna components [13]. One may construct a certain beam shape and point a finely focused beam in any direction by adjusting and synchronising the phases and amplitudes of the microwaves delivered from each antenna element of the array [17].

Power Transfer Through Microwaves Has Advantages

- Cloud cover and rain are not a barrier to microwaves operating at frequencies lower than 10 GHz [8].
- Microwaves are safer than lasers because they have a lower energy density.

Challenges

- Increasing the microwave beam-pointing control's precision.
- Boost microwave conversion efficiency from DC power (in space).
- Enhanced the microwave to DC power conversion efficiency (on the ground).
- Make the electronic modules lighter and smaller.

The losses that occur during the transmission and distribution of electrical power are one of the main problems in the power system [8–10]. The power generation grows and the power loss likewise increases as the demand rises daily. Power loss happens most frequently during transmission and distribution [9]. A rough estimate of 26% power loss occurs during transmission and distribution. The resistance of grid wires is the primary cause of power loss during transmission and distribution [10–13]. India's energy infrastructure has the largest transmission and distribution losses in the world, according to the World Resources Institute (WRI), at a staggering 27% [5]. This percentage has been estimated at 30%, 40%, and even more by a variety of Indian government bodies. Technical losses (inefficiencies in the grid) and theft are blamed for this [2].

By selecting an alternate power transmission method that offers significantly greater efficiency, lower transmission costs, and prevents power theft [40–42], the aforementioned issue can be resolved. One of the next technologies, microwave power transmission, may be the just replacement for effective power transfer.

William C. Brown, a pioneer in wireless power transmission technology, has conceived, constructed, and tested a device to illustrate how electricity may be sent via free space using microwaves [41]. Figure 5.3's functional block diagram may be used to describe the Wireless Power Transmission System idea.

The microwave power source on the transmission side generates microwave power, while electronic control circuits manage the amount of microwave power that is sent [43]. The wave guide ferrite circulator is associated with the microwave power source by way of the Coax to Waveguide Adaptor [46]. This adaptor protects the microwave source from the power that is reflected back to it. The impedances of both the transmitting antenna and the microwave source are brought into alignment by the tuner [47]. After then, the Directional Coupler will split the attenuated signals according to the direction in which the signal is being transmitted. The power that is transmitted from the transmitting antenna into empty space is received

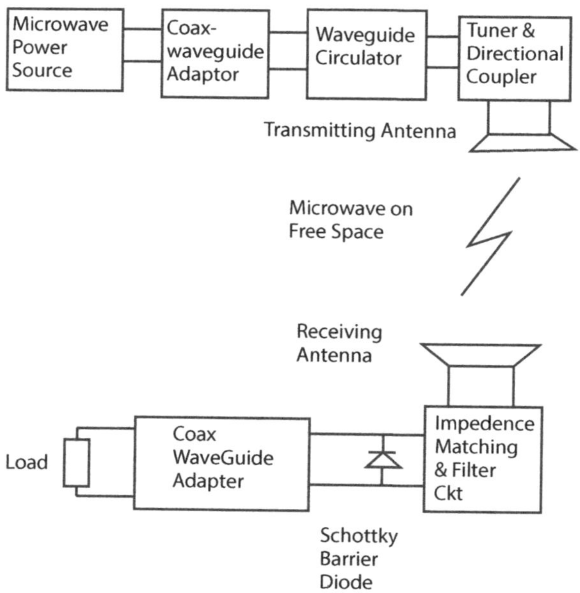

Figure 5.3 Functional block diagram of wireless power transmission [54].

in an equal amount by the receiving antenna [43]. At the receiving end of the device is a rectenna, which is responsible for receiving the power that is being sent and converting the microwave power into direct current power.

5.15 Electromagnetic Radiations

Electric and magnetic disturbances that oscillate or electrically charged particles moving through a vacuum or material create electromagnetic radiation, a kind of energy [21]. The combined wave travels in a direction that is perpendicular to both the magnetic and electric oscillating fields [24]. This is because the electric and magnetic fields, which are responsible for producing the disruption, are at right angles to one another. The emission of electrons results in the production of photons [26], which are packets of light energy that travel as quantized harmonic waves at the speed of light [29]. The electromagnetic spectrum is then utilised to classify this energy according to its wavelength in order to better understand it [31]. These electric and magnetic waves each have their own unique characteristics, including amplitude, wavelength, and frequency, and they move in a direction that is perpendicular to one another.

All electromagnetic radiation has the following general properties:

1. Electric vacuum may be traversed by electromagnetic radiation. Most other waves must pass through a material in order to propagate. For instance, in order to be heard, sound waves must pass through a gas, solid, or liquid.
2. The speed of light remains constant at all times. (Lightning speed: 2..99792458 x 108 m s-1)
3. Wavelengths are calculated as the separation between crests and troughs. It is typically represented by the Greek letter.

5.16 Conclusion

The purpose of this study was to highlight some significant characteristics of WSN with regard to application, design, and technology. When developing a WSN, it is necessary to take into consideration a variety of aspects including flexibility, power efficiency, fault tolerance, high fidelity, cheap cost, and quick deployment; nevertheless, application requirements are the most significant of these criteria. In the future, we anticipate that sensor networks will become an indispensable component of our lives as a result of applications in a wide variety of sectors.

References

1. Mohammed, S.S., Ramasamy, K., Shanmuganantham, T., Wireless power transmission–a next generation power transmission system. *Int. J. Comput. Appl.*, *1*, 13, 100–103, 2010.
2. García-Hernández, C.F., Ibarguengoytia-Gonzalez, P.H., García-Hernández, J., Pérez-Díaz, J.A., Wireless sensor networks and applications: A survey. *IJCSNS Int. J. Comput. Sci. Netw. Secur.*, *7*, 3, 264–273, 2007.
3. Ding, Z., Zhong, C., Ng, D.W.K., Peng, M., Suraweera, H.A., Schober, R., Poor, H.V., Application of smart antenna technologies in simultaneous wireless information and power transfer. *IEEE Commun. Mag.*, *53*, 4, 86–93, 2015.
4. Reddy, M.V., Hemanth, K.S., Mohan, C.V., Microwave power transmission–a next generation power transmission system. *IOSR J. Electr. Electron. Eng. (IOSR-JEEE)*, *4*, 5, 24–28, 2013.
5. Lee, C.K., Zhong, W.X., Hui, S.R., Recent progress in mid-range wireless power transfer, in: *2012 IEEE Energy Conversion Congress and Exposition (ECCE)*, 2012, September, IEEE, pp. 3819–3824.
6. Kalwar, K.A., Aamir, M., Mekhilef, S., Inductively coupled power transfer (ICPT) for electric vehicle charging–a review. *Renewable Sustainable Energy Rev.*, 47, 462–475, 2015.
7. Waffenschmidt, E. and Staring, T., Limitation of inductive power transfer for consumer applications, in: *2009 13th European Conference on Power Electronics and Applications*, 2009, September, IEEE, pp. 1–10.
8. Tilak, S., Abu-Ghazaleh, N.B., Heinzelman, W., A taxonomy of wireless micro-sensor network models. *ACM SIGMOBILE Mob. Comput. Commun. Rev.*, *6*, 2, 28–36, 2002.
9. Marins, J.L., Yun, X., Bachmann, E.R., McGhee, R.B., Zyda, M.J., An extended Kalman filter for quaternion-based orientation estimation using MARG sensors, in: *Proceedings 2001 IEEE/RSJ International Conference on Intelligent Robots and Systems. Expanding the Societal Role of Robotics in the the Next Millennium (Cat. No. 01CH37180)*, 2001, October, vol. 4, IEEE, pp. 2003–2011.
10. Hu, J.S. and Sun, K.C., A robust orientation estimation algorithm using MARG sensors. *IEEE Trans. Instrum. Meas.*, *64*, 3, 815–822, 2014.
11. Guoqing, Y., Zhanyou, S., Guochen, A., Application of digitally controlled potentiometers in the intelligent sensor's measure and control system,

in: *2007 8th International Conference on Electronic Measurement and Instruments*, 2007, August, IEEE, pp. 1–440.
12. Wu, Y., Liu, B., Wu, J., Zhao, L., Sun, T., Mao, Y., Wang, J., A transverse load sensor with ultra-sensitivity employing Vernier-effect improved parallel-structured fiber-optic Fabry-Perot interferometer. *IEEE Access, 7*, 120297–120303, 2019.
13. Gauglitz, G., Direct optical sensors: Principles and selected applications. *Anal. Bioanal. Chem.*, 381, *1*, 141–155, 2005.
14. Turner, R.C., Fuierer, P.A., Newnham, R.E., Shrout, T.R., Materials for high temperature acoustic and vibration sensors: A review. *Appl. Acoust.*, *41*, 4, 299–324, 1994.
15. Bakker, E. and Telting-Diaz, M., Electrochemical sensors. *Anal. Chem.*, *74*, 12, 2781–2800, 2002.
16. Fujita, H., A decade of MEMS and its future, in: *Proceedings IEEE the Tenth Annual International Workshop on Micro Electro Mechanical Systems. An Investigation of Micro Structures, Sensors, Actuators, Machines and Robots*, 1997, January, IEEE, pp. 1–7.
17. Wang, M., Ge, D., Zhang, L., Herder, J.L., Micro-scale realization of compliant mechanisms: Manufacturing processes and constituent materials—A review. *Chin. J. Mech. Eng.*, *34*, 1, 1–22, 2021.
18. Mehmood, Z., Haneef, I., Udrea, F., Material selection for optimum design of MEMS pressure sensors. *Microsyst. Technol.*, *26*, 9, 2751–2766, 2020.
19. Bernauer, F., Wassermann, J., Igel, H., Rotational sensors—A comparison of different sensor types. *J. Seismolog.*, *16*, 4, 595–602, 2012.
20. Algamili, A.S., Khir, M.H.M., Dennis, J.O., Ahmed, A.Y., Alabsi, S.S., Hashwan, S.S.B., Junaid, M.M., A review of actuation and sensing mechanisms in MEMS-based sensor devices. *Nanoscale Res. Lett.*, *16*, 1, 1–21, 2021.
21. Kashiri, N., Malzahn, J., Tsagarakis, N.G., On the sensor design of torque controlled actuators: A comparison study of strain gauge and encoder-based principles. *IEEE Rob. Autom. Lett.*, *2*, 2, 1186–1194, 2017.
22. Claeyssen, F., Lhermet, N., Le Letty, R., Bouchilloux, P., Actuators, transducers and motors based on giant magnetostrictive materials. *J. Alloys Compd.*, *258*, 1–2, 61–73, 1997.
23. Tzou, H.S., Lee, H.J., Arnold, S.M., Smart materials, precision sensors/actuators, smart structures, and structronic systems. *Mech. Adv. Mater. Struct.*, 11, 4–5, 367–393, 2004.
24. Bell, D.J., Lu, T.J., Fleck, N.A., Spearing, S.M., MEMS actuators and sensors: Observations on their performance and selection for purpose. *J. Micromech. Microeng.*, *15*, 7, S153, 2005.
25. Seferagić, A., Famaey, J., De Poorter, E., Hoebeke, J., Survey on wireless technology trade-offs for the industrial internet of things. *Sensors*, *20*, 2, 488, 2020.

26. Gunatilaka, D., High performance wireless sensor-actuator networks for industrial Internet of Things. Ph.D. Thesis, Washington University, St. Louis, MO, USA, 2019.
27. Perwej, Y., Omer, M.K., Sheta, O.E., Harb, H.A.M., Adrees, M.S., The future of Internet of Things (IoT) and its empowering technology. *Int. J. Eng. Sci.*, 9, 3, 20192, 2019.
28. Mazunga, F. and Nechibvute, A., Ultra-low power techniques in energy harvesting wireless sensor networks: Recent advances and issues. *Sci. Afr.*, *11*, e00720, 2021.
29. Wei, Q., Bai, K., Zhou, L., Hu, Z., Jin, Y., Li, J., A cluster-based energy optimization algorithm in wireless sensor networks with mobile sink. *Sensors*, *21*, 7, 2523, 2021.
30. Chen, Z., Sivaparthipan, C.B., Muthu, B., IoT based smart and intelligent smart city energy optimization. *Sustainable Energy Technol. Assess.*, *49*, 101724, 2022.
31. Jondhale, S.R., Maheswar, R., Lloret, J., Fundamentals of wireless sensor networks, in: *Received Signal Strength Based Target Localization and Tracking Using Wireless Sensor Networks*, pp. 1–19, Springer, Cham, 2022.
32. Karray, F., Jmal, M.W., Garcia-Ortiz, A., Abid, M., Obeid, A.M., A comprehensive survey on wireless sensor node hardware platforms. *Comput. Networks*, *144*, 89–110, 2018.
33. Yick, J., Mukherjee, B., Ghosal, D., Wireless sensor network survey. *Comput. Networks*, *52*, 12, 2292–2330, 2008.
34. Albreem, M.A., El-Saleh, A.A., Isa, M., Salah, W., Jusoh, M., Azizan, M.M., Ali, A., Green Internet of Things (IoT): An overview, in: *2017 IEEE 4th International Conference on Smart Instrumentation, Measurement and Application (ICSIMA)*, 2017, November, IEEE, pp. 1–6.
35. Albreem, M.A., Sheikh, A.M., Alsharif, M.H., Jusoh, M., Yasin, M.N.M., Green Internet of Things (GIoT): Applications, practices, awareness, and challenges. *IEEE Access*, *9*, 38833–38858, 2021.
36. Liu, L. and Mishra, A.R., Enabling technologies challenges of green Internet of Things (IoT) towards sustainable development in the era of Industry 4.0. *Technol. Econ. Dev. Econ.*, 1–32, 2022.
37. Ahmad, R., Asim, M.A., Khan, S.Z., Singh, B., Green IoT—Issues and challenges, in: *Proceedings of 2nd International Conference on Advanced Computing and Software Engineering (ICACSE)*, 2019, March.
38. Solanki, A. and Nayyar, A., Green Internet of Things (G-IoT): ICT technologies, principles, applications, projects, and challenges, in: *Handbook of Research on Big Data and the IoT*, pp. 379–405, IGI Global, Hershey, Pennsylvania, 2019.
39. Almalki, F., Alsamhi, S.H., Sahal, R., Hassan, J., Hawbani, A., Rajput, N.S., Breslin, J., Green IoT for eco-friendly and sustainable smart cities: Future directions and opportunities. *Mob. Netw. Appl.*, 1–25, 2021.
40. Maksimovic, M., Greening the future: Green Internet of Things (G-IoT) as a key technological enabler of sustainable development, in: *Internet of Things*

and Big Data Analytics Toward Next-Generation Intelligence, volume 21, pp. 283–313, Springer, Cham, 2018.
41. Pan, S.Y., Gao, M., Kim, H., Shah, K.J., Pei, S.L., Chiang, P.C., Advances and challenges in sustainable tourism toward a green economy. *Sci. Total Environ.*, *635*, 452–469, 2018.
42. Cavaliere, L.P.L., Khan, R., Rajest, S.S., Sundram, S., Jainani, K., Bagale, G., Regin, R., The impact of customer relationship management on customer satisfaction and retention: The mediation of service quality. *Turk. J. Physiother. Rehabil.*, *32*, 3, 22107–22121, 2021.
43. Fraga-Lamas, P., Lopes, S.I., Fernández-Caramés, T.M., Green IoT and edge AI as key technological enablers for a sustainable digital transition towards a smart circular economy: An industry 5.0 use case. *Sensors*, *21*, 17, 5745, 2021.
44. Said, O., Al-Makhadmeh, Z., Tolba, A.M.R., EMS: An energy management scheme for green IoT environments. *IEEE Access*, *8*, 44983–44998, 2020.
45. Zhu, C., Leung, V.C., Shu, L., Ngai, E.C.H., Green internet of things for smart world. *IEEE Access*, *3*, 2151–2162, 2015.
46. Nandyala, C.S. and Kim, H.K., Green IoT agriculture and healthcare application (GAHA). *Int. J. Smart Home*, *10*, 4, 289–300, 2016.
47. Bansal, M., Kumar, A., Virmani, A., Green IoT: Current scenario & future prospects. *J. Trends Comput. Sci. Smart Technol. (TCSST)*, *2*, 04, 173–180, 2020.
48. Bashar, D.A., Review on sustainable green Internet of Things and its application. *J. Sustain. Wireless Syst.*, *1*, 4, 256–264, 2020.
49. Sharma, P.K., Kumar, N., Park, J.H., Blockchain technology toward green IoT: Opportunities and challenges. *IEEE Network*, *34*, 4, 263–269, 2020.
50. Xie, L., Shi, Y., Hou, Y.T., Lou, A., Wireless power transfer and applications to sensor networks. *IEEE Wireless Commun.*, *20*, 4, 140–145, 2013.
51. Sidiku, M.B., Eronu, E.M., Ashigwuike, E.C., A review on wireless power transfer: Concepts, implementations, challenges, and mitigation scheme. *Niger. J. Technol.*, *39*, 4, 1206–1215, 2020.
52. Nguyen, M.T., Nguyen, C.V., Truong, L.H., Le, A.M., Quyen, T.V., Masaracchia, A., Teague, K.A., Electromagnetic field based WPT technologies for UAVS: A comprehensive survey. *Electronics*, 9, 3, 461, 2020.
53. https://circuitdigest.com/tutorial/different-types-of-wireless-power-transmission-technologies-and-their-working
54. Mohammed, S.S., Ramasamy, K., Shanmuganantham, T., Wireless power transmission–a next generation power transmission system. *Int. J. Comput. Appl.*, 1, 13, 100–103, 2010.
55. Dissanayake, T. D., Budgett, D.M., Hu, P., Bennet, L., Pyner, S., Booth, L., Amirapu, S., Wu, Y., Malpas, S.C., A novel low temperature transcutaneous energy transfer system suitable for high power implantable medical devices: Performance and validation in sheep. *Artif. Organs*, 34, 5, E160–E167, 2010.

6

Adaptive Energy Intelligence Using AI/ML Techniques

Gowthamani R.[1], Sasi Kala Rani K.[1], Manikandan M.[2] and Rohini M.[1]*

[1]*Department of Computer Science and Engineering, Sri Krishna College of Engineering and Technology, Coimbatore, Tamil Nadu, India*
[2]*School of Computing and Information Technology, REVA University, Bangalore, India*

Abstract
The world is developing at a quick pace through integrating physics and logic. The high boom of smart in anything (smart phones, smart cities, smart grid, smart industrial internet, smart homes and smart hospitals) and Internet of Things (IoT) enable more users to access data as well as computation power in real-time. Cyber physical system is more and more popular due to its integration services. However, there are several challenges to the improvement of the quality of service in CPS. The constraints of current cyber physical system are a subject of more prevalent concern. In response to these rising difficulties in energy production, new requirements and challenges spring up day by day, especially in energy distribution and optimization that can meet these challenges are sought after. In order to expand the efficiency and to decrease the quantity of energy to be used in cyber physical system for processing, numerous solutions need to be addressed. In this article, an elaborate work of computing machine learning algorithms has to be proposed for CPS is carried out and application areas in the Internet of Things are ascertained.

Keywords: Cyber physical system, machine learning, artificial intelligence, dynamic utilization

**Corresponding author*: gowthamanir@skcet.ac.in

Rathishchandra R. Gatti, Chandra Singh, Rajeev Agrawal and Felcy Jyothi Serrao (eds.) Self-Powered Cyber Physical Systems, (141–158) © 2023 Scrivener Publishing LLC

6.1 Introduction

The physical processes with computation are integrated inside the cyber physical systems (CPS). The physical processes controlled by embedded computer along with a network monitors, where the feedback loop of physical processes disturbs computation of CPS and vice versa. In the current physical world, time and simultaneous execution are inevitable. Achieving these two properties are very hard in the present world. In the IT revolution cyber physical system applications play a vital role. They include high self- reliance medical tools and systems, control the traffic and ensuring safety, highly progressive automotive systems, controlling of the processes, high energy conservation and saving, environmental control, instrumentation, the infrastructure control (resources like electric power, communications systems, water resources), distributed robotics (like telemedicine, tele conversation), defense, smart manufacturing system, and intelligence structures. This is very stress free to visualize the innovative abilities, such as generation of power and energy level connected with the power grid, where timing importance and security problems come to an end [1]. The important areas while seeing improvement and high progression of the cyber physical systems are considered as reducing the power needed by CPS by considering factors, like cost and the timing in addition to the enrichment of the designed products, which includes better product virtualization to enhance its design, care verification, and validation and improve the production. Hence, high flexibility with very low cost can be enabled with virtualization along with different levels of system development. The overall communication of designed product and this production system gives a significant position in Industry 4.0 development (known as advanced manufacturing, intelligence manufacturing, also the cyber physical production systems) perception. Later, development methods and system models in the design process of a system have a consideration, which relates to their importance as producer of transformation of current system evolutions, like mechatronics, health system, embedded system, instrumentation, and automation management systems, into a cyber physical system otherwise the universal integration combined with the Internet of Things (IoT). This perception leads to the evolution of mechanization, started in the mid of the year 1950s, which increased the aptitude to put on developments in computer technology and move toward an environment where computers are having an increasing response for the operations [2]. At present, computing systems use the combination of both cyber and physical systems, even though they are developed individually. Because of the current situation, the greatest performance factors, such as power problem, time efficiency, dynamic usage of different patterns, make an interconnection of both cyber and physical into a significant way. These are basic points that lead to very basic design of integration of both cyber and physical components at different levels [3]. This is considered the main aim to have a current research in usage of power in the area of cyber physical systems. CPSs are main systems which are integration of networking system, all the computation, and the physical processes [4–7], otherwise, in CPS physical components and software components intertwined deeply, where spatial and temporal operating scales [8], demonstrating different behavioral models, and can be interacting with one another in various ways that are modified along with context. There are some important characteristics in CPS: a) physical component cyber capability, b) automation system, c) combination of various temporal as well as spatial scales, and d) dynamic rescheduling [9]. Physical sensors and various cyber physical system, which includes the physical resources and various sensors for collecting data, are considered as part of CPS. And also need communication channel to transmit the data collected from sensor, which are mainly used for controlling and listening the cyber physical systems physical resources. On the other cyber side, all the calculations are worked out with the aim of exploiting use of renewable resources, and an appropriate decision is to be taken, with respect to physical resources in CPS. Accordingly, cyber physical systems change from the little, such as pace makers to very large scale like power grid [10].

Investigation on cyber physical systems broadly uses IoT sensors and persons (actors) for the purpose of monitoring and clear controlling the available physical elements in CPS, in which reliability of sensors and actor's lifetime are important role of the CPS system. The various sensors battery capacity and actor's battery capacity inadequateness act as a very notable task in hyper connected CPS. Communications within a limited physical space reduce the power of batteries faster than traditional sensor applications. Conservation of energy and the energy harvesting methods give growth for the lifetime of sensors. However, normal energy harvesting techniques cannot act as deterministic acceptable sources of energy due to their irregular accessibility. Newly, the combination of low-power sensors and radio frequency (RF) energy-harvesting capability is commercially available [11]. A computer-based algorithms are used to control and monitor the intelligent cyber physical system. In which, physical and software components are intensely tangled, can be eligible to work on various spatial and temporal scales, display individual and different behavior modalities, which can be interacted with one another in the habits that modify with framework [8]. The cyber physical system also includes transdisciplinary approaches, integrating basic concept of cybernetics, system design, mechatronics, and process science [12, 13]. Here the process control is also known as embedded systems. In these embedded systems, importance incline is to be huge on the computational factors, and limited on a powerful link in between computational and the physical elements. Cyber physical system is also the same to the Internet of Things (IoT), involving a similar architecture but cyber physical system displays advanced grouping and organization power between physical resources and computational elements [14].

Investigation on various fields like energy, computational power, and physical resource technologies gives the continuing fundamental modifications in supplying energy and energy system structure, which leads to the changes in energy power generation, distribution, and storage. This chapter tells the significance of maintaining efficient, reliable, and flexible electric power generation systems by using optimized machine learning algorithms to predict the low energy level. Investigation on various fields like energy, computational power, and physical resources technologies gives emphasis on intelligence power system. Nowadays, usage of energy vectors to realize the use of potential energy, combination of CPS and IoT sensors for optimizing energy efficiency, fossil and renewable energy system hybridization and applications that utilize carbon as energy is considered as current research areas in CPS.

6.2 Evolution of Cyber Physical System

This part delivers a summary of various types of existing systems and the changeover process of mechatronics to cyber physical system and the cloud based (IoT) systems. Mechatronics is also treated as an interdisciplinary area of engineering that goals to correlate and makes interconnection of various engineering, in which the interactions create the main role for the design of various product types. The process developer's interaction in different areas is often caught up to be considered insufficient accepting across various disciplines, and by considering inability in shared platforms to model the complex systems [15]. Therefore, several sub-systems are given by outside supplied users, there is a necessity of horizontal integrations inside organizations as well as need of vertical integration of sub system and main system suppliers of overall computational systems. In this point, Lee [16] mentioned that cyber physical system considered as the "integration of various computation and different physical processes." The monitoring of physical process is usually controlled by feedback loops with the help of embedded systems and monitors and vice versa.

Here, Figure 6.1 depicts the transition of mechatronic system to current cyber physical system and extended to IoT where x-axis and y-axis represent technologies and the system level, respectively. The promised aim of the revolution of cyber physical system is to enable the physical world can be controlled, monitored, and also influenced both ways like adaptively and intelligently. Most of the researchers show that the computing evolution of cyber physical computing to reach significance in the current systems. Here, the development of cyber physical system is mingled with technology, economic, and societal chances and the social challenges. Hence, there is a need for transdisciplinary theories, system models and methods, and supportive tools. Attaining this competence also gives the need for combining CPS with mechatronics, machine learning, and intelligence in engineering fields.

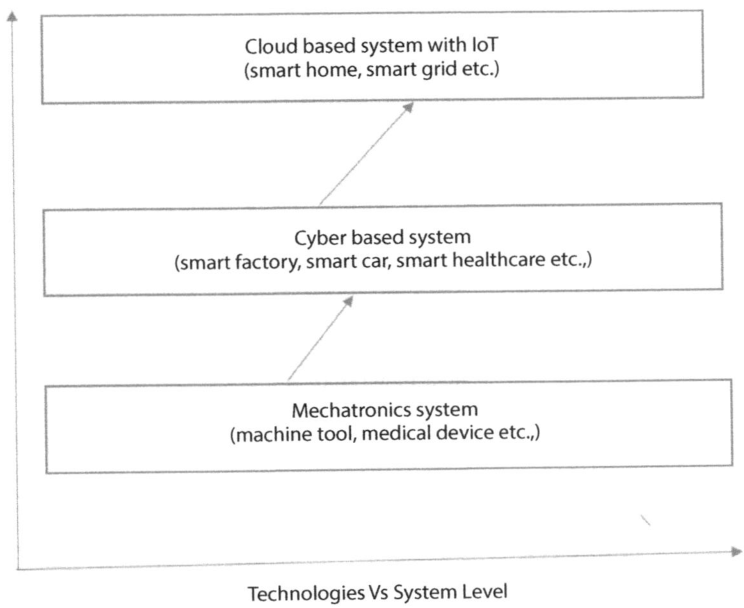

Figure 6.1 Transition of cyber physical system to Internet of Things.

6.3 Relationship With Internet of Things

A full-fledged CPS is intended for interacting components with a physical input, as well as output data instead of separate devices but it is not in a traditional embedded system. This format gives the ideas of how robotics and the sensor networks in addition to intelligence mechanisms provide a path for the importance of computational intelligence. Current improvements in engineering and science advance the physical link between various calculations and physical elements with the support of intelligent appliances, autonomy of the system, timing competence, system functionality, consistency, increasing the adaptability, safety, and system usability of cyber physical systems [17], which elaborate the growth of cyber physical systems in numerous directions, like energy building and health care system, etc. [18]. Figure 6.2 shows the relationship of CPS with IoT system. In the current growing field, Industry 4.0 is composed of CPS and IoT. The combination of IoT and CPS along with big data can be used in various

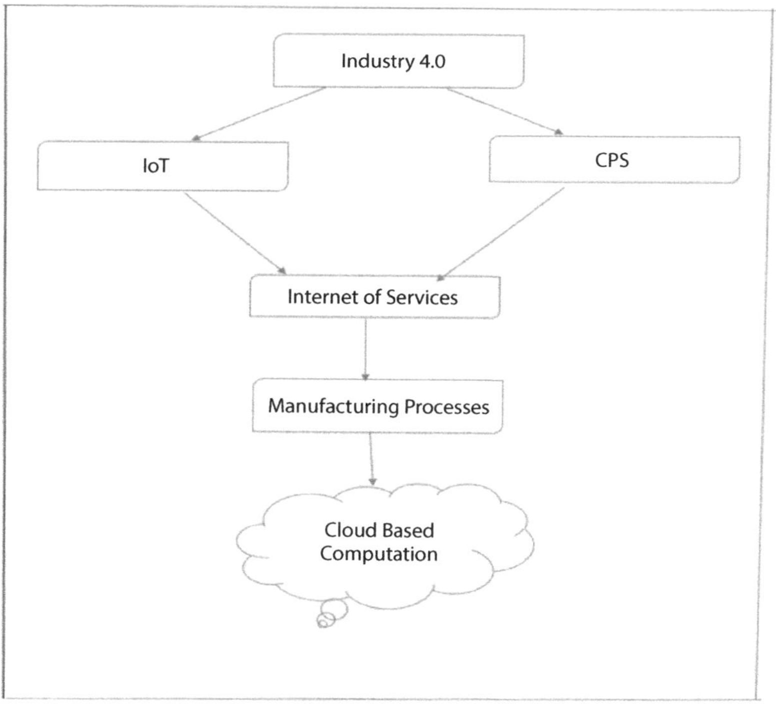

Figure 6.2 Relationship with IoT.

applications. In this chapter, the overview of how CPS is interrelated with IoT to sustain the energy level is discussed. Cyber physical system can be moved to cloud-based system to store the result of the model. CPS can be used enormously along with machine learning and artificial intelligence concept. Several Internet of services can also be done with CPS. Hence, IoT and CPS can be interrelated to make a system with accurate one and for processing vast amount of big data. In the manufacturing process, CPS can be combined and energy level is a crucial factor for better performance. To get enormous data, various sensors can be used and processed by CPS.

Most of the applications in the real world works with information as a rich environment, where information and communication can act as a basic driver for design process, where the integration of cyber physical systems and Internet of Things concept have been used by the designers. This leads to the designers to get new and unique simulation for expressing this situation. In the view of current industry, it is crucial while generating new products and system facilities with sound knowledge of information exchange specified by Internet of Things, hence there is a need for using smart systems to support resulting systems. It can be developed by developers with different functionality provided by various users in various pathways. Combination of this tells that IoT technologies assist to mix CPS and act as interface for CPS and users of the system.

Related applications of CPS are smart cities, energy efficient systems, intelligent buildings and environments, transportation infrastructure, and so on. Crosscutting IoT technologies gives vital role in allowing dependable IoT inside system engineering with CPS. This chapter focuses on basic engineering principles, such as machine learning, artificial intelligence, which reinforce the mixing of various components of cyber and physical system infrastructure and energy systems.

6.4 Challenges in Design and Integration of Cyber Physical Systems

The design of CPS needs a various disciplinary expansion process in which the person who designs should focus individual physical components and various computational components, along with integration of their components and also interaction between components. Challenges connected to CPS design are therefore measured as the view of the physical system process, various computation process and integration of the system respectively. Physical world directly linked with CPS system is considered as the main perspective, which are used to detect variations in this physical world system and realize the real time control of systems behavior [19]. While considering perspective of computation, CPS is considered as it is developed from embedded systems and also embedded systems are base for the CPS [20]. However, the link in CPS physical world is considered very stronger when comparing embedded systems. Hence, the various software component in a system is complex and large when comparing traditional embedded system. With respect to the viewpoint of main computation and physical processes integration, CPS design requires an integration inside the collaborative design of CPS. Physical components and computational components can interact with one another in various disciplines, leading to a variety of interruptions and several mismatches throughout the process of system design. Common understanding of communication is delayed due to a lack of specification and documentation interfaces between the various components.

Figure 6.3 depicts the CPS system integrated with the concepts generation, system modelling and integration of computation with respect to various disciplines. Earlier stage like conceptual design and modeling of system plays a vital role in system design and they enable foundations for the system [21].

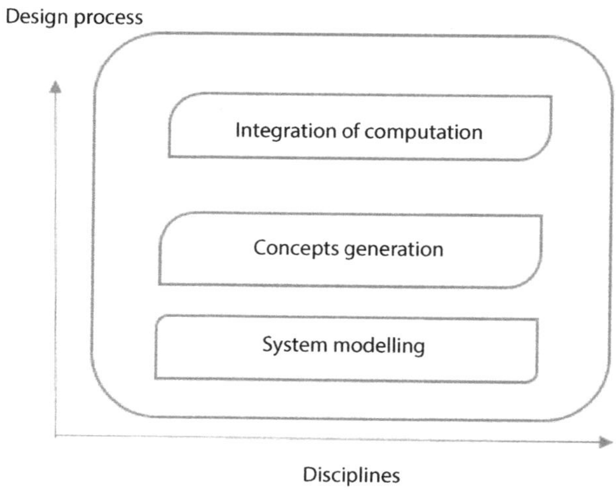

Figure 6.3 Integration with discipline and CPS.

6.5 Future Challenges and Promises

With respect to CPS advancement, several challenges are prominent when we adopt CPS in real world.

- **Modeling and design:** Hardware description languages and normal programming languages are very important for design of an entire system.
- **Security:** Failure and different attacks are main consideration while integrating cyber and physical components with one another.
- **Mobility:** In addition to static devices of CPS, in many cases we need to consider dynamic movement and device synchronization.
- **Distributed computation and network control:** CPS system needs to control overall computation and network. These systems get an overall synchronization action with CPS [25].
- **Scalability:** CPS are developed for various applications and it needs tools and strong foundations to develop the system to adopt the real world.
- **Energy:** CPS can be integrated to play an important role to connect physical devices. Therefore, energy needed for CPS system plays a vital role to develop the system.

6.6 Machine Learning Models

Machine learning is the way to designate a computer algorithm which is used to learn the system from available data. These algorithms [22] naturally use a sufficient value of input and less amount of features related to input for learning process. Currently, enormous ML techniques are proposed in estimation of consumption of energy level and various performance circumstances. In energy load forecasting, the machine learning becomes broadly used techniques in field of energy sector. Here, training samples are collected from past data, which can be used to form a corresponding network structure. To satisfy the accuracy of the system, the trained model can be used. Machine learning algorithms and models can be used for modeling of the system, designing the system and prediction of energy levels. There is a huge development in the field of ML models in developing energy systems. This part gives the idea of how machine learning model can be used in the field of energy prediction along with different view of models and also different applications [23]. ML models further classifies according to the energy type, ML model techniques and application types. This chapter further ends that ML algorithms can be used in an outstanding growth in accuracy, generalization capability of ML models in energy levels using hybrid ML algorithms. Hybrid model can combine two or more algorithms to predict the energy demand and therefore machine learning model can be used for energy consumption and sustainability of system [24].

6.7 Estimation of Building Energy Consumption

Machine learning models are used to find the relationship between various input features and target outputs (like energy performance) and also used as a black box where there is no need of information. The model can be trained to predict the output and sufficient amount of data used while predicting the output is wrong or not. This is also called as supervised learning. In supervised machine learning case, the final energy parameter is measured (engineering method) to train the model. Another model is called unsupervised learning where energy needed for the system is a vital factor and has an attention in analysis of energy level. The other name of unsupervised learning is unsupervised classification and focuses on unlabeled data and cluster these databases on hidden pattern and various similarities exists in the input features. Using unsupervised learning the bench marking of energy level is given as input to train the system and this is a baseline for calculating energy level. Therefore, unsupervised learning algorithm like clustering algorithm gives an accurate prediction while comparing traditional system. With the help of supervised or unsupervised machine learning, algorithm energy level is compared with less amount of time to self-power the system [26].

6.8 Development of Artificial Intelligence

In computer science, artificial intelligence (AI) is used to provide intelligence and uses various machine learning algorithm to recognize and extend the intelligence of human using various algorithms. AI also refers to intelligence of machine or having ability to duplicate the functionality of a human being. Also it has ability to solve and learn all types of problems.

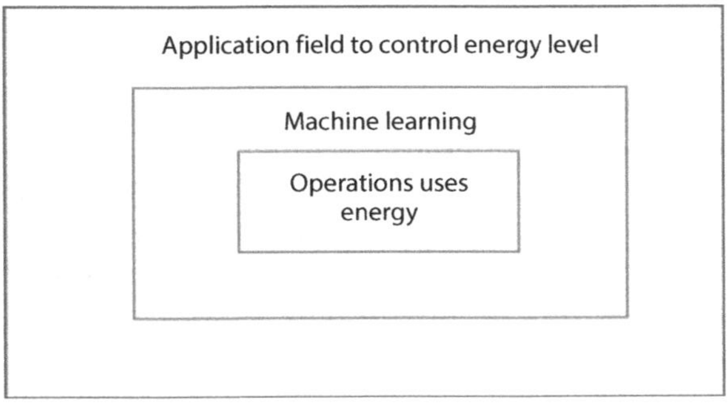

Figure 6.4 Applications of AI in self-power CPS.

In current computer field, this is also called as intelligent agents. For predicting energy level AI can be used along with ML algorithms, which can give efficient and accurate energy prediction techniques. AI architecture is combined with machine learning depicted in Figure 6.4. In last decades, which gives the better achievements in various research fields and they are divided into various forms based on application. They are artificial narrow intelligence (ANI) and artificial general intelligence (AGI). ANI used when the system is clear and boundary is defined. Which is used in many applications, gives better achievements and always better than human. AGI used as human brain which uses human intelligence. In many places, AI plays an important role to get better result and gradually clever into all aspects of human life. One among approach is to achieve AI is with the usage of machine learning algorithms [27]. Artificial neural networks act as an outline for various machine learning algorithms, which can be used to process complicated input data. The ANN can be used for several reasons, like regression, energy harvesting, and fitting the curve. The minimum complexity of ANN is considered as a main advantage of ANN model. In CPS system, ANN can be used to get the required data and can be processed by any of the ensemble algorithm to predict the energy level with accuracy.

6.9 Usage of AI/ML in Adaptive Energy Management

An energy system can be used to cluster the components and aimed to control the energy needed for the system. Energy systems also include various engineering systems, renewables and also alternatives. The improvement in the current energy system is deciding critical decision making processes for various applications to achieve the objective. The increasing utilization of data collectors in the system takes a large amount of data and energy. With the use of smart sensors, we can achieve a better level of production and consumption of energy. The available big data are used to create opportunity and good challenges for better decision making. Various ML models can be used to implement bulk data in several applications. Therefore, better prediction methods related to ML model give the extraction of various operational relations from the diverse observations like data driven model gains energy level in the real world. In the current world, in CPS energy systems the machine learning model plays an important role for predictive modeling of system, energy consumption, and various energy demand analysis with respect to accuracy, usefulness and also the system speed [28]. Several ML models provide a better knowledge on energy level functionality with respect to human interaction. The use of ML model along with CPS system can be used to check the boundary level and if it exceeds the boundary level, it will call self-powered CPS to get the energy.

6.10 Use of Hybrid/Ensemble Machine Learning Algorithm for Better Prediction

Ensemble model can be used to combine more than one machine learning algorithms and output of one algorithm can be given as input for another model, hence produce more accuracy while comparing single machine model in the real world. It is a technique which creates multiple models and then combine the result to produce better result. The model used to develop the system can be used with any machine learning algorithms like logistic regression, decision tree, SVM, etc. The input given to ensemble methods is called as base model. The ensemble model can be classified into voting, bagging, stacking and boosting. Based on the application, any of the ensemble model can be used to produce better accurate result. In ensemble machine learning model, support vector regression (SVR) machine learning method can be used along with any of the existing machine learning models like SVM, decision tree, LR (logistic regression). Here, in the proposed model, two algorithms along with AI technique can be used to predict the better result. In Multilayer perception, various neural network like backpropagation, probabilistic, convolutional, etc. can be used with ML models. Most MLP contains three basic layers, namely input layer, the hidden layer, and the output layer, and they are connected with one another. Here, prediction of energy level can be done with weight and activation function of MLP with the help of ensemble machine learning algorithm [29]. The combined data set given to ensemble model divided into two main parts, test and the training data. The training data can be used to train the system, where test data can be used to test the trained model is correct or not. The preprocessing step in machine learning algorithm can be used to smooth the system. Then the processed data can be used to establish a required training model. There are several parameters of CPS system can be given to train the system, so that energy level of the system can be compared with given criteria of the system. Ensemble model can predict the accuracy of energy level with the help of hybrid ML algorithms.

There are various ensemble algorithms can be used for better prediction. In max voting method, classification problem can be solved, to make a better prediction multiple data models are used for every data input. The prediction also known as vote. The majority result will give the final prediction. Like max voting technique, averaging technique can be used to predict the correct prediction based on multiple predictions done for each and every data point. Finally, average of all predictions can be used to find the resultant prediction. This averaging technique can be used in both regression and also calculating probabilities of classification problem. In weighted average technique, all the models assigned with different level of weights to find the better prediction. There are various advanced ensemble algorithms are there to give a better prediction, namely stacking, blending, bagging, and boosting. All the methods are used to predict the correct result with the help of multiple models like decision tree, SVM, KNN, etc. Before going into ensemble model, first the training data can be used to test the model. Then output of each model given to next model to boost up the prediction level. In case of boosting, each model tries to correct its previous model error. All the ensemble algorithms are using several steps to predict the accurate result. Bagging meta estimator is an another ensembling algorithm, which can be used for both classification problem and regression problem. Random forest is considered as best method, which uses bagging technique. In random forest, decision tree is considered as best estimator, which randomly selects input features and can be used to find best split in each and every node of given decision tree. Another technique, adaptive boosting is a simple algorithm, where it assigns weight for incorrect prediction and subsequent forthcoming model can be used to calculate the correct values [30].

The main aim of ML is considered as to train the stable models, which can be used for all aspects of the system. Ensemble learning is used to combine multiple machine learning models to get best and strong model. If any of the model predicts wrong prediction, other models can be used to correct their mistake. So hybrid approach can predict the correct value with the help of ML algorithms, which can be called as hybrid approach. This chapter gives overview of ML algorithms along with AI techniques can be used with CPS for better prediction of energy level. All the models trained the model and ensembles the final result with high accuracy. Ensemble modeling can be used to boost the prediction performance of the CPS model.

Figure 6.5 shows the flow of using ML/AI techniques to manage energy levels in dynamic manner, the dynamic machine learning algorithms bring improvement in CPS by optimizing the usage of power and energy delivery and also gives intelligence in the usage of energy needed by CPS. Cyber

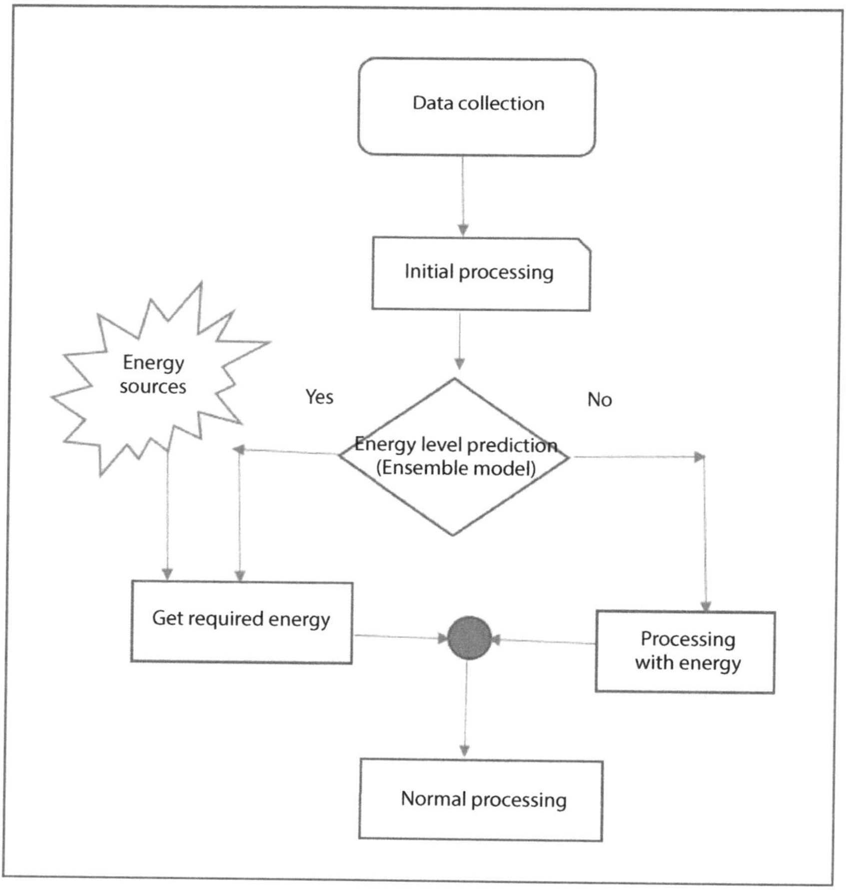

Figure 6.5 Framework of using ML/AI in energy management.

physical system can be used for embracing digital, analog, physical, and human components along with the function used through the integration of physics and logic functions. In recent years, there is an increasing interest in dynamic resource allocation, dynamic response and loading is a good place to provide high flexibility, which improves the reliability of energy systems in a cost effective way with the heuristic artificial intelligence based machine learning algorithms. A vast number of methods in artificial intelligence can be a handler for various challenges, pointing most optimal set of users to respond.

6.11 Conclusion

The smart in everywhere like sensors, grid, hospital, and multiple IoT technologies have empowered big data along with enormous experiments and chances to the CPS energy systems. The ML models can use big data in the creative nature can be used for known decision making and are considered as efficient models. The use of machine learning models in CPS energy systems can use ML algorithms, like ANN, SVM, decision trees, deep learning, ensemble techniques, and advanced hybrid ML models. It takes energy consumption prediction, dynamic utilization of energy, demand prediction of energy, and accurate cost prediction. Several machine learning models are proposed, in which hybrid models are better than the various conventional ML models. The development of renewable energy systems is closely dependent on the advancement of ML models. In the current world, dynamic energy systems have recently used in all places due to the performance improvement of CPS. Therefore, ML models have become indispensable with respect to energy systems. The chapter explains various aspects of CPS, ML algorithms, and AI techniques available to increase the energy level management. The hybrid model can be applied to increase the dynamic utilization of energy.

References

1. Lee, E.A., *Cyber physical systems: Design challenges*, Center for hybrid and embedded software systems, EECS University of California, Berkeley Berkeley, CA, USA, 2008.
2. Hehenberger, P., Vogel-Heuser, B., Bradley, D., Eynard, B., Tomiyama, T., Achiche, S., Design, modelling, simulation and integration of cyber physical systems: Methods and applications. *Comput. Ind.*, 82, 273–289, 2016.

3. Giese, H., Rumpe, B., Schätz, B., Sztipanovits, J., Science and engineering of cyber-physical systems. *Dagstuhl Rep.*, 1, 11, 1–22, 2012.
4. Conti, M., Das, S.K., Bisdikian, C., Kumar, M., Ni, L.M., Passarella, A., Roussos, G., Tröster, G., Tsudik, G., Zambonelli, F., Looking ahead in pervasive computing: Challenges and opportunities in the era of cyber–physical convergence. *Pervasive Mob. Comput.*, 8, 1, 2-21, 2012.
5. Sha, L., Gopalakrishnan, S., Liu, X., Wang, Q., Cyber-physical systems: A new frontier, in: *Machine Learning in Cyber Trust*, pp. 3–13, 2009.
6. Horváth, I. and Gerritsen, B., Cyber-physical systems: Concepts, technologies and implementation principles, in: *Tools and Methods of Competitive Engineering Symposium (TMCE)*, pp. 19–36, 2012.
7. Lee, E., Computing needs time. *Commun. ACM*, 52, 5, 70–79, 2009.
8. NSF, Cyber physical systems nsf10515, 2013, http://www.nsf.gov/pubs/ 2010/nsf10515/nsf10515.htm.
9. Miclea, L. *et al.*, About dependability in cyber-physical systems, in: *EWDTS*, pp. 17–21, 2011.
10. Khaitan, S.K. and McCalley, J.D., Design techniques and applications of cyberphysical systems: A survey. *IEEE Syst. J.*, 9, 2, 350–365, 2014.
11. Erol-Kantarci, M. and Jemison, W.D., Energy-harvesting low-power devices in cyber-physical systems, in: *Cyber-Physical Systems*, 2017.
12. Hancu, O., Maties, V., Balan, R., Stan, S., Mechatronic approach for design and control of a hydraulic 3-dof parallel robot. *The 18th International DAAAM Symposium, "Intelligent Manufacturing & Automation: Focus on Creativity, Responsibility and Ethics of Engineers*, 2007.
13. Suh, S.C., Tanik, U.J., Carbone, J.N., Eroglu, A. (Eds.), *Applied cyber-physical systems*. Springer New York, New York, NY, 2014.
14. Rad, C.-R., Hancu, O., Takacs, I.-A., Olteanu, G., Smart monitoring of potato crop: A cyber-physical system architecture model in the field of precision agriculture. *Conference Agriculture for Life, Life for Agriculture*, pp. 73–79, 2015.
15. Zheng, C., Le Duigou, J., Hehenberger, P., Bricogne, M., Eynard, B., Multidisciplinary integration during conceptual design process: A survey on design methods of cyber-physical systems, in: *DS 84: Proceedings of the DESIGN 2016 14th International Design Conference* (pp. 1625–1634), 2016.
16. Lee, E.A., Cyber physical systems: Design challenges. *Proc 11th IEEE Symposium on Object Oriented Real-Time Distributed Computing (ISORC)*, 2008.
17. Alippi, C., *Intelligence for embedded systems: A methodological approach*. Springer, 2014.
18. Wilson Boulevard, Arlington, Cyber-physical systems, Program Announcements & Information. The National Science Foundation, 2009.
19. Grimm, M., Anderl, R., Wang, Y., Conceptual approach for multidisciplinary cyber physical systems design and engineering. *Proc of TMCE 2014*, Budapest, Hungary, 2014.

20. Lee, E. and Seshia, S., *Introduction to embedded systems: A cyber physical systems approach*, Lee & Seshia, USA, 2011.
21. Komoto, H. and Tomiyama, T., A framework for computer-aided conceptual design and its application to system architecting of mechatronics products. *Comput.-Aided Des.*, 44, 931–946, 2012.
22. Ahmad, A.S., Hassan, M.Y., Abdullah, M.P., Rahman, H.A., Hussin, F., Abdullah, H., Saidur, R., A review on applications of ANN and SVM for building electrical energy consumption forecasting. *Renewable Sustainable Energy Rev.*, 33, 102–109, 2014, https://doi.org/10.1016/j.rser.2014.01.069.
23. Liu, R., Srivastava, A.K., Bakken, D.E., Askerman, A., Panciatici, P., Decentralized state estimation and remedial control action for minimum wind curtailment using distributed computing platform. *IEEE Trans. Ind. Appl.*, 53, 6, 5915, 2017, doi: 10.1109/TIA.2017.2740831. OSTI.
24. Sun, C.C., Liu, C.C., Xie, J., Cyber-physical system security of a power grid: State-of-the-art. *Electronics*, 5, 3, 40, 2016.
25. Baheti, R. and Gill, H., Cyber-physical systems. *The Impact of Control Technology*, 12, 1, 161–166, 2011.
26. Seyedzadeh, S., Rahimian, F.P., Glesk, I., Roper, M., *Machine learning for estimation of building energy consumption and performance: A review*, Springer, 2018.
27. Alam, A.G., Baek, C.I., Han, H., Prediction and analysis of building energy efficiency using artificial neural network and design of experiments. *Appl. Mech. Mater.*, 819, 541–545, 2016, https://doi.org/10.4028/www.scientific.net/AMM.819.541.
28. Mosavi, A., Salimi, M., Ardabili, S.F., Rabczuk, T., Shamshirband, S., Varkonyi-Koczy, A.R., State of the art of machine learning models in energy systems, a systematic review. *Energies*, 12, 1301, 2019, doi: 10.3390/en12071301.
29. Khan, P.W., Byun, Y.-C., Lee, S.-J., Kang, D.-H., Kang, J.-Y., Park, H.-S., Machine learning-based approach to predict energy consumption of renewable and nonrenewable power sources. *Energies*, 13, 4870, 2020, doi: 10.3390/en13184870.
30. https://www.analyticsvidhya.com/blog/2018/06/comprehensive-guide-for-ensemble-models

7

Renewable Energy Smart Grids for Electric Vehicles

Vishal H. Kanchan, Preethesh B., Hithesh Alen D'Costa, Sohan R. Alva and Rathishchandra Ramachandra Gatti*

Department of Mechanical Engineering, Sahyadri College of Engineering & Management, Mangalore, India

Abstract

The world is slowly transitioning from fossil fuel such as coal to renewable energy sources for power generation. Another significant transition is the slow transition from the use of internal combustion engine vehicles (ICEVs) to electric vehicles (EVs) on the power consumption side. While these two transitions are both economically and ecologically beneficial, they also come up with new challenges. From the power generation perspective, most of the renewables such as solar and wind energy sources are intermittent in nature thus adding complexity to steady supply of power to consumption side. From the power consumption perspective, the transition from ICEVs to EVs will create more dynamic electric loads to the utilities adding to more complexity in logistics of power delivery. Thus, electric grid has to face two major challenges—intermittent power generation and dynamic load power consumption. This necessitates the need for use of advanced smart grid technologies built using AI/ML algorithms for optimization of power generation and consumption logistics, block chain for transparencies in power transactions and IoT enabled edge computing for capturing data and analyzing the power transactions. Over the years, new concepts have emerged to merge the renewables and EVs to grids. The predominant concepts, such as grid to vehicles (G2V), vehicle to grid (V2G), virtual power plant (VPP) will be discussed to evaluate the feasibility of a super system of smart grid where renewables, EVs and the grid are seamlessly integrated. During such integration, such a smart grid will have multiple stakeholders, such as EV aggregators, utility service providers,

**Corresponding author*: gattirathish@gmail.com

Rathishchandra R. Gatti, Chandra Singh, Rajeev Agrawal and Felcy Jyothi Serrao (eds.) Self-Powered Cyber Physical Systems, (159–176) © 2023 Scrivener Publishing LLC

electrical transmission system operators (ETSOs), and electrical distribution system operators (EDSO). Some techno-business models that cater to these stakeholders will be discussed. The critical requirements and the feasibility challenges of V2G/G2V for EV subsystems, such as batteries and inverter technologies, precision metering and real-time pricing, fast charging capacities, battery life and frequency of charging will also be discussed.

Keywords: Electric vehicles, vehicle-to-grid, grid-to-vehicle, smart grids, renewable energy, virtual power plant, automotive cyber-physical-systems

7.1 Introduction

In both the transportation and power generating industries, fossil fuels constitute the world's major energy sources today. The dwindling supply of fossil fuels act1s as a wake-up call for these businesses to find other energy sources. In truth, the oil economy's future is not only unreliable, but also severely restricted, as global vehicle fleets see it. Furthermore, the combustion of fossil fuels emits greenhouse gases (GHGs), which result in global warming. As per the study [1], oil consumption in the transportation industry will have grown by 54% by 2035. Oil prices, according to the Energy Information Agency (EIA), will rise considerably over the next two decades. In this context, there have been a variety of efforts aimed at reducing oil use. In the transportation industry, electric vehicles (EVs) are a potential option that is fast increasing market share. According to economic analyses, internal combustion engine automobiles (ICEVs) will be phased out in the future in favor of electric vehicles (EVs). According to the Australian Energy Market Commission (AEMC) [2], by 2020, electric cars will account for less than 10% of new vehicle sales, and beyond 2020, they will contribute to a 15% to 40% rise in new light vehicle sales. Because EVs have a higher initial cost than ICEVs, a lot of effort will be required to meet future market predictions. The electrification of the transportation industry, on the other hand it looks to be one of the viable answers to global climate change, energy security, and geopolitical worries about fossil fuel supplies. Electric vehicles have the potential to serve as a distributed energy source for the electric grid. According to certain surveys, the majority of cars spend nearly 95% of their time parked, they can stay linked in this situation. Kempton [3] proposed the notion of vehicle to grid (V2G), which necessitates vehicles connecting to the grid and being prepared to dissipate the energy back into the system.

7.2 Integration of Electric Vehicles (EVs) into the Power Grid

Inclusion of a large number of electric automobiles into the utility grid is a significant issue that will need extensive research and monitoring in terms of financial implications, functioning, and management gains under optimal circumstances. Several recent studies have investigated the influence of electric cars on the power distribution network [4], whereas others have investigated the various application models for EV adoption in the power system [5]. The bulk of electric car charging systems, according to recent polls, are meant to be used at home. EV charging, on the other hand, is mostly expected to occur in public, commercial, or workplace charging stations [6]. As a result, it is expected that the impacts of EV charging will have a direct influence on the electric power distribution system. Overheated power transformers to the need for new power distribution equipment are just a few of the effects. The adoption of electric cars, on the other hand, has the potential to greatly improve the performance, efficiency, and power quality of the electric grid. This is achievable if the integration of a significant number of electric vehicles (EVs) is methodically managed and functionally organized to meet electric grid operational requirements [7]. Many techniques have been proposed in the literature to fully realize the benefits of incorporating big EV fleets into the power grid. The electric vehicle owner and the power business are the two foundations of this scenario. Most significantly, system participation can benefit both parties at the expense of sophisticated control, information technologies, and operational compromises. The EV aggregator is used directly in the most popular design, and it has lately piqued researchers' interest [8]. The aggregator is in charge of all essential operational activities, such as interacting with the distribution system operator (DSO), transmission system operator (TSO), and power system integrators. In most situations, the aggregator serves as a conduit between power market participants and customers. The virtual power plant (VPP) idea, in which automobiles are gathered and operated as a single dispersed source of energy, might potentially be used to achieve this EV integration [9].

7.3 EV Charging and Electric Grid Interaction

The charging of electric vehicles is among the most essential components of their use. There is a variety of charging levels for electric vehicles, each of which represents the vehicle's power and charging time. These numbers represent either sluggish or fast EV charging. Slow charging (up to 8 hours for PHEVs and 20 hours for BEVs) may be done at home or at work, whereas quick charging (typically 15 minutes to 1 hour) can be done at specialized charging stations in corporate or communal locations. The AC level 1 may be utilized in a domestic environment, but the AC level 2 is more suited to commercial and public settings such as offices, movie theatres and shopping malls. On the other hand, communal, residential, and business charging stations are all provided by DC-quick charging (DC levels 1–3) [10, 11]. The nominal ratings of the battery charger typically determine the charging power delivered, and in most studies conducted, the EV battery voltage is often restricted to less than or equal to 400VDC (DC bus voltage). Furthermore, depending on the battery storage size and charge level attributes (current and voltage ratings), the charging time that the EV spends fully charging its battery pack may vary. The issue of how to standardize the fast charging spectrum is a major topic of discussion. Fast charging, on the other hand, is necessary to charge an electric vehicle's battery in a matter of minutes. Global manufacturers collaborated with the Society of Automotive Engineers (SAE) to design a universal charging facility that included both AC and DC-fast charging choices. In a single entity connection, it combines AC single-phase charging, AC three-phase charging (AC-fast charging), and ultrafast DC charging (SAE combo standard). In addition, the CHAdeMO rapid charging standard developed by Tokyo Electric Power Company (TEPCO) is gaining traction in the electric vehicle market [11]. The adoption of electric automobiles as a dependable

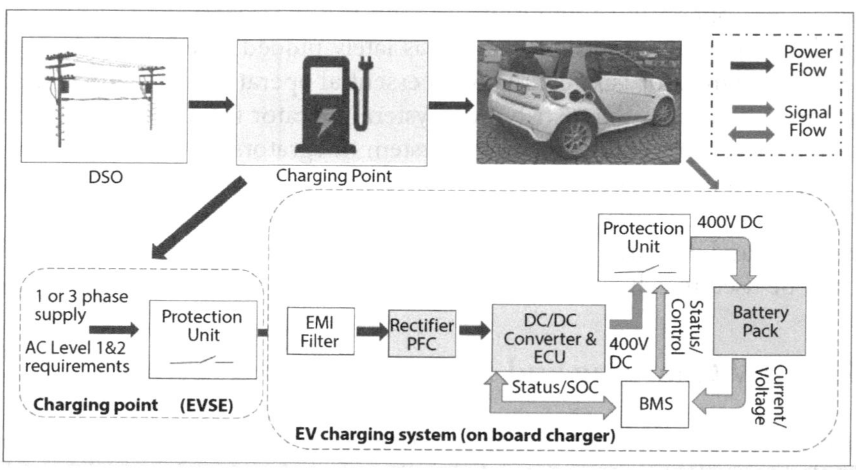

Figure 7.1 Configuration of EV charging at AC L1 and L2 setup [33].

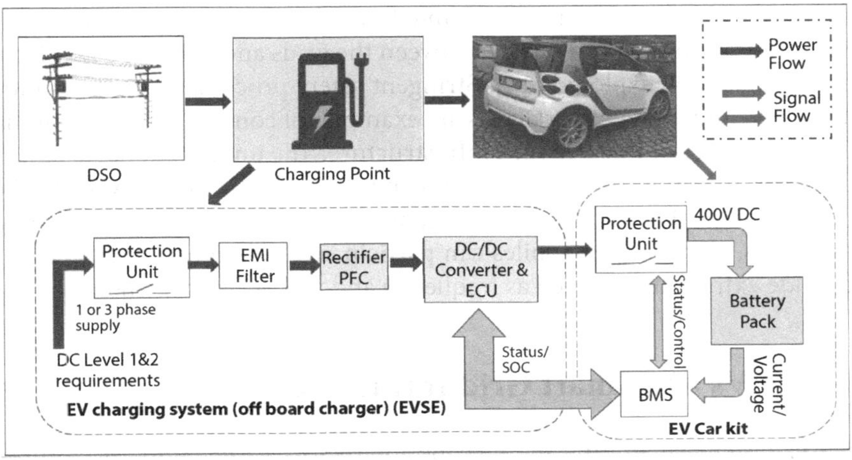

Figure 7.2 Configuration of EV charging at DC L1 and L2 setup [33].

mode of transportation will be aided by mimicking the fast ICEV refueling phenomena. Chaundhry and Bohn [12] discuss the V2G application utilizing DC fast charging level 1, which is limited to 36 kW, level 2, which is limited to 90 kW, and DC fast charging using the CHAdeMo standard, which is limited to 62.5 kW as in Figure 7.1. This is one of the demonstrations of the V2G's viability with DC quick charging infrastructure as shown in Figure 7.2. This study also looks into AC levels 1 and 2.

7.4 EVs with V2G System Architecture

Electric cars can be connected to electric grids and used for a variety of purposes, including active loads that consume grid power (while charging) or dynamic ESS that sends energy back into the grid. It's noteworthy that the segment is known as "vehicle to grid" (V2G). Because of the limited number of EVs accessible as commodities, their geographical position, and insufficient local battery capacity, V2G services are not feasible. To make the V2G idea a reality, a large number of electric vehicles are joined together in varied forms, based on the control systems and aims [13]. The pooling of EVs as a mono regulated dispersed source of energy could aid the electric grid in supervision and system management by participating in the energy market. V2G services can be offered via bidirectional or unidirectional power flow when an electric vehicle interacts with a smart grid. Electrical energy will flow from the vehicle to the grid (V2G) as well as from the grid to the EV while it is charging. In the literature, the cost and practicality of this two-way interface between the grids and collective EVs have been extensively explored [14]. Stringent safety procedures, such as anti-islanding, and high network costs are examples of constraints that limit the full advantages of this network infrastructure. The unidirectional arrangement, on the other hand, only allows energy to flow in one way, from the network to the EV (only for charging, not for discharging) [15]. According to studies, electric automobiles can participate in the power industry and provide extra features, such as frequency and voltage management.

7.5 EVs and Smart Grid Infrastructure

The electricity generating and distribution sectors are changing as far more distributed energy resources (DERs) enter the market. DERs have variable electricity generation and consumption in terms of time and place. As a result, energy conservation in the traditional power system has become increasingly complex and difficult. Energy grids are being created to increase the flexibility, efficiency, resiliency, and security of energy generation and delivery. The smart grid includes sophisticated connectivity, smart energy monitoring, and enhanced control. In the energy market, it can

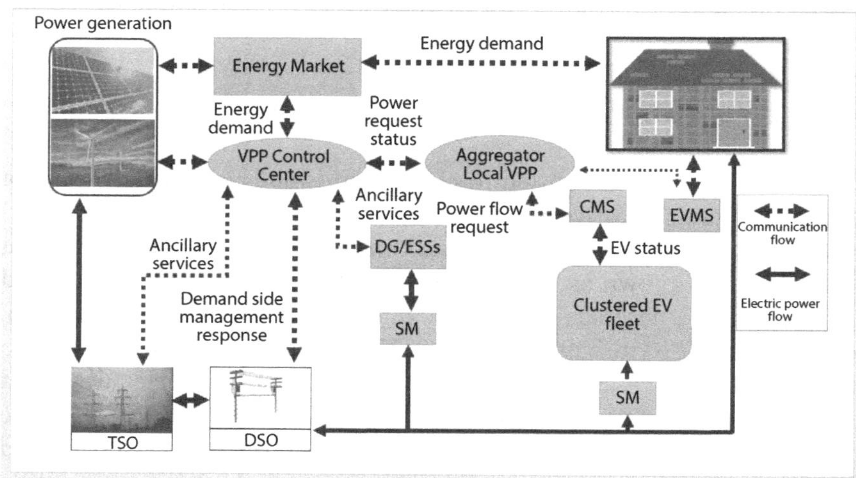

Figure 7.3 VPP realization and control in V2G context [33].

also offer EVs with a flexible and optimum deployment as active loads and potentially distributed, dispersed energy sources [16]. Researchers have attempted to assess and construct smart network system in order to stimulate the adoption of electric automobiles in the energy market. The construction of an integrated power network is being driven by the unification of methods and approaches in power transmission and networking as shown in Figure 7.3. Specifications and requirements for electric vehicle compatibility and smooth grid connectivity have previously been published [17]. Furthermore, the electric vehicle management system (EVM) of an EV collects and sends data to the GO or aggregator, and also conversely. A smart meter (SM) is a component of the EVM that enables real-time energy monitoring, connection, and administration. The impacts of EV charging that may be created in the smart grid environment through enhanced bidirectional data transmission are the subject of smart scheduling [18].

7.6 Renewable Energy Sources Integration With EVs

Renewable energy sources (RES) are being used in the electric grid more frequently, which is a positive indicator. The present power infrastructure has been affected by the unpredictability and asynchronous supply of power from numerous sources, notably wind and PV solar energy [19]. Based on available power sources, such as wind speed and solar radiation, the electricity generated by such RES might be extremely big (more than the power required) or very little (smaller than the power requirement) (less than the power demand). In a nutshell, these RES are time-variable, non-dispatchable, have minimal control, and have little power credit, which is problematic for designing power systems. Incorporating wind energy conversion systems (WECS) and photovoltaic solar systems into the electrical network is at an advanced stage and seems realistic, according to the majority of research [20]. However, balancing the power production from these RES on the grid using static energy storage systems (ESS) or controlled dispatch loads [21] might be a viable alternative. In the event of excessive or inadequate energy production, static power storage systems consume or provide electricity. Although that method necessitates a large upfront investment, it either delays or even raises the total cost of RES in the energy grid. Several scholars predict that the transportation industry will be electrified within the next decade, as previously stated.

7.6.1 PV Solar Energy With EVs

Photovoltaic solar energy-based power generation has already proven to be a viable option. Photovoltaic network panels are commonly used in combination to power the whole power grid. Since the popularity of electric automobiles rises, PV solar power is often likely to get affected for recharging and network assistance. A number of studies [22, 23] have shown that installing PV solar on parking lot roofs to charge EVs is a viable option. Furthermore, with these Photovoltaic solar networks, V2G interactions are feasible [24], and optimal production planning is achievable to minimize operational costs and improve network performance, as mentioned in [25]. To assess the power expenditures and emissions of an organizational charging station based on photovoltaic solar networks, Tulpule *et al.* [22] compare optimal charging techniques with uncontrolled charging systems. This study's daytime workplace EV charging behavior takes into consideration a variety of factors, including car parking costs and varied parking sites, to account for differences in solar insolation. According to the findings of this study, utilizing solar charging at work saves 0.6 tonnes of CO_2 per year, which equates to a 55% reduction in emissions when compared to using the home charge (night charging at home) plan. If the home charging system is employed, an optimum charging plan reduces CO2 emissions by 0.36 tonnes, which amounts to an 85% reduction in emissions. Because the expense of SMs and communications networks tends to increase the cost of home charging, a PV-based workplace power source appears to be the better alternative. [23] explores the effects of solar PV arrays put on a company parking lot's rooftop to provide charging facilities for commuters throughout the day. Solar electricity output (up to 12.6 kWh) is high throughout the summer, according to the study, and the most of the energy may be transmitted into the system (V2G) or utilized at work. With a somewhat longer payback period, this may be able to offset some of the early costs. However, during the winter, the generation (up to 3.78 kWh) is said to be adequate for recharging. Despite the fact that it may be important to demonstrate the practicalities, no study on how to balance the extra expenditure in the winter is offered. In addition, Traube *et al.* [24] model a bidirectional DC charger to connect the EV to the power source provided by the photovoltaic system. In addition, the load flow adjustment for the Photovoltaic inverter output is shown. This research examines three scenarios: charging electric cars without any special benefits (e.g., V2G), EVs providing network help when charging, and EVs providing network assistance without charging. On the 1.2-kW PV array, the findings show that such electric vehicle charger will adjust for high abrupt variations in photovoltaic power induced by clouding, which can reach 22.5% of the DC bus voltage per second.

7.6.2 Wind Energy With EVs

Wind energy conversion systems (WECSs) for energy production, as previously noted, are a popular and feasible alternative means of generating power. Several studies have looked at the synergy between WECSs and EVs in various scenarios to see how they affect the electric grid and how feasible they are [26, 27]. Lundand Kempton [26] examines the usage of electrical vehicles offering new functions and oversight in the US energy sector, with a focus on grid interaction with WECSs. Studies [27] investigate how much wind energy can be securely incorporated into an independent power system with EVs nearby. Electric cars are identified as players in main frequency regulation in this study, and their correlations during adaptive charging mode were then examined. Wind penetration in the isolated grid has increased from 41% to 59% thanks to Electrical vehicles and V2G services. All electric vehicles on the road, according to the research, have an interim charge and are willing to assist with grid stabilization. Pillai *et al.* [28] used the hourly Energy PLAN model and restricted duration-dynamic simulation settings to examine the ability of a confined electric grid (Danish island, Bornholm) to incorporate a substantial part of the wind energy system. The pooled EV batteries were employed for frequency control support in the V2G environment. The large fluctuations in wind power penetration produced the frequency instability. According to this study, a 16-MW aggregated EV battery storage capacity permitted 42 MW of penetration of wind energy into the electric grid (through V2G), but only 20 MW would be achievable without employing EV to offer V2G support. For hourly and limited time dynamic simulations, it was able to utilize 82% and 70% of wind energy output of the total output, respectively, with a pooled V2G capacity of 16 MW. The V2G was able to maintain frequency stability in all of the cases.

7.7 Application in Transport Sector

Setting the scene: the Hamburg container terminal Altenwerder
The Hamburg Container Terminal Altenwerder (CTA) (Figure 7.4), managed by Hamburger Hafen und Logistik AG, is the world's first climate-neutral container terminal (HHLA). The terminal has been outfitted with cutting-edge automation technologies and data processing systems since its debut in 2002. As a result, huge container ships can be emptied

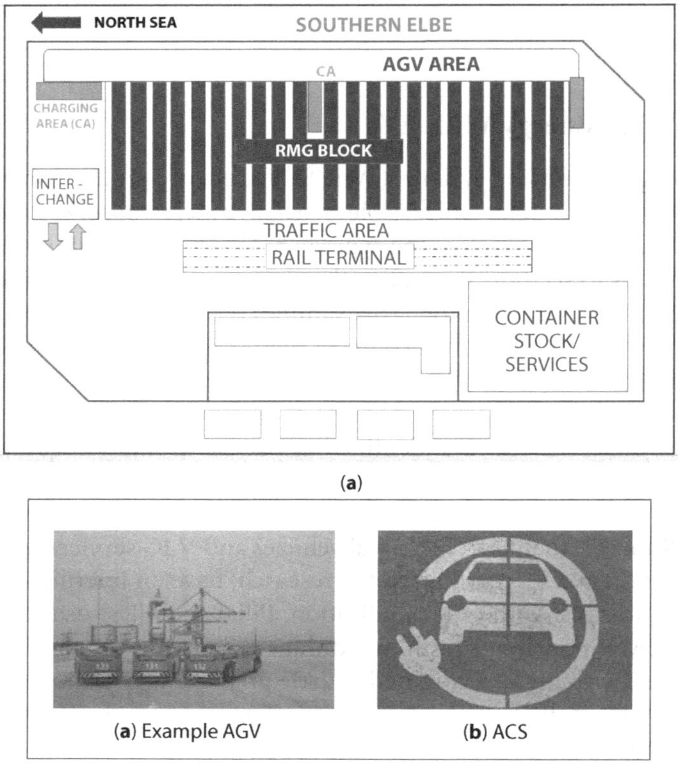

(b)

Figure 7.4 (a) Schematic layout of ACS [34]. (b) AGV and ACS [34].

and loaded more efficiently and with more automation. Because of its 1.4 km of quay wall and four docks, CTA has a compact configuration with less net transit routes. 14 container tower cranes manage the loading and unloading of containers in two stages: first, the vessel is loaded and discharged utilizing twin trolley tower cranes. After that, each cargo is placed into a functioning portal higher above, where it is prepared and inspected in readiness for future transport. The container is then received by a second trolley crane, which places it into 1 of the approximately 90 Automated Guided Vehicles (AGV) currently in operation (Figure 7.4). CTA's AGVs are being replaced with battery-electric systems, which were previously installed with diesel-electric systems. Thanks to European Regional Development Fund funding, CTA's AGV fleet would be fully transformed to fast-charging lithium-ion batteries by 2022. With a maximum charging and discharging rate of 220 kW and continuous use of all

18 automated power charging stations, the energy market could receive about 4 MW of load reduction and augmentation potential. Containers are transported from gantry cranes to storage using these trucks. They are entirely self-sufficient and do not require assistance from others. The AGV uses over 19,000 transponders embedded in the ground to calculate the quickest path. While allowing for other driving automobiles, the software calculates and manages. While allowing for other driving vehicles, software determines and regulates the shortest way to the target. Because AGVs are entirely self-contained, they will work their way to the ACSs and recharge as needed. The ACSs allow for fully automated and bi-directional charging of AGVs, giving power industries and transmission network services additional flexibility.

7.8 Application in Micro-Grid

Researchers investigated into how EVs interact with RESs on a micro grid to accomplish an efficiency-aware aim in [29–32]. Tushar *et al.* [29] propose a strategy for coordinating appliance and EV power consumption that is both efficient and effective. The goal is to use renewable energy sources as much as possible while reducing the quantity of power consumed from the grid. Only wind power, Photovoltaic solar electricity, a central controller, household items, and electric automobiles are considered in the micro grid. It may work in a stand-alone configuration or in conjunction with the macro grid. A mixed integer linear programming (MILP) issue is used to solve the scheduling problem. Electric vehicles may act as power storage device when renewable energy is plentiful, and they can also offer additional power when energy demand is high. They use a Markov chain state transition probability to forecast wind and solar electricity generation over the following 24 hours. Home appliances are categorized based on their tasks; some require exact scheduling while others require more flexibility. Electric cars are anticipated to enter the micro grid via a Poisson process. Charging EVs at home might take a long time since various EVs require varying amounts of energy. The IBM CPLEX MILP solver is used to solve the supplied model, which includes both continuous and discrete decision variables. Unlike Tushar *et al.* [29], Wu *et al.* [30] treat wind energy as a single resource for interacting with EVs, with the goal of improving wind power consumption efficiency in the micro grid. They suggest three coordinated energy-dissipating strategies for Electrical vehicles interacting with wind energy on the basis of control flexibility of EV charging and discharging behaviors. Their model takes into consideration instabilities in wind power output as well as individual vehicle SOC distributions. Valley scanning transmitting, controllable transmitting, and differential transmitting are the three strategies. Using a statistic model from a group of ICE cars, EV guiding trends are represented by departure time, arrival time, and driving distance. The normal distribution is used to simulate departure and arrival times. The logarithmic normal distribution is used to calculate the driving distance. The remaining available energy in an electric vehicle is referred to as SOC. The wind power-generating model is depicted in Section 7.6.2. To create evaluation situations, they use the Latin Hypercube Sampling method. The three techniques mentioned may be beneficial in increasing matching performance, according to the modeling statistics.

7.9 State-of-the-Art Review

Francis Mwasilu *et al.* (2014) examine and evaluates the most recent research and developments in the interplay between electric cars (EVs) with smart grids, as well as a model of the future electrical system. The smart grid's ultimate objective, as well as the eventual adoption of electric vehicles, poses a number of challenges in the field of electric network architecture, connectivity, and administration. A concept for incorporating Electric vehicles into the power network has been proposed after a thorough assessment of sophisticated remote monitoring and networking infrastructures. Several Electric vehicle quick charging approaches are also extensively explored in the context of the vehicle-to-grid (V2G) phenomenon in terms of the ability, consequences, and constraints. Sustainable resources (wind and solar photovoltaic) are being more integrated into the electrical grid. But, at the other end, their intermittent power generation poses distinct challenges in terms of power system network planning, operation, and administration. The adoption of electric automobiles in the power sector can act as a substitution for power grid abnormalities. A research study of sustainable power integration and the most current viable solution incorporating EVs is evaluated in this context, as well as an understanding of the prospective research gap that must be solved. The viability of the smart V2G system is also carefully assessed. The interactions of electric automobiles with the power grid as a potential power system concepts are investigated in this study, and each limitations of the study for potential solutions is identified.

The structure, advantages, and downsides of vehicle-to-grid technology are examined in Kang Miao *et al.* [35]. The essential evaluation approaches for achieving various vehicle to grid targets while meeting a range of constraints are also discussed in this study. The acceptance of electric vehicles as a replacement for traditional internal combustion engine automobiles has been aided by the power dilemma and communal concerns. The role of electric vehicles in the form of vehicle to grid technology has recently risen due to the expansion of the microgrid concept in the electric grid. Vehicle to grid technology allows bilateral energy transfer between electric vehicles and the electric network, allowing the power system to benefit from grid control, spinning reserve, peak load shaving, load levelling, and reactive power compensation, among other features. Developmental techniques are commonly used because connecting car to grid technology is a difficult single objective issue with numerous conflicting aims and limitations.

The issue of electric car charging in the context of transmission network, as well as the link to sustainable power, is the focus of this research in Liansheng Liu *et al.* [36]. The first half of the presentation is an overview of electric vehicles and sustainable power sources, with a focus on the various types of Electrical vehicles and techniques for estimating renewable energy. The present study is split into three areas based on the goals: price-conscious, performance-conscious, and emissions-conscious interactions between Electrical vehicles and sustainable power sources. The overview for each division includes an explanation of basic principles, a description of solutions, and a rating of various works. Finally, many key unsolved concerns with EVs and RESs are discussed, as well as some potential solutions. Among the most effective alternatives for decreasing pollution in transportation networks are electric vehicles (EVs). Electric vehicles (EVs) are growing increasingly popular, putting additional strain on the power infrastructure, which is a big issue. Integrating local energy generation, such as sustainable power sources, with power systems is one effective approach to lessen the impacts. Due to the erratic and retention characteristics of sustainable power systems, coordinating Electric Vehicle charging with other power station demands and sustainable production becomes extremely difficult.

We illustrate how a network of rechargeable battery mobility devices may be utilized to offer frequency retention resources in Stefanie Holly *et al.* [34]. To do so, we undertake a comprehensive study of the use case before recommending technology to address the issue. We give a high-level summary of the tasks and their sub-components, which include:

(a) future accessibility of Automated Guided Vehicles (AGVs) using an artificial neural network
(b) a heuristic method for calculating flexibility in the market
(c) modelling to determine if flexibility schedules are plausible,

(d) a system with several agents that monitors and controls the AGVs in real time and
(e) fleet flexibility is included in a virtual power plant.

In addition, we go through how they proceeded about analyzing the economics of delivering an operation-critical service in a technical environment with a lot of volatility and unpredictability.

7.10 Future Trends

Electric automobiles have been offered as a way to utilize renewable energy sources. Electric vehicles have been shown to provide peripheral grid services such as frequency and voltage control, excess power utilization, and voltage control assistance to improve production, safeguard the electric grid and lower energy network production costs. Enhanced connectivity, management, and monitoring techniques might be used to integrate EVs into the smart grid system. The microgrid can push for EV compatibility for grid support in this situation. Following up on that remarks, this research delves deep into the relationship between smart grids and electric automobiles. To promote the V2G framework's deployment over other energy storage technologies, more research and analysis is required. Modern technology in these areas must be utilized to detect issues and limits in order to accomplish near real-time connectivity and energy monitoring. Communication latency, reactive routing, and cyber security are all key considerations for the V2G transactions framework's effective and efficient deployment in the power grid context. Furthermore, recent research has looked into the possibility of a power grid based on V2G methods. The low penetration of electric automobiles with V2G capability is among the biggest obstacles to EV adoption in the energy industry. Other factors cited by automakers include EV technology's side effects, such as low prices and high efficiency power converters (for EV chargers). Battery wear from frequent charging and discharging cycles remains a problem for successful V2G operation with current battery technology. The research on lithium ion (LFP) batteries has generated some promising results. However, in order to ensure a high penetration of EVs, more thorough studies are needed that consider a variety of research topics, such as battery lifespan extension techniques and price–benefit analyses for its (batteries) implementation in V2G services. In addition, similar research on other battery systems, such as NiMH and other lithium ion compositions have not been published yet.

References

1. *International energy outlook 2011*, Energy Information Administration (EIA), Office of Integrated Analysis and Forecasting, U.S. Department of Energy, Washington, DC, 2011.
2. Australian Energy Market Commission (AEMC), Market arrangements for electric and natural gas vehicles. Approach paper, Sydney, 2011.
3. Kempton, W. and Letendre, S., Electric vehicles as a new power source for electric utilities. *J. Transp. Res. Part D*, 2, 3, 157–75, 1997.
4. Green II, R.C., Wang, L., Alam, M., The impact of plug-in hybrid electric vehicles on distribution networks: A review and outlook. *Renew. Sustain. Energy Rev.*, 15, 1, 544–53, 2011.
5. Shaaban, M.F., Atwa, M.Y., El-Saadany, E.F., PEVs modeling and impacts mitigation in distribution networks. *IEEE Trans. Power Syst.*, 28, 2, 1122–31, 2013.
6. Su, W., Smart grid operations integrated with plug-in electric vehicles and renewable energy resources [Ph.D. dissertation], Department of Electrical and Computer Engineering, North Carolina State University, North Carolina, 2013.
7. Pecas, L.J.A., Soares, F.J., Almeida, P.M.R., Integration of electric vehicles in the electric power system. *Proc. IEEE*, 99, 1, 168–83, 2011.
8. Bessa, R.J. and Matos, M.A., Economic and technical management of an aggregation agent for electric vehicles: A literature survey. *Euro. Trans. Electr. Power*, 22, 3, 334–50, 2012.
9. Pillai, J.R. and Bak-Jensen, B., Integration of vehicle to grid in the western Danish power system. *IEEE Trans. Sustain. Energy*, 2, 1, 12–9, 2011.
10. SAE Electric vehicle and plug-in hybrid electric vehicle conductive charge coupler. SAE standard J1772, 2012.
11. Yilmaz, M. and Krein, P.T., Review of battery charger topologies, charging power levels, and infrastructure for plug-in electric and hybrid vehicles. *IEEE Trans. Power Electron.*, 28, 5, 2151–69, 2013.
12. Chaundhry, H. and Bohn, T., A V2G application using fast charging and its impact on the grid, in: *IEEE Transportation Electrification Conference and Expo (ITEC)*, 2012.
13. Sortomme, E. and El-Sharkawi, M.A., Optimal scheduling of vehicle-to-grid energy and ancillary services. *IEEE Trans. Smart Grid*, 3, 1, 351–9, 2012.
14. Dashora, Y., Barnes, J.W., Pillai, R.S., Combs, T., Hilliard, M., Optimized energy management for large organizations utilizing an on-site PHEV fleet, storage devices and renewable electricity generation. *Energy Syst.*, 3, 133–51, 2012.
15. Sortomme, E. and El-Sharkawi, M.A., Optimal charging strategies for unidirectional vehicle-to-grid. *IEEE Trans. Smart Grid*, 2, 1, 131–8, 2011.

16. Su, W., Rahimi-Eichi, H., Zeng, W., Chow, M.-Y., A survey on the electrification of transportation in a smart grid environment. *IEEE Trans. Industr. Inform.*, 8, 1, 1–10, 2012.
17. Gungor, V.C. *et al.*, Smart grid technologies: Communication technologies and standards. *IEEE Trans. Industr. Inform.*, 7, 4, 529–39, 2011.
18. Guo, Q. *et al.*, Factor Analysis of the aggregated electric vehicle load based on data mining. *Energies*, 5, 2053–70, 2012.
19. International Electro technical Commission (IEC), Grid integration of large-capacity renewable energy sources and use of large-capacity electrical energy storage. White paper 3, 2012.
20. Dallinger, D., Gerda, S., Wietschel, M., Integration of intermittent renewable power supply using gridconnected vehicles – a 2030 case study for California and Germany. *Appl. Energy*, 104, 666–82, 2013.
21. Battke, B., Schmidt, T.S., Grosspietsch, D., Hoffmann, V.H., A review and probalistic model of lifecycle costs of stationary batteries in multiple applications. *Renew. Sustain. Energy Rev.*, 25, 240–50, 2013.
22. Tulpule, P., Marano, V., Yurkovich, S., Rizzoni, G., Economic and environmental impacts of a PV powered workplace parking garage charging station. *Appl. Energy*, 108, 323–32, 2013.
23. Birnnie III, D.P., Solar-to-vehicle (S2V) systems for powering commuters of the future. *J. Power Sources*, 186, 539–42, 2009.
24. Traube, J., Lu, F., Maksimovic, D., Photovoltaic power system with integrated electric vehicle DC charger and enhanced grid support, in: *Proceedings of the15th International Power Electronics and Motion Control Conference (EPE/PEMC)*, 2012.
25. Derakhshandeh, S.Y. *et al.*, Coordination of generation scheduling with PEVs charging in industrial microgrids. *IEEE Trans. Power Syst.*, 28, 3, 3451–61, 2013.
26. Lund, H. and Kempton, W., Integration of renewable energy into the transport and electricity sectors through V2G. *Energy Policy*, 36, 3578–87, 2008.
27. Lopes, J.A.P., Almeida, P.M.R., Soares, F.J., Using vehicle-to-grid to maximize the integration of intermittent renewable energy resources in islanded electric grids, in: *Proceedings of the International Conference on Clean Electrical Power*, pp. 290–5, 2009.
28. Pillai, R.J., Heussen, K., Østergaard, P.A., Comparative analysis of hourly and dynamic power balancing models for validating future energy scenarios. *Energy*, 36, 3233–43, 2011.
29. Tushar, M.H.K., Assi, C., Maier, M., Uddin, M.F., Smart micro grids: Optimal joint scheduling for electric vehicles and home appliances. *IEEE Trans. Smart Grid*, 5, 239–50, 2014.
30. Wu, T., Yang, Q., Bao, Z., Yan, W., Coordinated energy dispatching in micro grid with wind power generation and plug-in electric vehicles. *IEEE Trans. Smart Grid*, 4, 1453–63, 2013.

31. Strnad, I., Skrlec, D., Tomisa, T., A model for the efficient use of electricity produced from renewable energy sources for electric vehicle charging, in: *Proceedings of 2013 4th International Youth Conference on Energy (IYCE)*, pp. 1–8, 2013.
32. Zhu, L., Yu, F.R., Ning, B., Tang, T., Optimal charging control for electric vehicles in smart micro grids with renewable energy sources, in: *Proceedings of IEEE 75thconference on Vehicular Technology Conference (VTC Spring)*, pp. 1–5, 2012.
33. Mwasilu, F., Justo, J.J., Kim, E.-K., Do, T.D., Jung, J.-W., Electric vehicles and smart grid interaction: A review on vehicle to grid and renewable energy sources integration. *Renew. Sust. Energ. Rev.*, 34, 501–516, 2014.
34. Holly, S., Nieße, A., Tröschel, M., Hammer, L., Franzius, C., Dmitriyev, V., Dorfner J. *et al.*, Flexibility management and provision of balancing services with battery-electric automated guided vehicles in the Hamburg container terminal Altenwerder. *Energy Inform.*, 3, 1, 1–20, 2020.
35. Miao, T.K., Ramachandaramurthy, V.K., Yong, J.Y., Integration of electric vehicles in smart grid: A review on vehicle to grid technologies and optimization techniques. *Renew. Sustain. Energy Rev.*, 53, 720–732, 2016.
36. Liu, L., Kong, F., Liu, X., Peng, Y., Wang, Q., A review on electric vehicles interacting with renewable energy in smart grid. *Renew. Sustain. Energy Rev.*, 51, 648–661, 2015.

8

Recent Advances in Integrating Renewable Energy Micro-Grid Systems With Electric Vehicles

Hitesh Alen D'Costa, Sohan R. Alva, Vishal H. Kanchan, Preethesh B. and Rathishchandra R. Gatti*

Department of Mechanical Engineering, Sahyadri College of Engineering & Management, Mangalore, India

Abstract

Decentralizing power grids becomes increasingly essential due to rapid urbanization with increasing power loads across the nation. Over the years, microgrids have proved to be a promising alternative to the traditional centralized grids for serving remote locations where the power distribution costs from the centralized grid have become expensive and difficult to operate. On the other hand, due to advancements in power conversion technologies, harnessing renewable energy sources (RES), such as solar, wind, and wave power, has become practical and cost-competitive compared to their fossil fuel contenders such as coal or shale gas. Considering these two growing trends of microgrids and renewables, renewable energy microgrids prove to be very promising to meet the power demands of remote locations.

One more trend on the power consumption side is the transition of automotive consumer demand from Internal Combustion Engine Vehicles (ICEVs) to electrical vehicles (EVs). Although this transition is environmentally friendly and economical, it adds more dynamic loads to the grid on the demand side. While such fluctuating dynamic loads add complexity to a large-scale centralized grid, they can be handled with reduced logistic complexity to the localized microgrids. Hence, it is a favorable option to integrate local RES microgrids with the EV power supply. This chapter discusses the recent technological trends in systems and subsystems of RES microgrids and EVs that aid in seamless integration, considering the optimal load matching of power demand and supply.

**Corresponding author*: gattirathish@gmail.com

Rathishchandra R. Gatti, Chandra Singh, Rajeev Agrawal and Felcy Jyothi Serrao (eds.) Self-Powered Cyber Physical Systems, (177–192) © 2023 Scrivener Publishing LLC

Keywords: Micro-grids, decentralized grids, renewable energy, electric vehicle, intelligent transportation, electric vehicle charging, energy distribution

8.1 Introduction

Electric cars (EVs), particularly plug-in electric vehicles, are becoming increasingly popular due to their decreased dependence on oil and lower pollutants. By 2022, more than 35 million electric vehicles will be on the road worldwide [1]. This widespread adoption of EVs results in a significant increase in charging demand, putting a strain on the power infrastructure. One potential risk mitigation approach for the electrification of the infrastructure is to network power sources, such as renewable energy grids with fast charging points. One state-of-the-art deployment of such electrification can be found in Tesla's Solar city, networks of Photovoltaic grids with Tesla's EV charging stations. Using renewable energy sources to charge electric vehicles reduces CO2 emissions and even lowers charging prices. However, renewable energy sources, such as solar and wind fluctuate significantly over time because they rely on weather conditions. They are characterized by irregularity and unavoidability. Coordinating EV charging with other renewable energy and grid demand is exceedingly challenging due to these characteristics.

Micro Grid (MG) technologies, on the other hand, enable two-way communication between power consumers and the grid, allowing electric cars to communicate with the grid (EVs). The power infrastructure can handle many EVs while coordinating charging in the context of MG. Much recent research has focused on EVs interacting with RESs in the MG environment. The researchers carefully investigated and classified this study, which looks into these studies in depth. This assessment adds to or differs from prior EV integration assessments. Researchers are looking into saving operational expenses, power generation costs, billing costs, and ways to boost service provider profit margins. These studies examine how electric cars and renewable energy sources may be incorporated into the electric grid to reduce emissions further. As seen above, the interaction between RESs and EVs has been widely studied. This research aims to examine the various technologies utilized in the presence of MG. The goal of every study is crucial, and it necessitates various techniques. This chapter examines several interactions aims in light of this point.

This review article by Liansheng Lin *et al.* (2015) [3] provides an overview of EVs and RESs, focusing on the basic types of EVs and prediction techniques for RESs. The current research efforts are then classified into three groups based on the objectives: Interactions between EVs and RESs that are cost, efficiency, and emission-free.

Francis Mwasilu *et al.* (2014) [2] provides a comprehensive overview and assessment of the most recent research and advancements in the interaction of electric vehicles (EVs) with the smart grid, depicting the future electric power system model. A plan for integrating EVs into the electric grid is proposed after an extensive evaluation of sophisticated smart metering and communication infrastructures. In addition, the capabilities, effects, and limitations of various EV smart charging methods are thoroughly investigated in the context of the vehicle-to-grid (V2G) phenomenon. This study thoroughly investigates the interactions of electric vehicles with the power grids and identities, the issues and solutions for futuristic energy, and the infrastructure for viable solutions are identified.

Li Zhu *et al.* (2012) [5] proposes that we study charging strategies in smart microgrids with electric vehicles and renewable energy sources in this research. An ideal charging strategy is determined based on the states of renewable energy sources, battery states, and the number of charging EVs in order to maximize energy usage while meeting service availability restrictions. Plug-in electric vehicles are gaining popularity (EVs). Charging electric vehicles using smart microgrids powered by renewable energy sources is becoming increasingly common. Although some research on renewable energy sources and electric vehicles in smart microgrids has been done, the stochastic features and dynamic interplay between these two essential green solutions should be extensively examined. The optimal charging problem is modeled as a stochastic decision process. The findings demonstrate that the suggested method can increase EV service availability in microgrids powered by renewable energy sources.

8.2 Electric Vehicles and Renewable Energy Sources: A General Overview

This section covers a broad introduction to both EVs and RESs. In the case of electric vehicles, two sorts of plug-in electric vehicles (PEVs) that are now on the market are described in depth, as well as the chargers for the batteries that go with them. Wind and solar energy are the most researched RESs and the models used to estimate electricity output.

8.2.1 Electric Vehicles

The power converter is a key difference between conventional ICEVs and EVs. A conventional ICEV uses an internal combustion engine to convert burning fuel heat into kinetic motion. An EV, on the contrary, uses a motor to convert electric energy into kinetic motion. Therefore, an EV needs an electric power source to fuel its motor. There are three options, namely, (i) external power sources that can be plugged into the EVs, such as a fast charging station (ii) an on-chassis generator which can generate energy in the vehicle (Eg: PV rooftop on an EV) and (iii) on-chassis energy storage such as batteries. The most common option is the on-chassis energy storage EVs, generally referred to as battery electric vehicles (BEVs). In recent years, there has been increasing research on EVs that use external power sources, generally termed plug-in electric vehicles (PEVs).

8.2.2 Battery Electric Vehicles

BEVs are made of three primary subsystems—electric prime-mover (motor), electronic controller for controlling the kinetic motion, and rechargeable battery system. The electrical battery plays a crucial role in providing a regular supply of electric energy required for the kinetic motion of the EV.

Figure 8.1 depicts the typical architecture of BEVs with the prime mover, rechargeable battery, and controller. The prime mover or the motor converts electrical energy from the rechargeable battery into kinetic motion. This kinetic motion needs to be controlled by employing a controller. There are two typical BEV controllers: two-quadrant controllers and four-quadrant controllers. While a two-quadrant controller can only control forward and reverse motion, a four-quadrant controller has advanced motion controls with regenerative braking controls. Since most of the prime movers used in EVs are typically AC motors, they need an AC power source due to their

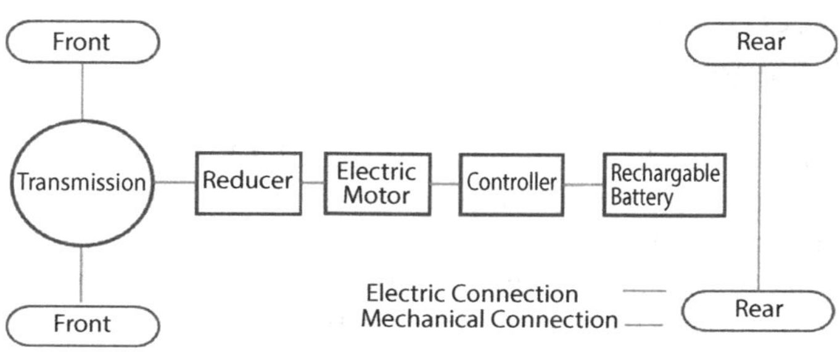

Figure 8.1 The construction of BEVs.

low cost. This necessitates including inverters that convert DC power from the batteries to AC power.

8.2.3 Parallel Hybrid Electric Vehicles

Based on the powertrains, hybrid electric vehicles (HEVs) can be classified into three fundamental types, namely series hybrid electric vehicles (SHEVs), parallel hybrid electric vehicles (PHEVs), and power split hybrid electric vehicles (PSHEVs). Among these, the most commonly deployed HEVs are PHEVs. PHEVs consist of both IC engine and an AC motor as the prime movers acting as dual propulsion systems. This dual propulsion can be operated in primarily two modes—electric charge draining (ECD) mode and electric charge sustaining (ECS) mode. Initially, the PHEV will operate in ECD mode draining the energy stored in the battery until the electric charge falls to a preset minimum threshold. Once the PHEV goes below this threshold, the ECS mode is triggered, wherein the IC engine is used for additional power. An optimal combination of ICEVs and battery power sources can increase the energy efficiency of the PHEV's drivetrain.

8.2.4 Battery Chargers for EVs

Because of their higher energy density capacity, longer battery life and lower ecological impact, lithium-ion batteries are used in most EVs as shown in Figure 8.2. Charging Li-ion batteries, on the other hand, takes work; careful control of the charger's voltage and current output is required. On the other hand, a significant voltage and current fluctuation might damage the battery. As a result, a better battery charger design has the potential to accelerate the development of PHEVs substantially. On-board or off-board PHEVs battery chargers are now available. An onboard charger's power output is usually limited due to weight, size, and cost considerations.

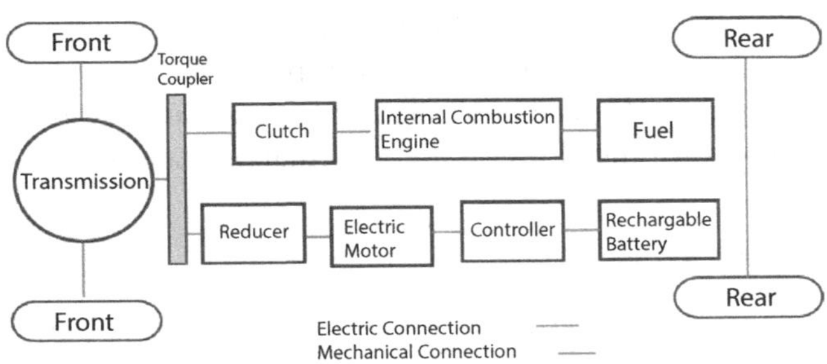

Figure 8.2 The construction of parallel hybrid EVs.

Meanwhile, an off-board charger is exempt from these constraints. There are two types of battery chargers for the PEVs based on the direction of charge flow: uni-directional chargers (UDCs) and bidirectional battery chargers (BBCs). In UBCs, the electrical power power flows from the charger to the battery. Contrary to this, in BBCs, power can flow from the charger to the battery and vice versa. This way, the EVs can send the new power back to the grid in BBCs. However, BBCs are expensive and complex to connect and require more hardware than UBCs. Hence, UBCs are widely adopted in level 1 and level 2 charging. On the other hand, numerous studies have been conducted assuming that bidirectional chargers are available and that vehicle-to-grid (V2G) technology has been realized. Electric cars can return power to the grid, according to the V2G technique [6].

8.2.5 Renewable Energy Sources

People are looking for more potential options as they become more aware of the environmental issues and the depletion of fossil fuels. RESs are considered a favorable alternative to the growing environmental concerns for their environmentally friendly qualities. The two most common types of RESs are solar energy and wind energy. Due to a large amount of viable power generation techniques available, tap them and various power generation techniques.

8.2.5.1 Wind Energy

Wind energy is becoming increasingly important for delivering sustainable energy. The global wind power capacity has increased by nearly 200,000 MW in the last 5 years, reaching 318,137 MW in 2013, as per Global Wind Energy Council. By 2030, the European Wind Energy Association wants to provide 23% of Europe's electricity demand using wind energy. It is critical to select advantageous wind farm locations in order to increase the efficiency of power generated by wind energy. As an initial step, the quantity and quality of wind are evaluated by assigning the wind speed. The use of testing equipment to measure wind speed at a specific vertical distance from sea-level (e.g., 15 m) typically much lower than the actual height of wind turbines.

8.2.5.2 Solar Energy

Solar energy can be harvested using concentrated solar power facilities or photovoltaic panels. CSP facilities concentrate sunlight using mirrors or lenses, resulting in an extreme temperature. When the Solar thermal heat energy is used to convert water into steam. The steam is then used to produce electric power by passing it through a steam generator to power traditional steam turbines or engines. In contrast, PV solar panels use a solid-state semiconductor to convert sunlight directly into energy. PV solar technology has progressed faster than CSP in recent years. According to a study published by the Solar Energy Industries Association, the solar sector in the United States built 4751 MW of new PV solar panels in 2013. Solar cells are the most basic component of PV solar panels and work well at low temperatures. The Energy efficiency of a PV changes only slightly as the temperature rises. For a temperature rise of 1°C, the diminishing sunlight rate is 0.2% to 0.5%. Solar cell manufacturers often use heat insulation to avoid significant temperature swings. While a result, as the ambient temperature changes, the solar variation is quite minimal. This explains why many academics consider solar cell efficiency to be constant. In addition to sunlight, the power generation of solar plants is based on availability, solar irradiation intensity and area of PV panels.

8.3 Microgrid

For small regional consumptions, a micro-grid is an appropriate power distribution system. A microgrid is a decentralized electric distribution facility that operates independently when disconnected from the microgrid [4]. Microgrids can be established for community-based infrastructure such as schools, hospitals as in-house grinders within industrial plants. They are especially useful for remote regions situated far away from the central macrogrid but have access to energy sources. A microgrid is a large set of (1) loader (sinks) that are networked to each other, (2) power sources, and (3) the power distribution network within a clearly defined boundary. Indeed, the concept of microgrids arose from the techno-commercial feasibility of distributed grid systems for generating and consumption. A microgrid is a collection of loads and circulated energy assets surrounded by clearly defined electrical boundaries and operate as a single well-regulated entity concerning the grid. The most significant feature is that it can connect to and disconnect from the grid, allowing it to function in both grid-connected and island mode in a coordinated and controlled manner. The microgrid can also be designed to fulfil the microgrid's unique requirements, such as boosting local resiliency, reducing the number of feeder losses, encouraging native voltages, and improving efficiency by using waste heat and uninterruptible power supply functions or voltage sag adjustment.

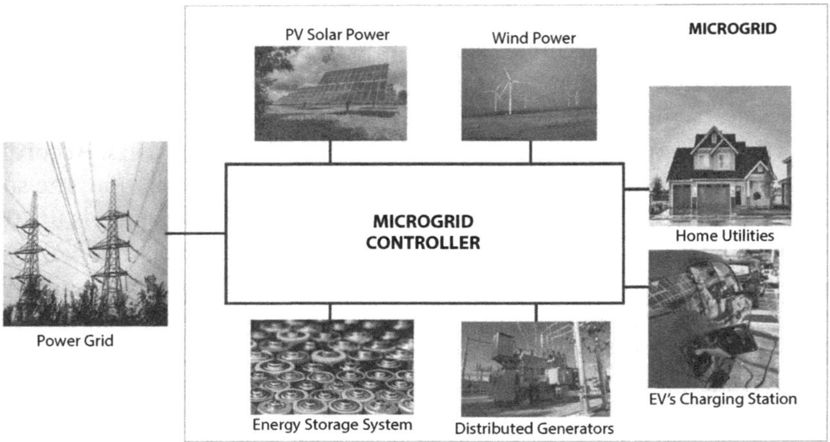

Figure 8.3 Typical microgrid layout.

Microgrids have scope for further integration of new energy sources, both renewable and non-renewable, thus having flexibility in both efficiency and scaling up of their capacity supplies of dispersed energy [7]. Furthermore, using local energy sources to support local loads reduces the distribution of energy losses thereby increasing the grid efficiency. Decisions must be made about whether electricity for transportation will be generated in a carbon-free manner or whether the current energy mix will be maintained—a calculation of the amount of electrical energy required throughout the world. If fossil-fuel-based transportation power is to be replaced, significant. Electrification is required for infrastructure transportation. As a result, it is reasonable to believe that it is possible to generate energy for transportation and other purposes using methods that do not emit carbon dioxide or reduce the use of fossil fuels. Almost all such operations will now include a blend of nuclear and other energy sources. Among RESs, solar energy is considered the primary source of all energy. When it comes to renewable that can be harnessed, wind, solar, and tidal energy play an important role, followed by other RESs. Figure 8.3 depicts a microgrid design and its applications.

8.3.1 Domestic Use

Domestic use as previously stated, microgrids can be used for various purposes, including residential construction, commercial buildings, and EV charging stations. Because residence appliances run on AC electricity, such as televisions, air conditioners, fans, washing machines, HVAC refrigerators, and microwave ovens, microgrid technologies allow these items to be readily operated while reducing the effect on distribution grid stations.

8.3.2 Industrial Use

Industrial use various innovative planning strategies are being advocated to extend and develop energy consumption in various contexts using the microgrid's bidirectional flow of electric power and information. One such expansion is the application of microgrids in various industries, such as petroleum. These businesses are frequently located in areas where natural reserves have been identified through seismographic surveys centralized industrial oil and gas unit has been established. The wellhead is the actual location from which oil and gas are extracted. It is located more than 40 km from the central facility and lacks the power to regulate process variables, such as pressure, temperature, flow, and level. Hence, adequate instrumentation is necessary for the regulation of process flow lines. As a result, microgrids may have advantages, particularly in the oil and gas industries.

8.3.3 Benefits of Microgrids

The key advantages of a microgrid are as follows:

1. Integration of electric vehicles.
2. Direct DC charging saves energy.
3. Integration of renewable energy sources.
4. Control and monitoring have been improved.
5. System dependability has improved.
6. Entrepreneurial possibilities are made more accessible.
7. Systems for storing energy.
8. It is possible to charge wirelessly.

Improved public health consumers will see lower electric costs, while utilities will see more income.

8.3.4 Locations of Microgrid

Over 30% of the energy usage-related radiation and about 32% of energy consumption is due to residential and commercial infrastructure. Solar energy and wind energy prove to be promising RES energy generation for such facilities due to reasonable costs and PV panels and turbines. That is affordable for such facilities. These RESs were primarily developed for community microgrids to facilitate the power of multiple EV charging stations and achieve grid dependence.

A microgrid that combines RES systems with EV infrastructure provides energy security, cost savings, and reliability. Charging points will serve as energy gateways, allowing electrical power transmission and energy storage in the vehicles. According to the world forecast scenario of microgrid revenue in the world market 2013 to 2020, the average scenario of microgrid revenue in the world market 2013 to 2020 is ($40 billion), while the revenue from the aggressive projection scenario is about $58,000 million.

8.4 Interactions Between Cost-Conscious EVs and RESs

This part looks at research initiatives to establish cost-conscious interactions between EVs and RESs. Depending on your point of view, cost-aware engagement may be viewed in various ways. Charging service providers entails lowering operating costs while increasing profit. For charging service users (EV owners), it shows a desire to reduce the cost of battery charging. It can include power generation, system lifetime costs, and energy transmission costs for the grid.

8.4.1 Operational Cost Reduction

Operational cost reduction charging companies are concerned about operating costs, which can be viewed from two angles. In one case, the interaction of EVs and RESs is considered in a specific environment, such as a microgrid. The topic of minimizing operational costs may be investigated from the standpoint of such a system. The charging utilities are primarily interested in lowering their operational costs. To minimize the operating costs, optimization techniques, such as nonlinear programming or mixed integer programming can be employed. Evolutionary algorithms (e.g., particle swarm optimization), control strategies (e.g., fuzzy control), and other methods can be used to address the issues described. Microgrids are a common application case for lowering system operating costs. Information and communication technologies can be used to control the component systems of a microgrid. The microgrid is a frequent application case for system running costs. Information and communication technologies may regulate power generators, energy storage devices, and loads in a microgrid. Wind power generators, PV solar power generators, distributed generators, battery energy storage systems (BESS), home EV charging appliances, and EV charging stations are among the subsystems that make up a microgrid, with the microgrid controller coordinating and scheduling power supply. A microgrid can connect to the electrical grid to exchange electricity. Several studies have been carried out to minimize the system operational cost of a microgrid with EVs and RESs.

The microgrid comprises renewable energy, utilities, battery switching stations, and transmission infrastructure. EVs and fast-charging stations will be charged when there is an excess of power generated by RESs. If there is a power outage, EV batteries and BSS can be used to generate electricity. The basic idea is to combine EVs, BSS, and loads to reduce the microgrid's operating costs. They use a price-based approach to coordinate EV and BSS charging in grid-connected mode. As a result, EV costs have decreased while BSS earnings have increased.

8.4.2 Lowering the Electricity Generation Cost

EV charging providers and users are affected by the cost of energy generation. Researchers have been studying the interplay between EVs and RESs to lower the power generation cost. The studies employed mathematical models such as optimum energy flow and the knapsack problem, as well as associated solutions such as dynamic programming and PSO. The top-level controller manages RES scheduling usage on an hourly basis. A middle-level controller schedules the charging demands of electric vehicles on a minute scale. The time-level controller manages EV charging in real-time by utilizing power grid frequency variation. The Knapsack problem is used to forecast conventional and wind energy use.

8.4.3 Growth in Profit or Benefit

Chargers for electric vehicles (EVs) want to increase their profit or gain from offering charging services. This goal may be achieved with the aid of renewable energy. It can be examined from the standpoint of maximizing investment returns or managing power delivery. Linear programming and stochastic programming are examples of models used in this area. The simplex approach and linearization method were used to solve the formulations.

8.4.4 Reduction in Charging Cost for EVs Owners

One of their main worries is the expense of charging. The cost of EV charging may be efficiently lowered if the fluctuation of RESs generation and EV charging needs can be planned correctly. Researchers have used quadratic programming, stochastic programming, and other techniques to formulate issues of minimizing charging costs. The unpredictability of renewable energy is extensively handled in their models. Meanwhile, many approaches, such as Lyapunov optimization and the graph search algorithm, have effectively tackled the optimization problem.

8.4.5 Other Cost-Conscious Efforts

The mentioned research activities are divided into four categories, with each part having the same cost-conscious goal. There are, however, other worthwhile projects with other objectives, such as decreasing system costs and lowering distribution costs. Similarly, mixed-integer nonlinear programming, linear programming, and other related activities are modeled and solved using logic control and genetic algorithms.

8.5 Interaction Between Efficiency-Conscious EVs and RESs

The efficient use of renewable energy may considerably cut fossil energy usage. This can have several advantages, such as lowering air pollution and improving customer cost performance. The interplay of EVs and RESs can help achieve this goal to some extent. The research initiatives mentioned in this section aim to accomplish three goals: increasing the deployment of RESs stations, improving energy distribution, and improving energy management. Optimization techniques, such as linear programming and convex programming, can be used to express efficiency-aware objectives. Optimization tools (e.g., IBM CPLEX) and evolutionary algorithms are currently available solutions for the challenges mentioned above in this area (e.g., genetic algorithm).

8.5.1 Microgrid Implementation

The research on EV-RES interaction in microgrids relevant to Operational efficiency goals has been completed, and an efficient scheduling system for coordinating appliance and electric vehicle use has been developed. The goal is to maximize the power generated by renewable energy sources while reducing the quantity of electricity used from the grid. Only wind power, PV solar electricity, a central controller, household appliances, and electric vehicles are considered in the microgrid. It may run in self-contained mode or in a manner that interacts with the microgrid. A mixed integer linear programming (MILP) issue is used to solve the scheduling problem. EVs may be considered energy storage when renewable energy is abundant; when electricity demand is high, EVs will offer additional power. They utilize a Markov chain state transition probability to forecast wind and solar electricity generation over the following 24 hours. Home appliances are classified according to their functions; for example, some demand precise scheduling while others require greater flexibility. The arrival of electric vehicles in a microgrid is thought to be a Poisson process. Charging EVs at home can take a long time since EVs with different amounts of needed energy are allocated randomly. The IBM CPLEX MILP solver is used to solve the specified model, which contains both continuous and discrete decision variables.

8.5.2 Increasing the Use of RESs

EV charging is a form of variable load that may be utilized to maximize the usage of renewable energy sources. They utilize an aggregator to manage the charging behavior of electric vehicles and meet demand from residential households at the same time. Only the behavior of full-time workers' electric vehicles is modeled using empirical driving characteristics. They get the demand for residential households from a German data set.

The electricity was generated using RESs and a typical generator. The sequence is only activated when the energy needed to charge the domestic application exceeds the amount of renewable energy available. As a result, conventional generation will be reduced if EV demand can be coordinated with available RES power.

8.5.3 Other Works With a Focus on Efficiency

Other works with a focus on efficiency according to Germany's weekly driving profiles. A method for optimizing individual electric vehicle charging to maximize the use of RESs. The charging duration for electric vehicles is calculated by identifying times when the most renewable energy is generated. The authors of this article believe that EVs can be notified when RES becomes available. In this way, EV owners can tailor their charging requirements to the highest renewable energy output. They examine 6466 driving profiles and categorize them into two groups, namely (full-time employees and retired). The average driving distance of EV owners differentiates the groups. Then they perform an uncoordinated benchmark simulation to determine the baseline level of RESs used in EVs. The relative proportion of RESs is calculated using data from the eastern half of Germany in 2011 on load and RESs generation. The time difference is 15 minutes.

Based on the data presented above, EV owners can choose time slots when the percentage of RESs generation is highest. A linear programming technique is used to model individual EV optimization. According to the simulation results, the key elements from driving profiles are Fast-charging location and overall driving distance covered for a week. Employee groups can increase wind share by at least 80%. Wind share, on the other hand, could be increased by at least 103%. The findings are exciting for academics, and the authors suggest that they expand their research in the future to include individual economic implications.

8.6 Open Problems

Three key outstanding challenges shared by various aims of EVs interacting with RESs are addressed in this section.

8.6.1 Grid Integration of RESs on a Large Scale

There is a considerable distinction between RES-generating facilities and traditional power plants. Consider a wind turbine, which uses power electronics to generate electricity. Synchronous generators are more commonly used in traditional power facilities. Integrating RESs into the electricity system, this disparity between RESs and traditional generators creates additional challenges. These issues can be seen from two perspectives. Voltage control, fault current, harmonic distortion, and flicker all affect the local power system. In the case of microgrid systems, the consequences include load-generation imbalances, reactive power generation, and decreased frequency management. The same difficulties arise when a PV solar power system interacts with a microgrid system. However, these issues will not threaten power infrastructure if RESs are only used on a limited scale. As RESs become more widely used, these issues will worsen. One potential path is to propose more realistic techniques to accomplish large-scale RESs integration into the grid.

8.6.2 The Use of EV Batteries in Conjunction With RESs

The use of EV batteries in conjunction with RESs The battery pack of an electric vehicle is used as a source of energy storage in the study presented in this paper. Suppose there is an excess of power generated by RESs, and buffered in EV batteries. When there is a power outage, the battery of an electric vehicle can be discharged to provide electricity. This, however, will extend the battery's charging-discharging cycle. Almost all-electric vehicles on the market today use Li-ion batteries. As the frequency of charging–discharge cycles increases, the power and capacity of a Li-ion battery will dwindle dramatically—several factors influence power, including the loss of lithium and an increase in dependability. The battery capacity will diminish proportionally to the depletion of active materials such as graphite and lithium. As a result, the usable life of the energy storage will be reduced. EV owners will carefully consider whether to participate in the integration of renewable energy sources. There are two approaches to dealing with this issue. One possibility is to tweak the balance between energy storage life and RES integration. Researchers can create a gaming strategy that will benefit electric vehicle owners. The other option is introducing new energy storage (battery) technology with a prolonged usable life.

8.6.3 V2G's Ability to Allow the Interaction of RESs

In the current market, however, the majority of EVs produced do not support V2G technology. The major objective of integrating electric cars into the transportation industry is to replace internal combustion engines (ICE). Some issues must be addressed in depth for V2G to be practical. To begin, EVs must be fitted with bidirectional power converters and sophisticated communications devices in order to use the V2G feature. Second, the charging station must have a smart charging point that can interact with the EV's control unit. Third, as the number of EVs grows, the grid operator's control approach will become increasingly complicated. Even if their EVs can support V2G, some owners may refrain from participating in the V2G application. The number of useable EVs cannot be precisely determined. Fourth, for V2G, EVs are mostly employed as energy storage or sources. The distribution of EVs cannot be predicted properly in advance due to their mobility. This may cause difficulty with the power distribution network's ability to schedule electricity among EVs, RESs, and charging stations. Finally, it is necessary to charge and discharge the battery of an electric vehicle regularly. This will shorten the battery's usable life cycle. To make V2G a reality, all of these issues must be resolved.

8.7 Conclusion

The transportation industry is being transformed by electric cars. Electric cars are expected to have various positive impacts, including less dependency on fossil fuels and lower carbon emissions. Meanwhile, charging a large number of electric vehicles has a significant impact on the power grid. A feasible option for reducing this impact is to use microgrid technology and local power generation, such as renewable energy. This review looks at all of the existing works in this genre. The related works are divided into three categories: cost-conscious charging, efficiency-conscious charging, and emissions-conscious charging. For each area of activities, an overview of the focused issues and solutions is provided. The results of their evaluations and comparisons are also shown. Finally, several key unanswered questions and their solutions are explored. This evaluation is intended to help fellow students and instructors better understand the present state-of-the-art and contribute more to this subject.

References

1. Tanaka, N., *Technology roadmap: Electric and plug-in hybrid electric vehicles*, International Energy Agency. Technical Report, 2011. https://www.iea.org/reports/electric-and-plug-in-hybrid-electric-vehicles-roadmap
2. Mwasilu, F., Justo, J.J., Kim, E.-K., Do, T.D., Jung, J.-W., Electric vehicles and smart grid interaction: A review on vehicle to grid and renewable energy sources integration. *Renew. Sustain. Energy Rev.*, 34, 501–16, 2014.
3. Liu, L., Kong, F., Liu, X., Peng, Y., Wang, Q., A review on electric vehicles interacting with renewable energy in smart grid, 2015.
4. Richardson, D.B., Electric vehicles and the electric grid: A review of modeling approaches, impacts, and renewable energy integration. *Renew. Sustain. Energy Rev.*, 19, 247–54, 2013.
5. Zhu, L., Richard Yu, F., Ning, B., Tang, T., Optimal charging control for electric vehicles in smart microgrids with renewable energy sources, 2012.
6. Su, W., Eichi, H., Zeng, W., Chow, M.-Y., A survey on the electrification of transportation in a smart grid environment. *IEEE Trans. Ind. Inf.*, 8, 1–10, 2012.
7. Bukhari, S.A.A.S., Cao, W.-P., Soomro, T.A., Guanhao, D., Future of microgrids with distributed generation and electric vehicles, 2017.

9

Overview of Fast Charging Technologies of Electric Vehicles

Sohan R. Alva, Vishal H. Kanchan, Preethesh B., Hithesh Alen D'Costa and Rathishchandra Ramachandra Gatti*

Department of Mechanical Engineering, Sahyadri College of Engineering & Management, Mangalore, India

Abstract

Advances in new fast chargeable batteries, as well as fast-charging infrastructure, have accelerated the growth of the electric vehicle market. So far, slow charging of electric vehicles has been one major hurdle for people to migrate from fossil fuel vehicles to electric vehicles. Recharge times must be comparable to the time it takes to fill a gasoline tank to acquire widespread consumer acceptance. Electric vehicle (EV) battery charging durations must be decreased to an average of seven minutes to contend with fossil fuel-powered transportation. This chapter examines EV fast-charging technologies and their effects on EV power train systems such as battery systems and power electronic systems. The chapter also discusses recent novel approaches and the leveraging of emerging technologies such as wireless networks, AI/ML, and energy storage materials. For systems-level design and power conversion topologies to advance state-of-the-art fast charging. The feasibility of different levels of EV charging, namely, Level 1 (5 kW or less), level 2 (5–50 kW) and level 3 (Ultra-fast charging > 50 kW), as per the Department of Energy, US, is discussed.

As the charging rates increase, new issues emerge in the power train systems. Typical issues include heating of energy storage elements causing both component failures and safety concerns, increased size of electrical conductors with high amperage rating, power demand exceeding rated power and issues from electrical grid utilities supplying the necessary electrical power for fast-charging stations. These challenges are discussed along with the state-of-the-art solutions and possible future solutions. Finally, the chapter focuses on the impact and leverages of fast charging technologies on self-poweredautomotive cyber-physical systems.

**Corresponding author*: gattirathish@gmail.com

Rathishchandra R. Gatti, Chandra Singh, Rajeev Agrawal and Felcy Jyothi Serrao (eds.) Self-Powered Cyber Physical Systems, (193–212) © 2023 Scrivener Publishing LLC

***Keywords*:** Electric vehicle, fast-charging, battery technologies, energy storage, fast-charging stations, automotive CPS

9.1 Introduction

The transport system in the world largely depends upon fossil fuels. Over the years, the usage of fossil fuels hasreached exorbitant volumes, which has threatened the future availability of fossil fuels. One of the aspects of environmental pollution is the release of harmful gases produced by combustible vehicles. As a solution, there arose a need for an environment-friendly transport system. Plug-in electric vehicles were a better solutionto reduce emissions and ecological damage.

As a result of research work done in the field of electric-operated vehicles, many improvements and new concepts have emerged in this field. Electric cars can be charged by plugging into a charging point and drawing power from the grid. They use rechargeable batteries to store electricity. A public charging station or a home charger can charge electric vehicles. Nevertheless, a major problem in electric vehicles is charging the battery in minimum time and storing the charge for longer.

9.2 Different Levels of Charging Electric Vehicles

There are different levels of charging electric vehicles:

Level I—less than 5 kW, AC voltage of 120 V [1]
Level II—5 kW and 50 kW, AC voltages of 120 and 240 V [1]
Level III—greater than 50 kW, DC voltage up to 600 V [1]

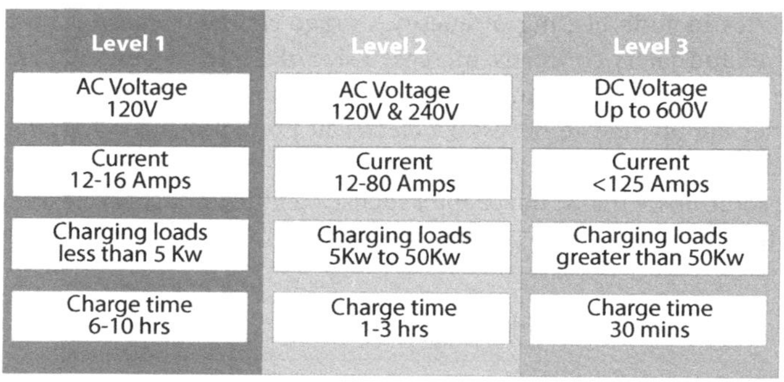

Figure 9.1 Different charging levels of EV.

9.2.1 Level I

Electric vehicles (EVs) can be connected to the most commonly grounded electrical socket. It is also called "home charging," comprising an onboard charger for electric vehicles that harnesses energy from the domestic grid AC supply [2]. Slow charging distinguishes AC level 1 charging, which battery providers provide, so it lacks a standardized charger. A 10-kWh battery takes 3 to 20 hours to charge using a standardized charger [3]. Most "adaptable" batteries are used for 2 W/3 W, and some permitted 4 W EVs of India employ them [3]. This is the optimum charging range for fixed batteries as they are air-cooled. The battery charging rate must be low to manage heat or thermal dissipation.

9.2.2 Level II

This method uses dedicated alternating current (AC) typically found in private and public infrastructure as Electric vehicle supply equipment (EVSE). On the other end, electric vehicles have built-in onboard chargers that accept energy from these EVSEs [2]. Level II is sometimes known as "AC Fast charging." Typically, 3.6, 7, and 10 kWh AC chargers with compatible connectors are used for AC level 2 fast charging. In 2021, level 2 fast chargers will be included in all-Indian BHARAT 2 AC and other country-specific standards for most of the 4WD launches. With this calibre, a 10-kWh battery top-up takes 90 minutes, and a full 30-kWh cycle takes around 4 hours. Liquid cooling batteries are used by global vehicle original equipment manufacturers (OEMs) for 2/3/4 wheelers to provide quick charging. Fast charging can shorten the battery's life if the cells are not kept at optimal temperatures. OEMs have previously proven that batteries with improved cooling capability can now run for 1.5 million kilometers before losing their capacity [3].

9.2.3 Level III

This method makes use of dedicated direct current (DC). In this, the EVSEs allow EVs to receive electrical supply from an off-board charger of government and privately owned EV infrastructure. The conductive charging system architecture supplies energy via an off-board charger [2]. DC & Level 3 Fast charging technology bypasses the vehicle's onboard charging circuit and delivers direct current to the batteries. As the cooling and battery management system becomes even more important to establish an equilibrium between cooling costs and charging competence, various players are studying combinations of AC and DC to find a balance between cooling costs and charging efficiency. Using high-speed FAST charging, a 40-kWh battery may be charged to 80% capacity in 30 minutes [3]. Figure 9.1 shows different charging levels of EVs.

9.2.4 DC vs AC

Level 3 Charging or DC Fast Charge are more technical phrases for fast or rapid charging. DC Fast charge stations are high-capacity commercial installations located only in public areas. Direct Current (DC) is used to charge batteries in fast charging. Electric vehicles and plug-in hybrids feature a built-in converter, or rectifier, to convert AC to DC. The charging speed is partly determined by the converter's ability to convert AC to DC [4]. Alternating current (AC) is used to charge electric vehicles at home.

9.2.5 Fast Charging

Fast charging attempts to recharge electric car batteries simultaneously as combustion vehicle refuelling. Fast charging takes roughly 20 minutes to charge up to 80% capacity. Fast charging is also called level 3, rapid, or quick charging. Table 9.1 lists the type of charge (along with their state

Table 9.1 DC charging power levels.

Charge type	Power level, kW		
	Small cars	Medium cars	Heavy duty
Quick charge, 70% SOC, 60 min	20	35	75
Rapid charge, 60% SOC, 15 min	60	125	250
Fast charge, 100% SOC, 10 min	125	250	500
Hybrid (Plug-in), 30 min	10	20	40

of charge [SOC]) and charges power levels for different vehicles. The California Air Resources Board (CARB) establishes a certification criterion for fast charging in the zero emissions vehicle (ZEV) mandate program as a 10-minute charge that permits the vehicle to go 100 miles [1].

Despite the advantages, some setbacks of fast charging must be addressed. One of the impacts of fast charging is the battery capacity of electric vehicles, which declines with frequent fast charging. A high KVA-rated current is passed to the EV battery during fast charging, adversely impacting the batteries and the power management circuits.

9.3 State-of-the-Art Fast-Charging Implementation

Among fast charging technologies, extreme fast charging (XFC) technology is proven to satisfy the present and future refill demands of electric mobility. Tu [5] discusses how XFC technology can benefit EVs if implemented with new power electronics converter elements. XFC stations need to beequipped with new solid-state transformers (SSTs) compared to the conventional line-frequency transformers for fast charging. Medium voltage SSTs are beneficial for XFC stations for charging electric vehicles.

A queuing math model for estimating the future deployment of charging infrastructure was developed by Gnann *et al.* [6]. Sweden and Norway's charging big data set was used as the basis for the model development. The model explained the future energy infrastructure demands for fast charging and how comparable the investments made on fast charging stations were to the fossil and other fuels. Theratio of 1000 batteries charged per fast charge point for a power rating of 150 kW. The payback excess was approximately 0.05 to 0.15 €/kWh per charging station. Certain power requirements, such as battery capacity and power rating increase day by day and thus pose a challenge for the design of fast charging infrastructure. More advanced models are required to predict the infrastructure required for fast charging to meet the future energy needs of the growing EV market.

Collin *et al.* [7] provide an overview of EV rapid charging methods and their implications for battery systems, including heat management and limits. The articleprovides potential new techniques and prospects for power electronic converter topologies and systems-level research to enhance the state-of-the-art in fast charging.

Bryden *et al.* [8] discuss two models for developing fast charging facilities—one for forecasting the peak time slots during which fast charging is utilized for EVs and the other to predict the power capacity requirements of fast charging facilities. These two models were predicted using comparative studies on fueling vehicles at gasoline stations. The data such as start time, finish time, and distance travelled were collected from these fossil fuel vehicles, assuming them as EVs instead of the comparative approach. Peak fuel usage was observed in the evenings for lengthy trips between 3:00 PM and 7:00 PM. It was alsoestimated that an average EV equivalent would require about 400 kW of fast charging power, assuming 3 miles/kWh efficiency of the EV and the trip requirement of 1200 mph for 80% of lengthy trips.

Wolbertus and Van den Hoed [9] aim to examine current charging trends at fast charging stations and the function of fast charging compared to other charging alternatives. These trends are investigated concerning the cars' technological capabilities, and it is discovered that as battery capacity increases, the demand for quick charging diminishes. However, there are signs that rapid charging is seen as more convenient for vehicles with high charge capabilities, as they are used more frequently. If chargingcapabilities rise in the future, such data suggest a higher proportion for rapid charging. According to a geographical study's results, most rapid charging occurs at a significant distance from home, implying that most charging takes place "on the road." Some fast-charging sessions are convenientlylocated near home, especially for people who need access to private home charging. This indicates that quick charging has some future promise in cities with much on-street parking.

Botsford and Szczepanek [1] discuss the challenges facing the future widespread usage of electric cars (EVs) in terms of battery charging infrastructure, both rapid and slow. The following are specifically discussed:

- Definitions of charge schemes: What is rapid charging, and how does it work? What does it mean to chargeslowly?
- Requirements for infrastructure and grid impacts: connectors, charger/vehicle communications, time-of-use power prices, and grid upgrades/synergies are all factors to be considered.
- For the key rival lithium battery chemistries, the trade-offs between cycle life and charge methods, Vehicle cost and charging time, and power density vs energy density must be considered.

Sbordone *et al.* [10] offer an overview of the various types of electric vehicle charging stations and storage solutions enabling the integration of EV charging stations into smart grids based on current international European standards. After that, the real-world implementation of an electric vehicle rapid charging station with an energy storage system (ESS) is detailed. This prototype was created, implemented, and is now accessible in ENEA laboratories in Italy. The prototype system has been put through its paces to evaluate its capabilities in conjunction with a smart grid accessible at the same ENEA facility, which includes a smart metering system.

Morrissey, Weldon, and O'Mahony [11] studied household event data on the frequency of fast charging in domestic households. They found that, most often, fast charging was done during the evening when there is peak grid power usage. This can create additional loads to the grid; hence, incentivization to encourage off-peak period usage was recommended in their study. They also found that car parks with the highest usage frequency would be the best avenues for placing the fast-charging facilities. This indicates thatpublic and private fast-charging car park facilities may shortly have good market potential.

9.4 DC Fast-Charging Structure

A high-power DC charger transforms an incoming 3-phase AC supply to the DC voltage required by the vehiclebeing charged using an AC-DC and DC-DC converter. A data transfer channel is also added to give information about the vehicle and the charging condition of the battery. Owner data and Vehicle informationare protected by one final element: a secure data

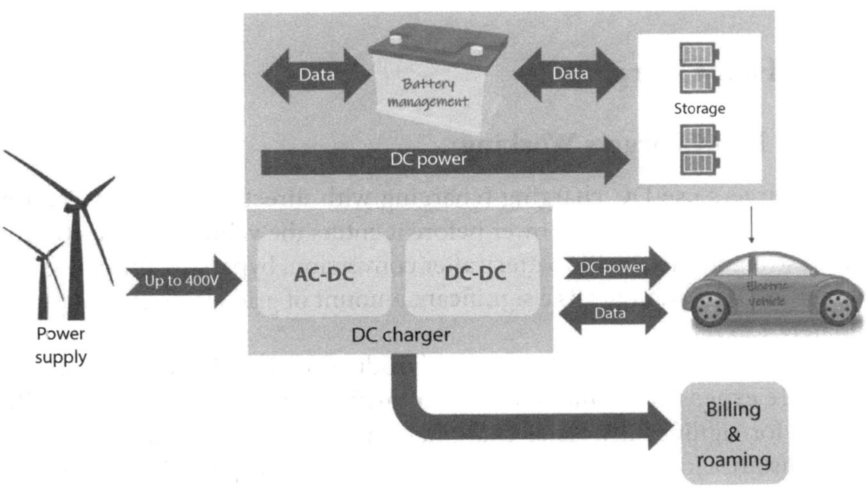

Figure 9.2 DC fast charging structure.

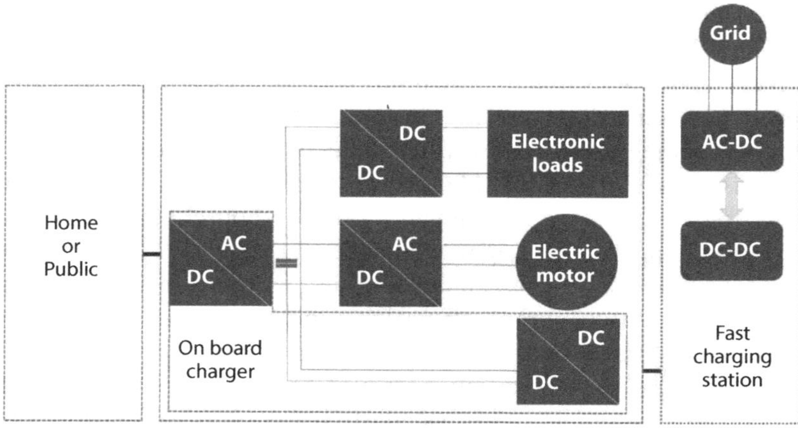

Figure 9.3 Charging system configuration for electric vehicle.

connection for billing purposes. Figure 9.2 shows DC fast charging structure. Figure 9.3 shows the charging system configuration for the electric vehicle.

To allow electricity to flow from the mains to the EV, three functions must be performed in general. Throughout the charging process, the supply voltage must be consistently higher than the primary EV electrical system voltage. As EV electrical system operates on direct current, the alternating current supplied by the mains must be converted or rectified to direct current. The EV must have a physical connection to the power source [12].

9.5 Fast Chargers

9.5.1 Fast Chargers Working

Fast chargers use DC charging (charging with direct current) for electric vehicles; they convert the power before it enters the vehicle. The electricity is fed straight into the EV battery after conversion, bypassing the car's converter. A DC system needs a significant amount of grid electricity (approximately 125 A).

As a result, its operating expenses (production, installation, and operation) are expensive, resulting in higher tariffs. However, because it usually allows for significantly faster charging, it is the ideal charging option for long-distance journeys (for cars that support DC charging). These chargers can be found more near highways than in residential or office areas.

Level 1 DC fast chargers include a 48 V/72 V output voltage, 10 kW–15 kW power outputs, and a maximumcurrent of up to 200 amps.

Level 2 DC fast chargers offer output voltages up to 1000 volts and power outputs ranging from 30 kW to150 kW.

9.5.2 DC Plug Connectors

Electric vehicle manufacturers use around four DC fast charging connectors [13].

SAE Combo Charging System (CCS)—Used by automobile manufacturers such as BMW and Volkswagen. CHAdeMO—Nissan, as well as other Japanese manufacturers, such as Mitsubishi. Supercharger—Tesla standard connector.

GB/T—BYD and other Chinese enterprises. This standard is also used by Mahindra and Tata electric vehicles.

9.5.3 EV Fast-Charging Infrastructure

Electric vehicle supply equipment (EVSE), also known as EV charging infrastructure, is an important partof a healthy EV ecosystem that necessitates a careful design and specialized electrical infrastructure at various levels of the distribution grid. Electric car charging stations are a collection of power electronics that securely deliver regulated electricity from the grid to vehicle batteries. They are usually placed on the wall or a pedestal. Chargers of various sorts deliver varied current and voltage levels to meet the needs of different vehicles' batteries. The wattage of electric car chargers ranges from 500 watts (W) to 500 kilowatts (kW). Faster charging speeds are expected to be possible as battery chemistry improves.

DC chargers or DC fast charger's potential to give higher charging rates. They also direct current to thevehicle battery and bypass the onboard converter. Almost all DC fast charging stations are commercial operations owned and controlled by a charge network operator. DC Fast charging requires dedicated grid infrastructure as well as some level of design and planning phases. DC fast charging stations,at the very least, requires the installation of dedicated three-phase power supply equipment that uses substantially more current than AC charging options.

9.6 Today's Situation and Future Needs

In some countries in Europe, it is estimated that there is at least one fast charging station with a power ratingof 150 kW and above for 1000 EVs [4].

This ratio of fast charging stations to the number of EVs may improve shortly. The current payback benefit of approximately 0.05-0.15 €/kWh may also improve over time by economies of scale. Fast charging mainly depends on grid factors and consumer factors. On the consumer end, fast charging depends on battery capacities and EV power rating. The grid factors include the peak demands, grid capacity, number of fast charging stations to each grid, other competing grid loads, and preferences for the loads, to name a few. Extensive research is required to collate the data on energy usage, frequency of use and frequency of EV charging, battery capacities used inthe EVs of different kinds, and demographic data of EV ownership. Suitable AI algorithms can be used to devise fast charging requirements for a specific location based on these parameters.

9.7 Fast-Charging Point Power Requirements

One of the financial requirements to be considered during the design and installation of fast charging points (FCPs) is that they must be made future-proof—meaning that once they are installed, they need not be replaced and can have only minor upgrades for the next X years, where X is the time where is economically viable capital returns. Ideally, there must be mutual information exchange between the EV OEMs and the FCP OEMs to ensure that corresponding FCP upgrades meet the EV power upgrades. One of the ways to do this is by billing norms that can mandate EV power upgrades. Most FCPs that are currently used have 50 KW capacity but may have to upgrade to 350 KW to meet future EV consumption. Two typical FCP protocols, Combined Charging Systems (CCS) and CHAdeMO, suggest this 350-kW capacity upgrade for the near future.

The USABC is working on battery goals for a mass market, commercially viable full-battery electric car. The USABC's goal for fast charging is to get an 80% charge in 15 minutes. If the EV reaches 3 miles per kWh, this equates to a charging rate of 320 kW or 16 miles per minute for a 100-kWh EV. The power needs for charging electric cars on lengthy excursions are discussed in this section (as explained in Bryden *et al.* [6]). The stopping time (TB) is chosen to be less than one hour since lengthier stops may signal that the driver has arrived at his destination and is thus ignored. A 1.5-hour journey (t1) followed by another 1.5-hour trip (t2) is the baseline situation for trip lengths (t2). It is worth noting that these 1.5-hour journey lengths are unconnected to the previous long-distance definition, which is a trip in which the EV range is exceeded, with a base case of 200 miles, because travelling 200 miles in 1.5 hours is unachievable. In this, 1.5-hour trips were chosen because they are long enough for the driver to be on a long-distance trip while still being short enough to give enough data points for the study. While travelling long distances, people stop for short amounts of time, with 60% pauses lasting less than 12 minutes. To put charging power into perspective, if an EV achieves 3 miles per kWh, then 10 miles per minute is 200 kW. To meet 80% of travels, the charging rate must be 20 miles per minute or 400 kW. To meet 80% of travels, the charging rate must be between 18 and 23 miles per minute, or 360 to 460 kW [8].

9.8 Recent Technologies in Fast Charging, Machine Learning, and Artificial Intelligence

The technologies used to build and use EV charging infrastructure can substantially impact energysystems. Machine learning and artificial intelligence have recently advanced rapidly, spurring innovation in electric vehicle technologies.

9.8.1 Machine Learning

Using computational intelligence for charge scheduling during long-distance travel reduces the long charging time. Machine learning is used to

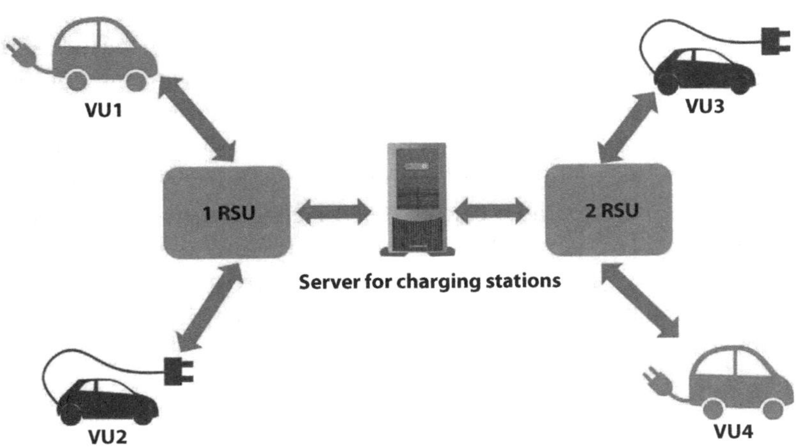

Figure 9.4 Charging station selection scenario.

schedule an electric car's charging in a charging station, allowingthe driver to stay within driving range and plan their route with the least time spent waiting for the vehicle to be recharged [14].

The charging station server, roadside unit (RSU), and vehicle unit are shown in Figure 9.4 as part of the charging station choices (VU). The RSU is in charge of communication between the charging station unit and the VU. Zigbee will be deployed as a communication mechanism on the VU, charging station unit, and RSU. It connects up to 255 nodes per network using the 2.4 GHz radio band [15]. The charging station server may be configured using a machine learning technique to choose the closest fast charging point with the least waiting time.

9.8.2 Artificial Intelligence

Artificial intelligence (AI) could aid in developing long-lasting and fast-charging electric vehicle (EV) batteries. Artificial intelligence optimizes EV battery recharging while also extending the battery's life. AL has also been used to help researchers better understand current batteries by enhancing battery management systems and developing detailed mathematical models to replicate their performance.

9.8.3 Energy Storage Materials

Due to surging demands for fast charging, there is tremendous focus on research for finding the best and most economical options in energy storage and conversion systems. Several battery technologies have been explored, such as lithium-based, sodium-based, magnesium-based, zinc-based, and supercapacitors [16]. Among lithium-based batteries, lithium-ion batteries (LIBs) are much more favourable compared to lithium-sulfur batteries (LSBs) and lithium-metal batteries (LMBs), although they are expensive. LIBs have a high energy density, high reliability, high-rate capability, and high voltage stability even at high voltages, making them suitable candidates for EVs that need fast charging capabilities [17]. Due to the expensive LIBs and the geographically limited sources of Lithium, battery manufacturers are exploring ways to increase the power capacities of other cheaper alternatives such as Sodium-ion batteries.

9.9 Effect of Fast Charging on EV Powertrain Systems

9.9.1 Battery Technology Gap and Lithium Plating

One of the commonly occurring issues in LIBs during fast charging in cold temperatures is lithium plating. Due to cold temperatures, during fast charging, there are multiple layers of deposition of Lithium on the surfaces of the battery anode elements, as depicted in Figure 9.5. This not only decreases the battery's useful life but also decreases safety. This is one of the main reasons why current LIBs are made of fluid electrolytes, metal oxide.

Cathodes and graphite anodes are deemed unsuitable for fast charging unless they work with decreased electrochemical performance. Hence, if at all LIBs must be used for fast charging, the ways to detect lithium plating and its systematic elimination are required not only to maintain the battery performance but also to ensure the overall safety of the battery operations in the EV. One of the non-destructive ways to detect lithium plating, especially when there are a multitude of LIB cells is to measure the variation in the cell thickness [18].

The discovery of new anode materials with enhanced lithium-ion diffusion coefficients is recognized as a step forward in efficiently tackling the challenge of quick charging for electric cars. According to researchers, red phosphorous is the most promising anode for satisfying the dual criteria of high energy density and quick charging performance to the greatest degree feasible [18].

9.9.2 Thermal Management Systems

The ambient conditions, predominantly temperature and humidity, seriously affect the batteries' overall performance. All batteries are based on some exothermic electrochemical processes, generating much heat.

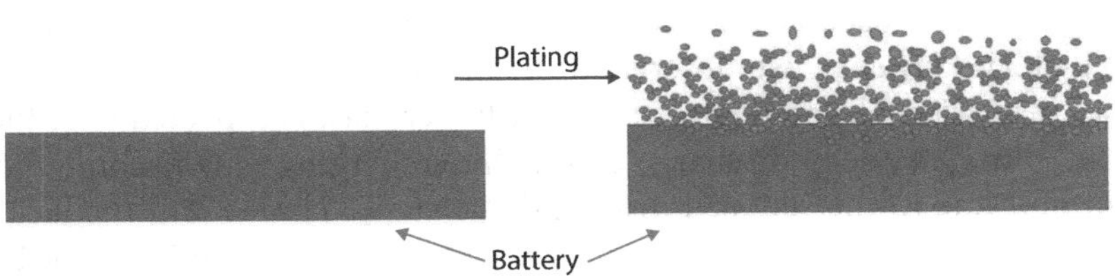

Figure 9.5 Lithium plating.

Hence, proper cooling and insulation are required to keep the battery performance at steady operating levels and ensure no overheating occurs in the EV that may heat or burn the other EV components in the vicinity. Another major issue with the thermal management of the batteries is the ambient temperature conditions. If the ambient temperature is too hot, it adds up to the exothermal heat of the battery. It may cause severe conditions such as enlargement of the battery cells and pressure buildup, causing the gases to escape and leading to severe failure modes such as battery explosion. Thermal management of batteries becomes increasingly risky as we transition from normal charging mode to fast charging mode. Since fast charging is a prerequisite for future EVs, thermal management of batteries will play a crucial role in the safety of EVs and next to the energy performance of the EV. An intelligent battery management system (BMS) employed in modern EVs has adaptive current control to the batteries based on ambient temperature conditions. The BMS actuates either cooling or heating systems to keep the temperature steady, suited for optimal battery performance and safety. The other factors that the BMS will also require are the fast charging time and the current charging rating that will influence the battery temperature [18].

9.9.3 Battery Cycle Life

Mechanisms negatively impact the battery's cycle life and are initiated when the battery is overcharged, undercharged, or deeply depleted. As a result, charging must continuously be monitored, and any deviation from recommended charging schedules will result in capacity loss and battery cycle life degradation. Within this context, it is claimed in this study that it is how the user charges and discharges the battery, rather than the fast-charging technology, that may wear out the battery. The batteries can be fast-charged continuously for many depths of discharge up to 100% state of charge (SOC) under a more advanced fast-charging regime. As a result, regular charging is not used at all. The onboard computer continuously controls the battery. However, user intervention is possible. Therefore, undercharge cases may arise. The ability to charge the battery fast is expected to reduce, if not eliminate, deep discharge cases. Undercharging may become more common due to the reduced driving distance per charge and the lack of deep discharge, resulting in more frequent charging interruptions. The battery's cycle life is linearly reduced when it is undercharged. Fast charging hurts the battery's cycle life, whether utilized partially (in conjunction with normal charging) or exclusively. When compared to regular charging, the average kilometer distance per charge is decreased due to battery degradation [18].

Fast charging, when done correctly, does not harm the battery, according to this study's findings. Mechanisms activated when a battery is charged quickly and damaging to the battery's cycle life can also be triggered when normal charging is done [10]. As a result, the study's attention has switched to crucial deviations from normal charging, such as deep drain and undercharge, which are more likely to occur when a user can charge the battery quickly [12].

9.10 Grid Impacts Caused by EV Charging

As a result of EV adoption, there have been concerns about the impact of EV rapid charging on the electrical grid. Large EV fleets connecting to the power grid for charging can result in a wide range of charging rates, and the dynamic behaviour of EVs further complicates the possible impacts. The following are the various grid impacts caused by EV charging [19].

9.10.1 Impact on Load Profile

EVs pose a major load to the existing electric grids, and if EVs are to be adopted as mainstream transportation systems in any region, then it is required to upgrade the grid capacities to meet these loads. Most studies prove that homeowners typically charge their vehicles once they return from work, ranging from 3 PM to 7 PM in the evenings. This is a peak load time zone where other loads from domestic household appliances are also active. Unless there are schemes such as incentives such as discounted prices during off-peak periods, it is easier to combat this logistical p issue of load distribution.

9.10.2 Impact on Grid Components

Due to increased load conditions on the existing grids by the adoption of EVs, the power transmission components such as cables and transformers might have overheating hazards due to overloads. Redesign of the grids with due consideration of overloading, especially at peak hours, need to be done by using transmission line components of higher rated capacities.

9.10.3 Impact on Power Losses

When an urban or rural region is electrified for the adoption of EVs, the grid's loading capacity, power supply, and usage will increase. This will also increase the power losses as the electrification will have more power sinks (or consumption points) compared to the earlier infrastructure before EV adoption. However, these power losses are inevitable and may be considered proportionalto the increase in grid capacity.

9.10.4 Impact on Voltage Profile

FCPs are generally characterized as power points with high voltage requirements that may come outside the regulated voltage norms. Introducing several FCPs across the region may increase the risks of voltage dips. These potential variations in the voltages must be considered suitable premises while designing or redesigning EV-enabled grids.

9.10.5 Harmonic Impact

Power electronics control the fast-charging operation of an electric vehicle charger. The power grid may experience power quality concerns due to switching these power electronics components. The harmonic problem is the main concern, as significant harmonic distortion can cause system components to de-rate.

9.11 Fast-Charging Technologies on the Self-Powered Automotive Cyber-Physical Systems

Autonomous electric vehicles can be considered deep cyber-physical systems (CPS) with complex interconnections between the physical EV with the intelligent layers that include the IoT sensor layers and the applications. However, self-powered autonomous vehicles remain a distant concept regarding fast charging. There can be autonomous cars that can be solar-powered but not at the fast charging rates of, say, 20 minutes. Also, having energy generation embedded in the EV increases its size and may disturbaerodynamics. Hence, self-powering of EV CPS is possible by deploying suitable charging infrastructure. Another trend in autonomous vehicles is independent charging, where the EV charges at the FCP without human intervention. This can be possible in two ways, namely, conductive (robotic) chargers or inductive (wireless) charging pads [20]. One of the best examples of conducive charging is Tesla's snake charger. Despite the creep factor and system complexity, the capacity to establish conductive connections under controlled, supercharging-type conditions can provide a quick charge and restore the autonomous vehicle to action. However, such robotic chargers are expensive to be deployed and maintain at various locations.

Another great option, as mentioned above, is the wireless charging pad. In this method, the EV can be charged just by parking it near the wireless charging pad. The charging takes place wirelessly without any physical connection. The charging pad can be built within an enclosed air-tight shell or underneath the parking floor. This wireless method is maintenance friendly since there are no physical connections. However, this type of charging is not favorable for fast charging.

Another concept known as emotional charge can be developed if the wireless charging pods are available throughout the road so that charging occurs when the vehicle is moving. However, such deployment of charging pods throughout the road may be a very expensive option.

9.12 Conclusions

As electric vehicles become more common, there is a rising demand for more efficient charging methods. Charging the vehicles using fast chargers available at fast charging stations (DC fast charging stations) is a more feasible answer. This chapter gives an overview of fast charging, fast chargers, fast charging infrastructure and various other technologies used in fast charging. It also informs about the effects of fast charging on EV powertrains and its impacts on the power grid. This chapter concludes by reviewing today's situation and future needs of fast charging infrastructures and overviews charging technologies on self-empowered automotive CPS. As fast charging and battery technologies continue to evolve and improve shortly, it can be anticipated that the charging time will drop even further.

References

1. Botsford, C. and Szczepanek, A., Fast charging vs slow charging: Pros and cons for the new age of electric vehicles. *24th Int. Batter. Hybrid Fuel Cell Electr. Veh. Symp. Exhib. 2009*, EVS 24, vol. 3, pp. 1811–1819, 2009.

2. Dericioğlu, Ç., Yirik, E., Ünal, E., Cuma, M.U., Onur, B., Tümay, M., A review of charging technologies for commercial electric vehicles. *Int. J. Adv. Automot. Technol.*, 2, 1, 61–70, 2018, doi: 10.15659/ijaat.18.01.892.
3. https://www.financialexpress.com/auto/industry/electric-cars-ev-charging-station-charging-infrastructure-EV-adoption-in-India-electric-scooters/2014692/.
4. https://evglobe.com/what-is-fast-charging-for-electric-cars/.
5. Tu, H., Feng, H., Srdic, S., Lukic, S., Extreme fast charging of electric vehicles: A technology overview. *IEEE Trans. Transp. Electr.*, 5, 4, 861–878, 2019, DOI: 10.1109/TTE.2019.2958709.
6. Gnann, T., Funke, S., Jakobsson, N., Plötz, P., Sprei, F., Bennehag, A., Fast charging infrastructure for electric vehicles: Today's situation and future needs. *Transp. Res. Part D Transp. Environ.*, 62, March, 314–329, 2018, DOI: 10.1016/j.trd.2018.03.004.
7. Collin, R., Miao, Y., Yokochi, A., Enjeti, P., Von Jouanne, A., Advanced electric vehicle fast-charging technologies. *Energies*, 12, 10, 1839, 2019, doi: 10.3390/en12101839.
8. Bryden, T.S., Hilton, G., Cruden, A., Holton, T., Electric vehicle fast charging station usage and power requirements. *Energy*, 152, 322–332, 2018, DOI: 10.1016/j.energy.2018.03.149.
9. Wolbertus, R. and van den Hoed, R., Fast charging systems for electric passenger vehicles. *World Electr. Veh. J.*, 11, 4, 1–11, 2020, DOI: 10.3390/wevj11040073.
10. Sbordone, D., Bertini, I., Di Pietra, B., Falvo, M.C., Genovese, A., Martirano, L., EV fast charging stations and energy storage technologies: A real implementation in the smart microgrid paradigm. *Electr. Power Syst. Res.*, 120, 96–108, 2015, doi: 10.1016/j.epsr.2014.07.033.
11. Morrissey, P., Weldon, P., O'Mahony, M., Future standard and fast charging infrastructure planning: An analysis of electric vehicle charging behaviour. *Energy Policy*, 89, 2016, 257–270, 2016, doi: 10.1016/j.enpol.2015.12.001.
12. Rantik, M., Life cycle assessment of five batteries for electric vehicles under different charging regimes, 1999.
13. https://www.pluginindia.com/blogs/bharat-ev-specifications-for-ac-and-dc-charging-everything-you- need-to-know.
14. Brenna, M., Foiadelli, F., Leone, C., Longo, M., Electric vehicles charging technology review and optimal size estimation. *J. Electr. Eng. Technol.*, 15, 6, 2539–2552, 2020, DOI: 10.1007/s42835-020-00547-x.
15. Vanitha, V., Resmi, R., Reddy, K.N.S.V., Machine learning-based charge scheduling of electric vehicles with minimum waiting time. *Comput. Intell.*, 37, 3, 1047–1055, March, 1–9, 2020, DOI: 10.1111/coin.12333.
16. Ma, J.M. and Li, Y.T., Editorial for advanced energy storage and conversion materials and technologies. *Rare Met.*, 40, 2, 246–248, 2021, DOI: 10.1007/s12598-020-01654-4.

17. Wu, Y. *et al.*, An exploration of new energy storage system: High energy density, high safety, and fast charging lithium ion battery. *Adv. Funct. Mater.*, 29, 1, 1–7, 2019, DOI: 10.1002/adfm.201805978.
18. https://www.ionenergy.co/resources/blogs/fast-charging-for-electric-vehicles.
19. Yong, J.Y., Ramachandaramurthy, V.K., Tan, K.M., Mithulananthan, N., A review on the state-of-the-art technologies of an electric vehicle, its impacts and prospects. *Renew. Sustain. Energy Rev.*, 49, 365–385, 2015, DOI: 10.1016/j.rser.2015.04.130.
20. https://www.pluglesspower.com/learn/solve-last-mile-vehicle-autonomy/.

10

A Survey of VANET Routing Attacks and Defense Mechanisms in Intelligent Transportation System

Allam Balaram[1]*, P. Chandana[2], Shaik Abdul Nabi[3] and M. SilpaRaj[4]

[1]*Department of Computer Science and Engineering, MLR Institute of Technology, Hyderabad, Telangana, India*
[2]*Dept. of CSE (AI&ML), Institute of Aeronautical Engineering, Hyderabad, Telangana, India*
[3]*Department of Computer Science and Engineering, AVN Institute of Engineering and Technology, Hyderabad, India*
[4]*Department of Emerging Technologies (Cyber Security), CVR College of Engineering, Ibrahimpatnam, Rangareddy, Telangana, India*

Abstract

Due to the tremendous growth in the vehicle industry, Vehicular Ad hoc NETworks (VANETs) have become important in wireless communication networks. High-speed vehicular nodes make the network highly dynamic. Besides, VANETs consist of several vehicles and extend the network scalability over the entire road area. Those mobility and scalability features create challenges in the design of the VANET routing protocol. Moreover, those features have made the routing protocols vulnerable to various routing attacks. The malicious vehicles in VANET make challenges in the design of an efficient and secure communication protocol. This survey summarizes the architecture of VANETs and discusses their security, the impact of attacks on routing efficiency, and challenges. The survey analyzes routing attacks on various VANET applications and the existing defence systems against various routing attacks in VANETs.

Moreover, it comprehensively covers the anonymous routing schemes, which help to hide the vehicular identity from others. Various intrusion detection systems are also discussed. Briefly, the survey covers most information about VANET and its security systems.

**Corresponding author*: drbalaramallam@gmail.com

Rathishchandra R. Gatti, Chandra Singh, Rajeev Agrawal and Felcy Jyothi Serrao (eds.) Self-Powered Cyber Physical Systems, (213–226) © 2023 Scrivener Publishing LLC

***Keywords*:** VANETs, VANET mobility, VANET routing, routing attacks

10.1 Introduction

Recently, research on Vehicular Ad hoc NETworks (VANETs) plays a major role in the Intelligent Transportation System (ITS) [1]. The idea of an ITS is to develop an advanced application to serve innovative services related to modes of transport management for the better, safer and smarter use of transport systems using networks. Examples of these technologies include emergency services during an accident, enforcing traffic laws, signs and speed limitations changes based on the conditions. The technologies include in Intelligent Transport systems most basic systems (traffic surveillance, car navigation, sign representations, speed cameras to monitor accidents and vehicles) to more advanced services like parking guidance and weather information. The VANET is a kind of MANETs with road routes. The VANET consists of On-Board Units (OBUs) & Road Side Units (RSUs). Communication in VANETs is classified into different forms of communication like Vehicle-to-Vehicle (V2V) and Vehicle-to-Infrastructure (V2I) communications [2]. The major functionality involved in VANETs is establishing operational communication among neighbors and RSU to obtain important information for making decisions in driving [3]. Unique features of the vehicles are scalability, high mobility, and volatility. The vehicular nodes are much more dynamic. For example, drivers drive a car at various speeds and change their locations frequently. The vehicles with high mobility make the network topology highly dynamic. Besides, the road topology consists of several vehicles [4]. These challenging characteristics make the VANET weak against various network attacks and necessitate the design of secure VANETs [5]. The VANETs face several security issues connected to privacy and authentication [6–10]. Moreover, the malicious vehicles incur several security issues in the design of routing protocols over VANETs [6]. For example, 1) the attackers may forward the packets to the non-existing path that causes considerable delay. 2) The attackers may introduce routing loops, resulting in network congestion and communication medium contention. 3) The attackers may act cooperatively to prevent the source node from discovering an optimal path to the destination. To solve such issues, several defence mechanisms have been developed in VANETs.

10.2 Attacks in VANET

The network layer attackers perform malicious activities in terms of modification, interception, injection, and deletion in VANETs. Attacks may be launched at any routing phase, such as route discovery, route maintenance, and data transmission. The survey divides the VANET attacks differently based on the application of VANET. The VANET attacks can be grouped under normal V2V communication, safety application, and infotainment applications because the routing attacks affect the normal operations in VANET like infotainment, comfort and applications [7].

10.2.1 Attack on V2V Communication

Developing a secure VANET communication is important since V2V communication is affected by various threats like eavesdropping attacks, certificate replication and location privacy attacks.

(i) Certificate replication attack: The malicious vehicles exploit false certificates and keys of other vehicles for authenticating the vehicular nodes, and it attempts to create ambiguity for the trusted authority.
(ii) Eavesdropping attack: The attacker uses a valid certificate of others and attempts to collect private data of vehicles, like location and user ID.
(iii) Privacy attack: It is a type of attack, which includes various forms of malicious activities on privacy-preserving schemes, including vehicle tracking. The attacker obtains information like position, ID, and cryptography keys of other vehicles and uses them to affect the VANET communication [7].

10.2.2 Various Attacks on Safety Applications

Various attacks on VANET safety communication are mostly related to channel allocation. These are listed below.

(i) Denial of service attacks: The malicious vehicle broadcasts multiple messages, which block all communication initiated by the legitimate vehicles. Several attackers implement this kind of attack parallel in a distributed way, and thus it is called a Distributed DoS [8].
(ii) Jamming attack: The impact of the jamming attack on VANET security applications is high. The untrustworthy jamming vehicle consumes heavy power by transmitting the messages in the equivalent frequency range and performing frequent alert injections [9] to affect the broadcast communication in VANET.
(iii) Platooning attack: The platooning attacker nodes are located in the same zone. It aims to disturb legitimate vehicle communication and isolate the trusted nodes from the network operation [7].
(iv) Betrayal attack occurs when a vehicle in a trusted source enters a malicious region node and receives false messages from it [7].

10.2.3 Attack on Infotainment Applications

The infotainment applications in VANET include the safety and comfort of the passengers. The attacks on those applications are listed below.

(i) Illusion attack: The malicious vehicle sends fake warning messages to the neighbors, creating false illusions about the nearby vehicles.
(ii) Inject message attack: The attackers send replays of the same or new messages or modify the original messages by injecting the malicious packets into the network to disturb the VANET communication.

10.3 Impacts of Attacks on VANET Routing

The misbehaving nodes cause a severe threat to the routing in VANET [2, 10]. The misbehaving nodes are hard to detect. The misbehaving node deviates the packet from the normal routing process. It may be due to several reasons. Enforcement of cooperation among nodes considerably reduces the misbehaving nodes. The misbehavior of a node may be either intentional or non-intentional.

10.4 Nonintentional Misbehavior

When a node becomes faulty, it may deviate from the routing process unintentionally or accidentally. A node may misbehave non-intentionally as it runs out of battery. Moreover, a node can be considered faulty when its software fails. The faulty nodes effectively reduce the node density. A packet may be lost or corrupted due to wireless transmission problems [13]. Radio inference, the hidden terminal problem, and high bit rate error are some of the non-intentional causes of the packet drop. Contention and a buffer overflow in the MAC layer cause the higher layer to drop packets. These are the packet drops due to non-intentional misbehavior, which happens naturally depending on the network environment.

- **Inactive Nodes:** An important step in routing protocols is route discovery and packet forwarding. Inactive nodes in the network do not involve in these two steps. Inactive nodes include resource-constrained (transmission power, bandwidth), lazy, and misconfigured nodes. Inactive nodes are concerned only with the network layer. Such nodes are not considered as source and destination. The density of the network gets reduced by the number of inactive nodes. The inactive nodes moderately affect the wireless routing process, whereas the selfish nodes have an extremely destructive impact on the wireless routing process.

10.5 Intentional Misbehavior

A node may attempt to misbehave intentionally by taking advantage of the routing process. In most attack scenarios, intentionally misbehaving nodes involve packet dropping. Some solutions mitigate the such type of packet dropping. The security mechanisms must be able to distinguish both intentional and non-intentional misbehaving nodes. The misbehaving node mainly interrupts the goal of routing protocols.

Moreover, it degrades the routing performance, which may even result in network failure. The performance degradation is noticed with the metrics like throughput, packet delivery ratio, and end-to-end delay. A network may contain single misbehaving nodes or collaborative misbehaving nodes. These misbehaving nodes purposely cause more traffic in the network and drop the packets to degrade the network performance. For a secured VANET environment, it is necessary to mitigate the impacts of such attacks, to improve the routing process. The main objectives of the attacker are not to forward the intended packets and may even modify the packet content by adding or removing some fields. These nodes degrade the network performance by not participating in the routing process. The misbehaving nodes are broadly classified into selfish and malicious nodes [11].

- **Malicious Nodes:** Malicious nodes aim at reducing network resources. A malicious node may exist in several forms. A malicious node may compromise a legitimate node and drop the packets to degrade the network performance. Many forms of malicious nodes generate a different class of attacks to collapse the network. Cryptography techniques are ineffective in eliminating these attacks.
- **Selfish Nodes:** Selfish nodes hesitate to participate in the routing process. The selfish node participates in the route discovery process but does not participate in packet transmission activity. It remains selfish as it need not waste its resources for other nodes. This class of nodes does not cooperate with the other nodes and discards the packets purposely without forwarding the data packet to the intended destination [11].

10.6 Defence Mechanism of Routing Attacks in VANET Routing

The defence mechanisms have some provisions for identifying the attacker nodes in a network [12, 13]. After identifying the attacker nodes, it is essential to inform its identity to all other network nodes. It assists other nodes in the network select alternate paths without malicious nodes for routing [14, 15]. The attacks can be avoided or prohibited via two proactive and reactive approaches. The proactive defence mechanism prevents the attacker in the initial stage by applying cryptographic techniques. The reactive defence mechanism considers attacker prevention during the data transmission phase.

- **Proactive Mechanism:** This is the first defence line that uses conventional cryptography to ensure authentication and encryption schemes. The proactive mechanism pre-deploys security keys and initializes the security parameters. The digital signature ensures data integrity and authentication. Hashing technique is applied to enhance integrity in data transmission. Strong symmetric encryption can be used for secure communication. Several proactive defence mechanisms have been developed for VANET communication [16, 17].

- **Reactive Mechanism:** Misuse and the misbehavior detection system are the second lines of defence. It detects misuse based on the behaviour and patterns of well-known attacks. Intrusion and node misbehavior detection are reactive mechanisms to combat intrusion and misbehavior. The Sybil, DoS, and compromised RSU attacks can be detected using the misbehavior detection scheme [18, 19]. In Biswas *et al.* and Fasbender *et al.* [20, 21], the certificate replication attack is detected, but they lack in detecting the coordinated attacks in VANET. The reactive mechanisms from [22] to [27] for VANETs are listed under the cryptography category in the following Table 10.1.

Table 10.1 Defense mechanisms in VANET.

Proposed approach or protocol	**Attack**	**Category**	**Description**
P^2DAP [18]	Sybil	Misbehavior detection	Detects Sybil attacks at low overhead and delay, but an attacker can compromise RSB
Data-centric misbehavior detection [19]	Sybil and Compromised RSU	Misbehavior detection	Low computation and communication costs. However, the results do not hold good for the group mobility model
An identity-based authentication scheme [20]	Certificate replication attack	Identity-based authentication	It can detect the Certificate replication attack alone, but it fails in detecting the collaborative attacks with Certificate replication attack

(Continued)

Table 10.1 Defense mechanisms in VANET. (*Continued*)

Proposed approach or protocol	Attack	Category	Description
Variable and Scalable Security [21]	DoS and certificate replication attack	Non-Disclosure Method	It is efficient to detect the attacks related to privacy and authentication but fails in supporting the data confidentiality
Verification based authentication scheme [22]	Replay, Bogus, DoS, and Certificate replication attack	Cryptography based detection	It detects the attackers successfully, but with high overhead
VANSec [23]	Eavesdropping, Location Tracking, and DoS	Cryptography based detection	Privacy-related issues are not focused
Security Protection [24]	Integrity Related attacks	Elliptic Curve Cryptography	It focuses on integrity, reliability, and confidentiality
A two-level privacy-preserving [25]	DoS and Sybil	Pseudonym based security	It can detect the DoS and Sybil attacks alone, but it fails in detecting the collaborative attacks
Privacy-Preserving Authentication Protocol [26]	Privacy and Security related attacks	Hybrid Cryptography	It shows a significant reduction in routing overhead
Enhancing Security Protection [27]	DoS and certificate replication attack	Elliptic Curve Cryptography	It increases the control overhead

10.7 Intrusion Detection Techniques in VANETs

Intrusion detection techniques can find malicious nodes that have crossed the border of the network. An intrusion detection system involves three stages. The first stage collects the data, the second stage analyzes the collected data, and finally, the last stage efforts to prevent the attack's impact on the system. IDS can be deployed as monitoring nodes as a special node. The special node deployment varies with the architecture of IDS and protocol [28]. There are three types of IDSs are used, such as specification-based intrusion detection (SBID), knowledge-based intrusion detection (KBID), and anomaly-based intrusion detection (ABID).

- **Anomaly-based Intrusion Detection (ABID):** ABID system identifies the profile of nodes that deviate from the normal profile and defines such activity as anomalous. It is also known as behavior-based intrusion detection. The model of the normal behaviour is compared with the current behavior of the network to identify intrusion in the network. ABID includes two phases, namely training and testing operations. The training phase models the normal behaviour of the network or users. Testing involves comparing the normal network behaviour with the current network behaviour. This technique uses statistical [29] or mathematical approaches to identify the deviations between the two models. The table [4] reviewed the IDS solutions [30, 31] that follow the anomaly-based IDS.
- **Knowledge-based intrusion detection:** KBID systems uphold a knowledge base with a pre-defined pattern of known attacks and identify the nodes with those patterns to detect them. KBID system identifies the negative patterns and triggers an alarm against the attackers [32]. KBID depends on knowledge about attacks.
- **Specification-Based system:** This system expresses a set of conditions that a program or protocol must satisfy. The protocol is announced as an intrusion if it does not satisfy natural operation conditions [32]. If there is any mismatch between the predefined set of conditions and the current routing scenario, it detects the attacker. In the following Table 10.1, the IDS solutions [33–35] follow the specification-based IDS.

10.8 Anonymous Routing in VANETs

The VANET should meet the security requirements like integrity, non-repudiation, authentication, and confidentiality to protect the network against routing attackers. Apart from these basic security requirements, it is important to preserve sensitive information of the vehicle's owner, such as location privacy and identity, in contrast to illegal tracing and user profiling [36, 43]. On the other hand, traceability is essential in the case of law enforcement. Moreover, the network authorities need privilege revocation after detecting malicious activities during network access. The following Table 10.1 lists some of the anonymous routing protocols in VANET.

Moreover, they describe their benefits and defects also. The following lists anonymous routing protocols for detecting malicious vehicles and compromised RSUs [37, 38, 44]. Some solutions, such as [39–42] described in the table, are proposed for detecting the replay and Sybil attacks in VANETs.

10.9 Challenges and Future Directions

There are several advantages to VANET. Apart from VANET's advantages, there are several issues in developing secure VANET communication. These challenges include time constraints, network scale, node mobility, and volatility [45].

Time Constraints: The major requirement in an efficient VANET communication protocol is to deliver the data packets within an acceptable time limit. Safety-related VANET communication requires stringent deadlines [46]. However, it takes work to implement the security mechanism since it consequently increases the delivery time of packets. In the future, the VANET secure routing protocols for safety-related applications must consider the time constraints in its design.

Network Scalability: The scalability denotes the network area and the number of vehicles connected [47]. Consequently, installing the security mechanism is complicated because the increased number of vehicles creates challenges in distributing public and private keys.

Node Mobility: High-speed vehicular nodes in VANET make challenges in the security design. It may increase the network overhead while applying the security mechanism in the VANET protocol. Therefore, mobility management is essential in a secure protocol for the significant interplay between the sender and receiver [48].

Volatility: As a part of the high mobility of vehicles, connections between vehicles may be lost. The personal contacts in the secure VANET routing protocols need long-life passwords, which reduces communication security [47].

The challenges above need to be solved to design efficient and secure VANET routing in the future for the intelligent transportation system.

10.10 Conclusion

As VANETs possess unique characteristics, this research area impacted wireless communications. The major critical issues in VANETs are routing security and information privacy. It has been reviewed as an art of identifying routing attacks on various VANET applications. Several defence mechanisms for VANET communication are reviewed. This survey has identified the benefits and defects in the existing defence mechanisms. It has discussed various IDS techniques in VANET. It lists the defence mechanisms for routing attacks with the use of IDS. Moreover, it has discussed the anonymous routing protocols developed for VANET. Finally, this survey has provided the challenges and future directions in VANET security.

References

1. Zeadally, S., Hunt, R., Chen, Y.S., Irwin, A., Hassan, A., Vehicular ad hoc networks (VANETS): Status, results, and challenges. *Telecommun. Syst.*, 50, 217–241, 2012.
2. Mishra, B., Nayak, P., Behera, S., Security in vehicular adhoc networks: A survey. *Proceedings of International Conference on Communication, Computing & Security*, pp. 590–595, 2011.
3. Moharram Mohammed, A. and Daraiseh Ahmad, A., Toward secure vehicular ad-hoc networks: A survey. *IETE Tech. Rev.*, 29, 1085, 80–89, 2012.
4. Raya, M. and Hubaux, J.-P., Securing vehicular ad hoc networks. *J. Comput. Secur.*, 15, 1, 39–68, 2007.
5. Engoulou, R.G., Bellaïche, M., Pierre, S., Quintero, A., VANET security surveys. *Comput. Commun.*, 44, 1–13, 2014.
6. Zeadally, S., Hunt, R., Chen, Y.-S., Irwin, A., Hassan, A., Vehicular ad hoc networks (VANETS): Status, results, and challenges. *Telecommun. Syst.*, 50, 4, 217–241, 2012.
7. Qu, F., Wu, Z., Wang, F., Cho, W., A security and privacy review of VANETs. *IEEE Trans. Intell. Transp. Syst.*, 16, 6, 2985–2996, 2015.
8. Mishra, R., Singh, A., Kumar, R., VANET security: Issues, challenges and solutions, in: *Proceedings of the International Conference on Electrical, Electronics, and Optimization Techniques (ICEEOT)*, pp. 1050–1055, 2016.
9. Cui, J., Wei, L., Zhang, J., Xu, Y., Zhong, H., An efficient message-authentication scheme based on edge computing for vehicular ad hoc networks. *IEEE Trans. Intell. Transp. Syst.*, 20, 5, 1621–1632, 2019.
10. Akalu, R., Privacy, consent and vehicular ad hoc networks (VANETs). *Comput. Law Secur. Rev.*, 34, 1, 37–46, 2018.

11. Malhi, A.K., Batra, S., Pannu, H.S., Security of vehicular ad-hoc networks: A comprehensive survey. *Comput. Secur.*, 89, 101664, 2020.
12. Fonseca, E. and Festag, A., A survey of existing approaches for secure ad hoc routing and their applicability to VANETS, vol. 28, NEC Network Laboratories, pp. 1–28, 2006.
13. Sirola, P., Joshi, A., Purohit, K.C., An analytical study of routing attacks in vehicular ad-hoc networks (VANETs). *Int. J. Comput. Sci. Eng. (IJCSE)*, 3, 4, 210–218, 2014.
14. Saini, M. and Singh, H., VANET its characteristics attacks and routing techniques: A survey. *Int. J. Sci. Res.*, 5, 5, 1595–1599, 2016.
15. Korba, A.A., Nafaa, M., Salim, G., Survey of routing attacks and countermeasures in mobile ad hoc networks, in: *IEEE UKSim 15th International Conference in Computer Modelling and Simulation*, pp. 693–698, 2013.
16. Bariah, L., Shehada, D., Salahat, E., Yeun, C.Y., Recent advances in VANET security: A survey, in: *IEEE 82nd Vehicular Technology Conference (VTC2015-Fall)*, pp. 1–7, 2015.
17. Chaubey, N.K., Security analysis of vehicular ad hoc networks (VANETs): A comprehensive study. *Int. J. Secur. Appl.*, 10, 5, 261–274, 2016.
18. Zhou, T., Choudhury, R.R., Ning, P., Chakrabarty, K., P2DAP – sybil attacks detection in vehicular ad hoc networks. *IEEE J. Sel. Areas Commun.*, 29, 3, 582–594, 2011.
19. Ruj, S., Cavenaghi, M.A., Huang, Z., Nayak, A., Stojmenovic, I., On data-centric misbehaviour detection in VANETs, in: *Vehicular Technology Conference*, pp. 1–5, 2011.
20. Biswas, S., Mišić, J., Mišić, V., An identity-based authentication scheme for safety messages in wave-enabled vanets. *Int. J. Parallel Emergent Distrib. Syst.*, 27, 6, 541–562, 2012.
21. Fasbender, Kesdogan, D., Kubitz, O., Variable and scalable security: Protection of location information in mobile IP. *IEEE VTS, 46th Vehicular Technology Conference*, USA, 1996.
22. Celes, A.A. and Edna Elizabeth, N., Verification based authentication scheme for bogus attacks in VANETs for secure communication, in: *IEEE International Conference on Communication and Signal Processing (ICCSP)*, pp. 0388–0392, 2018.
23. Ahmed, S., Ur Rehman, M., Ishtiaq, A., Khan, S., Ali, A., Begum, S., VANSec: Attack-resistant VANET security algorithm in terms of trust computation error and normalized routing overhead. *J. Sens.*, 2018, 6576841, 17, 2018.
24. Kumar, S. and Kumar, N., Enhancing security protection using elliptic curve encryption and digital signature in vehicular ad-hoc network. *Int. J. Eng. Technol.*, 7, 4, 837, 2018.
25. Rajput, U., Abbas, F., Eun, H., Hussain, R., Oh, H., A two level privacy preserving pseudonymous authentication protocol for VANET, in: *IEEE 11th International Conference on Wireless and Mobile Computing, Networking and Communications (WiMob)*, pp. 643–650, 2015.

26. Tahir, F., Nasir, S., Khalid, Z., Privacy-preserving authentication protocol based on hybrid cryptography for VANETs, in: *IEEE International Conference on Applied and Engineering Mathematics (ICAEM)*, pp. 80–85, 2019.
27. Ravi, K. and Kulkarni, S.A., A secure message authentication scheme for VANET using ECDSA, in: *Fourth International Conference on IEEE Computing, Communications and Networking Technologies (ICCCNT)*, pp. 1–6, 2013.
28. Erritali, M. and El Ouahidi, B., A survey on VANET intrusion detection systems, in: *Proceedings of the 2013 International Conference on Systems, Control, Signal Processing and Informatics*, pp. 16–19, 2013.
29. Erritali, M. and El Ouahidi, B., A review and classification of various VANET intrusion detection systems, in: *IEEE National Security Days (JNS3)*, pp. 1–6, 2013.
30. Hortelano, J., Ruiz, J.C., Manzoni, P., Evaluating the usefulness of watchdogs for intrusion detection in VANETs, in: *IEEE International Conference on Communications Workshops*, pp. 1–5, 2010.
31. Ghaleb, F.A., Saeed, F., Al-Sarem, M., Al-Rimy, B.A.S., Boulila, W., Eljialy, A.E.M., Aloufi, K., Alazab, M., Misbehavior-aware on-demand collaborative intrusion detection system using distributed ensemble learning for VANET. *Electronics*, 9, 9, 1411, 2020.
32. Loukas, G., Karapistoli, E., Panaousis, E., Sarigiannidis, P., Bezemskij, A., Vuong, T., A taxonomy and survey of cyber-physical intrusion detection approaches for vehicles. *Ad Hoc Netw.*, 84, 124–147, 2019.
33. Buchgger, S. and Le Boudec, J., Performance analysis of the CONFIDANT protocol, in: *Proc IEEE/ACM Workshop on Mobil Ad Hoc Networking and Computing (MobiHoc '02)*, Lausanne, Switzerland, pp. 226–336, 2002.
34. Zhang, Y. and Lee, W., Intrusion detection in wireless ad hoc networks. *6th Int'l. Conf. Mobile Comp. and Net.*, pp. 275–283, 2000.
35. Sen, J., An intrusion detection architecture for clustered wireless ad hoc networks. *Second International Conference on Computational Intelligence, Communication Systems and Networks*, 2010.
36. Blaze, M., Ioannidis, J., Keromytis, A.D., Malkin, T., Rubin, A., WAR: Wireless anonymous routing, in: *International Workshop on Security Protocols*, Springer, pp. 218–232, 2003.
37. Sun, J. *et al.*, An identity-based security system for user privacy in vehicular ad hoc networks. *IEEE Trans. Parallel Distrib. Syst.*, 21, 9, 1227–1239, 2010.
38. Caballero-Gil, P., Caballero-Gil, C., Molina-Gil, J., Hernández-Goya, C., Flexible authentication in vehicular ad hoc networks. *IEEE 15th Asia-Pacific Conference on Communications*, pp. 576–879, 2009.
39. Zhang, L. *et al.*, A scalable robust authentication protocol for secure vehicular communications. *IEEE Trans. Veh. Technol.*, 59, 4, 1606–1617, 2010.
40. Zhang, C. *et al.*, An efficient message authentication scheme for vehicular communications. *IEEE Trans. Veh. Technol.*, 57, 6, 3357–3368, 2008.

41. Ying, B., Makrakis, D., Mouftah, H.T., Efficient privacy preservation protocol using self-certified signature for VANETS, in: *ICWMC The Seventh International Conference on Wireless and Mobile Communications*, pp. 301–306, 2011.
42. Lu, H., Li, J., Guizani, M., A novel ID-based authentication framework with adaptive privacy preservation for VANETs. *IEEE Conference in Computing, Communications and Applications (ComComAp)*, pp. 345–350, 2012.
43. Sampigethaya, K., Li, M., Huang, L., Poovendran, R., AMOEBA: Robust location privacy scheme for VANET. *IEEE J. Sel. Areas Commun.*, 25, 8, 1569–1589, 2007.
44. Tan, Z., A privacy-preserving mutual authentication protocol for vehicle ad hoc networks. *J. Converg. Inf. Technol.*, 5, 7, 180–186, 2010.
45. Yang, J. and Fei, Z., Broadcasting with prediction and selective forwarding in vehicular networks. *Int. J. Distrib. Sens. Netw.*, 2013, 9, 2013.
46. Chen, W., Guha, R.K., Taek, J.K., Lee, J., Hsu, I.Y., A survey and challenges in routing and data dissemination in vehicular ad-hoc networks, in: *Proceedings of the IEEE International Conference on Vehicular Electronics and Safety (ICVES '08)*, pp. 328–333, 2008.
47. Kim, D., Cano, J.C., Wang, W., de Rango, F., Hua, K., Data dissemination in vehicular environments. *Int. J. Distrib. Sens. Netw.*, 2013, 2, 2013.
48. Gerlach, M., Festag, A., Leinmller, T., Goldacker, G., Harsch, C., Security Architecture for vehicular communication, in: *Proceedings of the 5th International Workshop on Intelligent Transportation (WIT '07)*, 2007.

11

ANN-Based Cracking Model for Flexible Pavement in the Urban Roads

Athiappan K.[1], Kandasamy A.[2], Karthik C.[3]* and Rajalakshmi M.[4]

[1]Department of Civil Engineering, Thiagarajar College of Engineering College, Madurai, Tamil Nadu, India
[2]Saveetha School of Engineering, SIMATS Deemed University, Chennai, Tamil Nadu, India
[3]Department of Robotics and Automation, Jyothi Engineering College, Thrissur, Kerala, India
[4]Department of Biomedical Engineering, Sethu Institute of Technology, Madurai, Tamil Nadu, India

Abstract

A novel technique is presented in this paper aimed at predicting the distress on pavement using a data-driven methodology. The main focus of this study is to manage the difficulties and ensure safe and comfortable roadway use. Diagnosing deterioration types and using proper maintenance techniques is critical, especially in the early phases. A model is developed for predicting the distresses using the ANN technique where the factors influencing the progression of cracking were drainage, modified structural number and equivalent standard axle load are considered. The model developed with ANN can be utilized for the optimum maintenance of urban roads, especially in metropolitan cities. The method is verified with extensive testing by performing it, and its analysis helps us to ensure its efficiency and adaptability.

***Keywords*:** Flexible pavement, crack prediction, artificial neural network, ANNs, sensitive analysis, artificial intelligence

**Corresponding author*: karthikc@jecc.ac.in

Rathishchandra R. Gatti, Chandra Singh, Rajeev Agrawal and Felcy Jyothi Serrao (eds.) Self-Powered Cyber Physical Systems, (227–242) © 2023 Scrivener Publishing LLC

11.1 Introduction

To efficiently construct flexible pavement, the researchers focused on a new methodology in the middle of the 20th century. Then they focused their research on the maintenance and rehabilitation of the constructed pavement at the end of the 20th and the start of the 21st century [1]. The pavement will get deteriorate very rapidly beyond the point of effective maintenance. In order to attain optimum pavement maintenance, highway engineers are using software developed for pavement management systems, such as HDM-4 are highly effective [5]. Cracking is an important parameter used to quantify bituminous pavement deterioration; the parameters, such as fatigue and ageing, are the principal factors for forming cracks in flexible pavement layers. The propagation of cracking is accelerated through embrittlement, which weakens the different layers of pavement resulting from ageing and water ingress. There are two types of cracking: Thermal cracking in the structural direction caused by load and age/environment. The bituminous pavement is triggered by structural cracking when the pavement attains 0.5% of the crack concerning the carriageway's total surface area. Thermal cracking in transverse is caused by temperature fluctuations or freeze–thaw cycles and so happens only in particular climates. Using performance models to predict the progression of cracking would be immensely helpful for deciding timely maintenance strategies. The current maintenance practices are ad-hoc, subjective and based on Engineer's experience. Given the fund's constraints and the need for judicious spending of available resources, maintenance planning and budgeting must be done based on scientific methods. Only recently, there were scientific and engineering tools available for local conditions to highway professionals. Under the financial constraints, efforts are underway to build road management and development systems to strengthen the quality road network. These tools are critical for determining the highway sector's funding needs, assessing alternative maintenance options, and prioritizing the work program. The prediction models presently available in the literature are location specific with the prevailing conditions and cannot be transferable to other locations with different conditions. It is also found that no models are available for predicting the performance of urban roads. Hence, in this study, deterioration prediction model was developed exclusively for urban roads. Because cracking is one of the most common problems in metropolitan roadways, it was determined to create crack propagation, prediction models. The following are the study's goals to examine the cracking state of chosen city roads at varied traffic, age, and discharge levels and model development and evaluation for forecasting crack advancement in city roads.

11.2 Literature Review

Pavement surface failure is broadly classified into disintegration cracks, deformation, and surface defects, according to IRC: 82-1982 [4]. Crack is found to be the most common failure on urban roads. The common types of cracks are hairline cracks, alligator cracks, longitudinal cracks, edge cracks, shrinkage cracks, and reflection cracks. Sanaa *et al.* [9] developed the model in the form given in the equation.

$$N_f = \beta_{f1}\, k_1\, (\sigma)^{\beta_{f2} k_2} (E)^{-\beta_{f3} k_3}, \tag{11.1}$$

where N_f = ratio of cracking to the number of repetitionσ = critical location tensile strain = material stiffness, k_1, k_2, k_3 = laboratory calibration parameter, β = calibration factor. The predictions matched the AASHO measurements in the fall season better than those in the spring. Moshe and Dinesh [7], in their study, have determined the cost-effective maintenance actions for infrastructure facilities by developing a performance deterioration model. The measurement model for cracks given in the equation was

$$CRx = 1.503[3.562\ AGE/SNC + 3.897\ ESAL/SNC + 1.654\ CP/SNC + e_1] + e_2, \tag{11.2}$$

where CR_x, cracking area concerning total surface area; AGE, age of pavement; SNC, structural number; CP, cumulative precipitation; ESAL, equivalent standard axles in cumulative number; e_1, e_2, error in the model (assumed to zero), The regression value of the model is $R^2 = 0.62$. V.K. Sood and Sharma [13] found out that the model developed based on Indian conditions would be cost-effective for transportation planning and implementation of pavement materials. Yongqi and Mtcalf [15] investigated the slab and developed a simplified fatigue crack initiation prediction model. The initial fracture equation shows a strong relationship between the load applied and the critical tensile stress developed. The occurrence and progression of a crack in the slabs are closely linked to the accumulation of persistent deformation. A sudden rise in structural and flexible distortion characterizes cracking. Pradeep *et al.* [8] developed the model by considering the structural condition index for existing pavement. Hsiang *et al.* [2] have developed a pavement design methodology because of the LTPP database's consistency, it monitors the data of pavements combined with more than 2,400 bitumen and Cement concrete pavement sample sections in North America from 1987. Samer *et al.* [11] have developed models for initiating and progressing infrastructure distress using 763 observations. The following data, such as cumulative traffic, pavement structural stability, environmental factor and pavement surface condition, are monitoring data collected from 116 test sections. The residuals are then introduced as an extra independent variable in the crack progression model, which is subsequently calculated. The artificial neural network (ANN) methodology was used to construct the model, which processes information in the same manner that biological nervous systems, such as the brain, do. It is a technique for nonlinear statistical data modeling. This can be used to represent complex input-output interactions or detect data patterns. The "input" layer modules attached to the output layer module have hidden modules in the ANN statistical analysis as shown in Figure 11.3. The basic information is fed into the network as input units. BPNNs have been recommended for modeling time-varying processes by Rumelhart and McClelland [6]. They use a vector containing previous values as the input. Hornik *et al.* [10] found that BPNN, including one hidden layer and suitable activating variables, can approximate any nonlinear continuous function.

11.3 Methodology

Figure 11.1 depicts the approach used in the research. The project was broken down into six stages. The identification of the pavement section was the first step, as shown in Figure 11.2. The Chennai Corporation was contacted in the second stage to collect service records for all of the sections over a 4-year period. The third stage was the measurement of cracking on study stretches. The fourth stage was developing a model for crack progression using the ANN technique. In the fifth stage, the model was validated using the data collected from in-service pavements. Finally, conclusions are drawn and a model for adoption is recommended.

- Criteria for Selection of Study Stretches

The study stretches are selected considering the following four parameters: age, traffic, drainage conditions, and surface, as shown in Table 11.1. Usually, the roads in Chennai are renewed once in 4 years in the case of

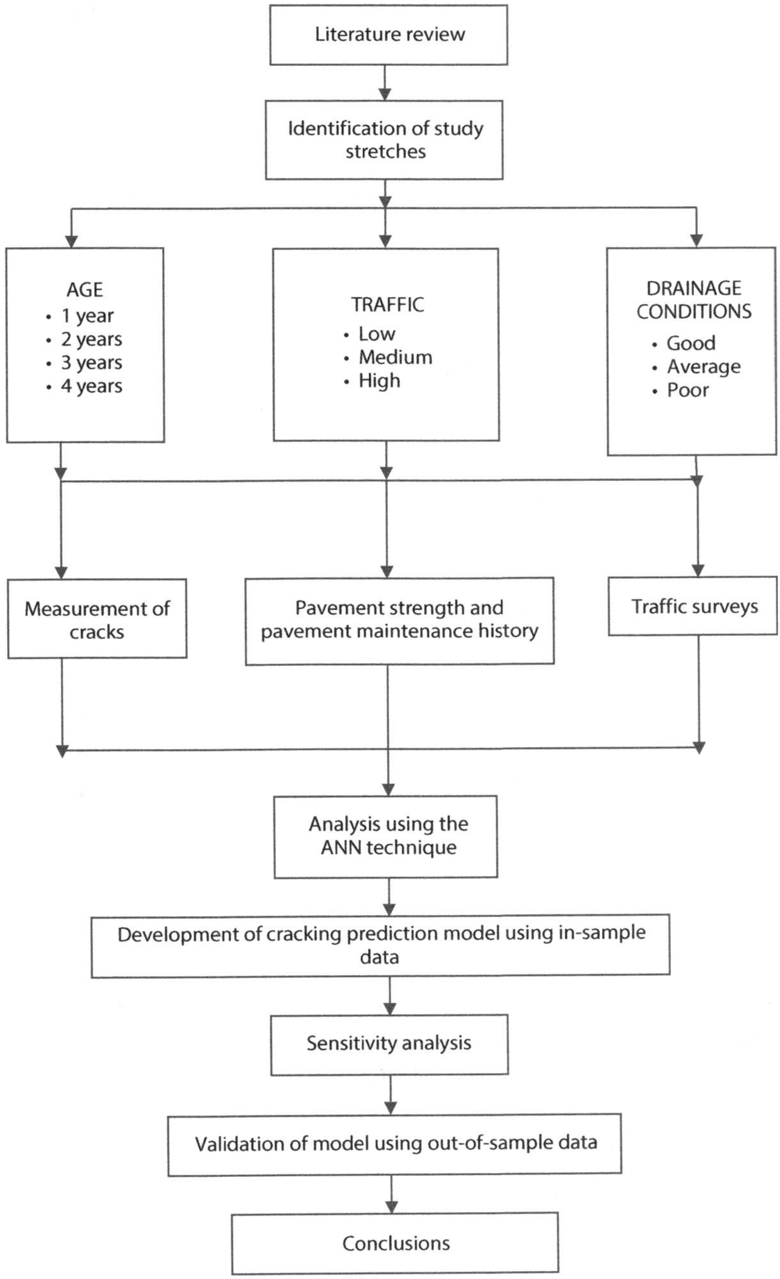

Figure 11.1 Flowchart showing methodology.

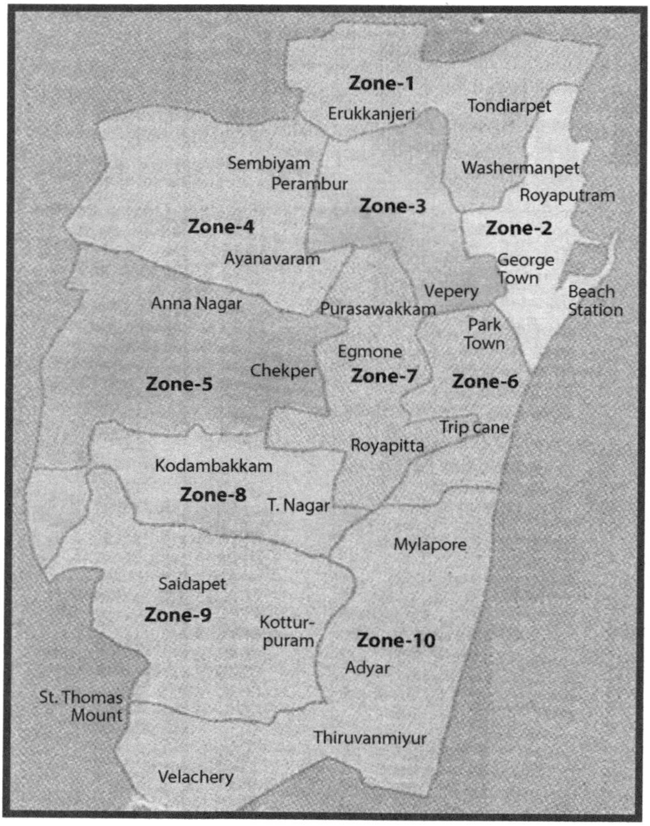

Figure 11.2 Zonal map of Chennai Municipal Corporation.

Table 11.1 Chennai road network.

<table>
<tr><th colspan="2">Description</th><th>Road length in km</th></tr>
<tr><td>Bus Route Roads (bituminous)</td><td rowspan="3">Roads maintained by Chennai Corporation</td><td>222</td></tr>
<tr><td>Interior Roads (bituminous)</td><td>2244</td></tr>
<tr><td>Interior roads (concrete)</td><td>231</td></tr>
<tr><td colspan="2">Roads maintained by Highways</td><td>150</td></tr>
<tr><td colspan="2">Total</td><td>2847</td></tr>
</table>

Source: (Chennai Corporation)

interior roads and once in 3 years in the case of bus route roads. Hence, in this study, SDBC surfacing with ages ranging from 1 to 4 years (1, 2, 3, and 4 years) were selected. For each age level, test sections were identified, covering three traffic levels (low, medium, and high) and drainage conditions as good, average, and poor. A total of 90 test sections were identified with various combinations of age, traffic, and drainage levels.

The traffic classifications in IRC 37-2001 were utilized in this study: low means less than 150 CVPD, medium means 150 to 1500 CVPD, and high means greater than 1500 CVPD. The drainage conditions are classified in this study on a subjective basis as good means the rainwater does not stagnate and flows freely into the drainage. The rainwater stagnates to some extent and flows into the drainage. The rainwater stagnates and does not flow into the drainage. The drainage coefficients were adopted as given in the HDM-4 manual. The coefficients adopted are shown in Table 11.2.

Other criteria considered in this study for selecting test sections are straight stretches on plain terrain. In terms of drainage, crust, surface quality, and subgrade, it stretches in a homogeneous longitudinal and transverse orientation.

- Traffic Volume Count Survey

A traffic volume count survey was conducted, and the number of commercial vehicles was segregated. The number of equivalent standard axle loads (ESAL) was calculated by multiplying CVPD by the vehicle damage factor [3].

- Measurement of Cracks

Surface cracks were measured on all the sections as a percentage area of cracking for the total area. For crack measurements, the deteriorated area

Table 11.2 Drainage coefficients.

Type of drain	Drainage coefficient
Good	1
Average	2
Poor	3

Source: (HDM-4 Manual) [12]

and all types of cracks combined were measured as a shape equal to a square or rectangle using a measuring scale. In the case of a single transverse or longitudinal crack, it was measured by considering the length of the crack and the minimum width of 30 cm to calculate the area of the crack. In each segment, the cracking is estimated as a percentage of the total area of the segment considered [16].

- Evaluation of Test Section

On each test segment, 0.9 × 0.9 m test pits were excavated. The trenches were excavated along the wheel route at a range of 0.5 to 1 m from the margin of the pavement, based on the roadway width. The pits were dug down to the subgrade level. Layer by layer, the pavement was excavated. During the excavating operation, the thickness of every layer and the type of material utilized was visibly recorded. On the soil layer, the relevant tests were performed.

- sieve analysis
- plasticity indices
- the dry density of subgrade soil
- moisture content necessary for compaction.
- soaked CBR

11.4 Structural Number

The structural number was used to estimate the structural strength of the pavement. It was determined by following parameters: subgrade strength, depth of every layer offered, and the great content utilized in each layer. The way to calculate the structural number is provided here [14],

$$SN = a_1 \times t_1 + a_2 \times t_2 + a_3 \times t_3 + \ldots\ldots\ldots\ldots\ldots + a_n \times t_n \quad (11.3)$$

where

$a_1, a_2 \ldots a_n$ are material coefficient used in each pavement layers,
$t_1, t_2 \ldots t_n$ are the thickness of each layer in inches.

The subgrade strength has an impact on the pavement's strength as well. As a result, the abovementioned structural number is changed to account for the subgrade impact (Paterson, 1987). The strength co-efficient suggested for different pavement materials used in India are given in Table 11.3 [16].

Table 11.3 Strength coefficient of different layers.

S. no.	Type of layer	Thickness (mm)	Co-efficient
1	Bituminous Concrete (BC)	40	0.3
2	Bituminous Concrete (BC)	25	0.28
3	Dense bituminous macadam (DBM)	-	0.28
4	Semi Dense Bituminous Concrete (SDBC)	25	0.25
5	Bituminous macadam (BM)	-	0.18
6	Premix carpet (PC)	25	0.18
7	Granular Sub-base	-	0.11
8	Water-bound macadam/Lime cement stabilized	-	0.14

Source: (CRRI 1994) [16]

11.5 Modeling Methodology

The sequential steps involved in modeling are shown in Figure 11.3.

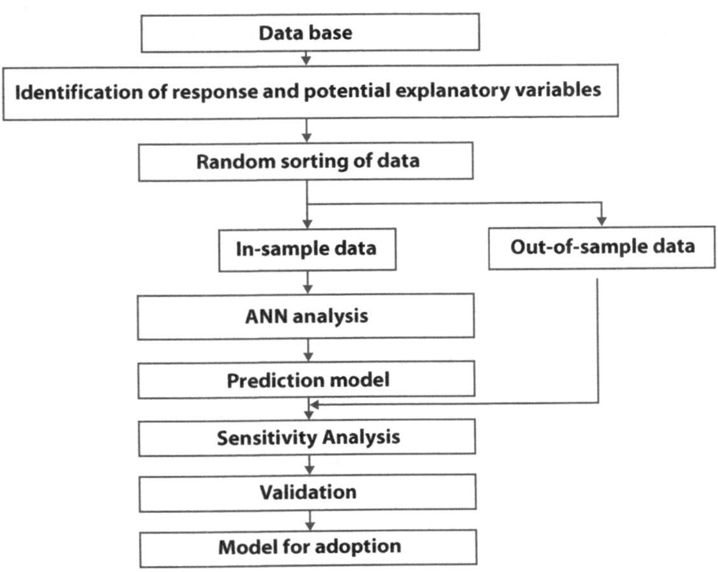

Figure 11.3 Modeling methodology.

- Laboratory Studies

The portions' layer thicknesses were tested physically, and subgrade soil samples were taken from the test pits. The granular size distribution, plasticity index, plastic limit, shear density, optimal moisture content, and California Bearing Ratio of the subgrade soil were all examined in the lab. MSN was estimated by considering the CBR value of the subgrade and thickness of the pavement.

- Development of Model

In developing a cracking model for flexible urban pavement, the technique utilized was Artificial Neural Network. The multilayered network with one hidden layer and one output layer was adopted in this architecture. As an input activity, the raw data is fed into the network. The weights determined each hidden unit activity on the links between the input and hidden units, as well as the actions of the input units. The activities of the hidden units and the weights between the hidden and output units determine the output unit's behaviour, as shown in Figure 11.4. The software MATLAB was used to write the architecture of the artificial neural network having six neurons as a hidden layer in which the weights from the input layers are stored. The logsig was used as a transfer function for these weights to be transformed to the hidden layer. For the hidden layer to the output, the purelin transfer function was utilized. The program was run with about 70% of the data for training the network, 25% for model validation and 5% for testing. Figure 11.5 represents the training of data using MATLAB.

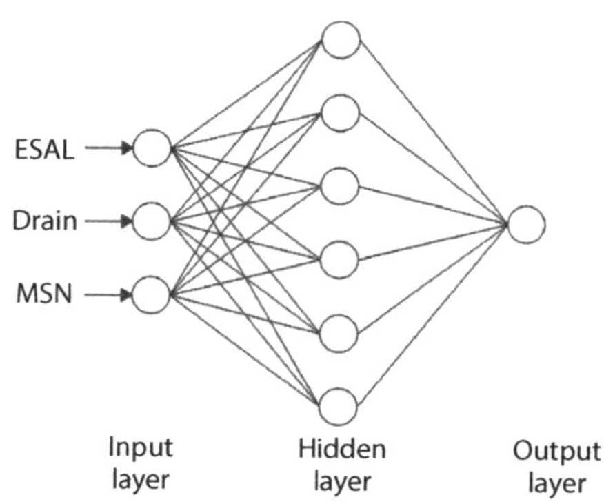

Figure 11.4 Network architecture.

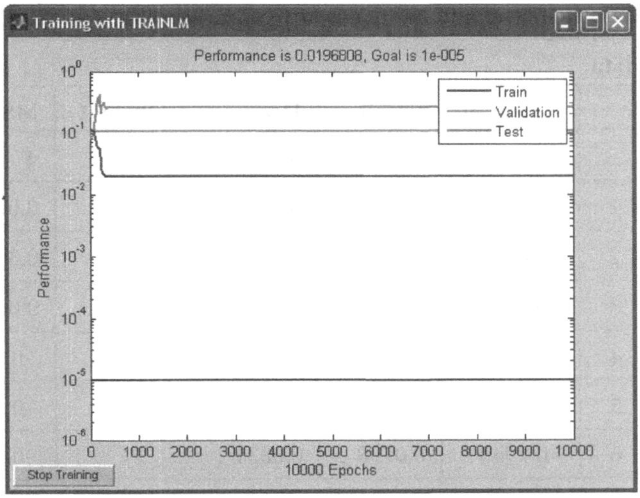

Figure 11.5 Training of data.

The output model from the network is represented by the equation given below.

$$\text{Cracking} = W_2\left[\frac{1}{(1+\exp(\text{-}n))}(W_1P+b_1)\right]+b_2 \tag{11.4}$$

where

W_1 = Weight matrix from the input layer to the hidden layer
P = Input matrix
ESAL = Equivalent Standard Axle Load in MSA
DC = Drainage coefficient
MSN = Modified Structural Number
W_2 = Ratio of Weight matrix between the hidden layer and output layer
b1 = hidden layer bias
b2 = output layer bias
n = Elements from the matrix (W_1P+b_1)

The model was developed with a mean square error of 0.0196, which is satisfactory. The weight matrices that come as the program's output are given in Tables 11.4 and 11.5.

Table 11.4 Input layer and hidden layer weighted matrix.

Input data					
		Constant	**Email**	**Drainage coefficient**	**MSN**
		1	2	3	4
Hidden Layer	1	−0.6916	−0.0071	0.2361	0.0763
	2	−0.6582	−0.0396	0.2344	0.3000
	3	−0.6931	−0.0083	0.2358	0.0703
	4	−0.7873	0.0729	0.0989	−0.1991
	5	−0.7643	0.0391	0.1477	−0.1347
	6	−0.7940	0.0798	0.0831	−0.2189

Source: (Output from MATLAB)

Table 11.5 Hidden layer and output layer weighted matrix.

	Hidden layer						
Output	**Constant**	1	2	3	4	5	6
	−1.6959	0.7718	0.7948	0.7714	0.7770	0.7704	0.7795

Source: (Output from MATLAB)

11.6 Model Validation

The validation and testing of the model were carried out with 25% and 5 % of the data. The predicted values and the variation are shown in Figure 11.6.

11.7 Sensitivity Analysis

The sensitivity of the cracking concerning the inputs is determined through sensitivity analysis. The median values of all the inputs are selected, and only one input variable is altered from its minimum to the maximum value in the range. The range of values of independent variables used in this study is given in Table 11.6.

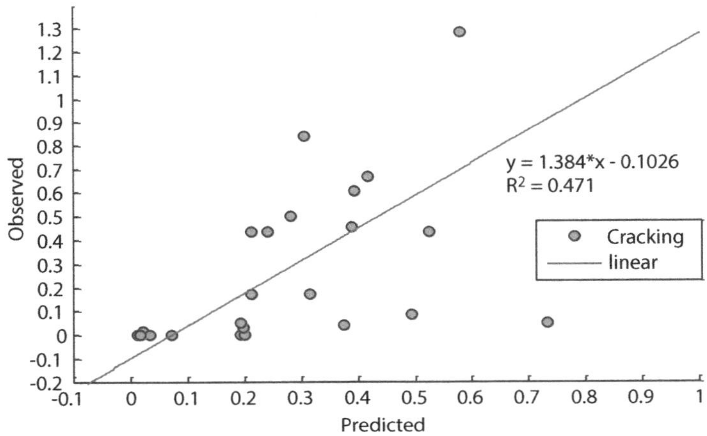

Figure 11.6 Predicted vs observed cracking.

Table 11.6 Range of independent variables.

Sl no.	Independent variable	Unit	Range	Median
1	Email	Msa	2–22	12
2	Drain	-	1–3	2
3	MSN	-	2–8	5

- Variation of cracking with ESAL

The increase in the ESAL has shown a steady increase in the percentage of cracking. From Figure 11.7, the effect of ESAL can be considered sensitive for cracking.

- Variation of cracking with drainage coefficient

The increase in the drainage coefficient has increased the percentage of cracks. Figure 11.8 shows that the crack has increased considerably for average and poor drainage.

- Variation of cracking with MSN

The increase in the MSN value has reduced the percentage of cracking, which is acceptable to be sensitive to cracking. The decreasing trend of cracking with the increase in the value of MSN is shown in Figure 11.9.

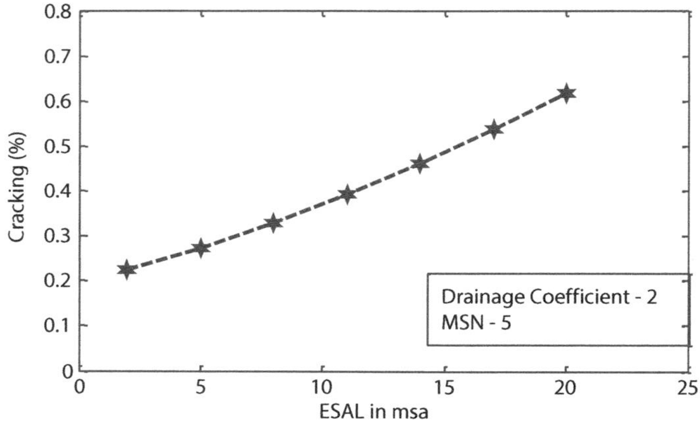

Figure 11.7 Variation of cracking with ESAL.

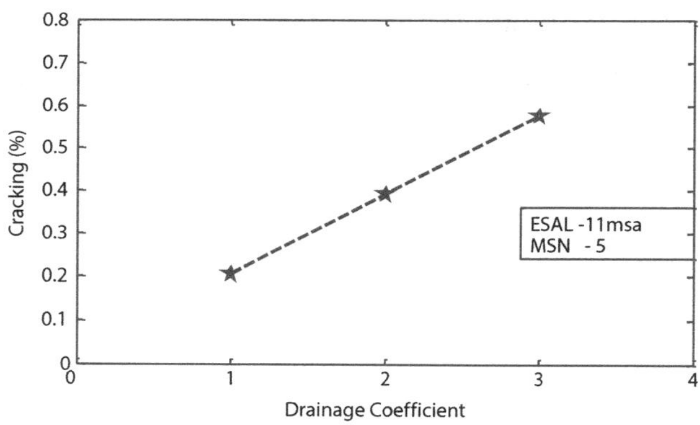

Figure 11.8 Variation of cracking with drainage coefficient.

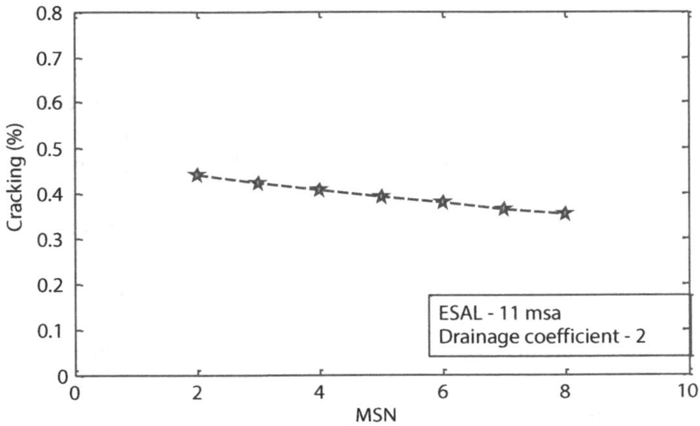

Figure 11.9 Variation of cracking with MSN.

11.8 Conclusions

The following conclusions are drawn from the laboratory and field studies: models developed for distress using the ANN technique are highly satisfactory. Factors influencing the progression of cracking are drainage, modified structural number, and equivalent standard axle load. Cracking increases with poor drainage conditions and decreases with an increase in pavement strength. The model developed in this study can be utilized for the optimum maintenance of urban roads.

11.9 Limitations

This model can be utilized only on the pavement with similar characteristics of data collected for model development.

11.10 Future Scope of Study

The researchers can develop a model that applies to all types of flexible pavement and all prevailing environmental conditions.

References

1. Guignier, F. and Madanat, S., Optimization of infrastracture systems maintenance and improvement policies. *J. Infrastruct. Syst.*, 5, 4, 124–134, 1999.
2. Ker, H.-W., Lee, Y.-H., Wu, P.-H., Development of fatigue cracking prediction models using long-term pavement performance database. *ASCE, J. Transp. Eng.*, 134, 11, 477–482, 2008.
3. IRC: 37-2001, Guidelines for the design of flexible pavements (second revision), Indian Roads Congress, New Delhi, 2001.
4. IRC: 82-1982, Code of practice for maintenance of bituminous surfaces of highways, Indian Roads Congress, New Delhi, 1982.
5. Hein, D. and Croteau, J.M., The impact of preventive maintenance programs on the condition of roadway networks paper, in: *Coordinating Pavement and Maintenance Management Session of the 2004 Annual Conference of the Transportation Association of Canada, Québec City, Québec*, 2004.
6. Rumelhart, D.E. and McClelland, J.L., Parallel distributed processing: Explorations in the microstructure of cognition, vol. 1, Foundations, MIT Press, Cambridge, MA, 1986.
7. Ben-Akiva, M. and Gopinath, D., Modeling infrastructure performance and user costs. *ASCE, J. Infrastruct. Syst.*, 1, 33–43, 1995.
8. Agarwal, P.K., Das, A., Chakroborty, P., Simple model for structural evaluation of asphalt concrete pavements at the network level. *ASCE, J. Infrast. Syst.*, 12, 1, 41–49, 2006.
9. Masad, S., Little, D., Masad, E., Analysis of flexible pavement response and performance using isotropic and anisotropic material properties. *J. Transp. Eng.*, 132, 4, 342–349, 2006.
10. Hornik, K., Stinchcombe, M., White, H., Multilayer feedforward networks are universal approximators. *Neural Newt.*, 2, 5, 359–366, 1989.
11. Madanat, S., Bulusu, S., Mahmoud, A., Estimation of infrastructure distress initiation and progression models. *ASCE, J. Infrastruct. Syst.*, 1, 3, 146–150, 1995.
12. Jaya, L.U., Study of HDM-4: An application to Indian conditions. Master of Technology Dissertation, Indian Institute of Technology, Bombay, India, 2004.
13. Sood, V.K. and Sharma, B.M., Development of pavement deterioration models for Indian conditions. *Journal of Indian Roads Congress*, No. 57-3, 1996.
14. Paterson, W.D.O., Road deterioration and maintenance effects, a World Bank Publication, 1987.
15. Li, Y. and Metcalf, J.B., Crack initiation model from asphalt slab test. *ASCE, J. Mater. Civ. Eng.*, 14, 4, 303–310, August 1, 2002.
16. Lou, Z., Gunaratne, M., Lu, J.J., Dietrich, B., Application of neural network model to forecast short-term pavement crack condition: Florida case study. *ASCE, J. Infrastruct. Syst.*, 7, 4, 166–171, 2001.

12

A Review of Autonomous Vehicles

Joyston J. D'Costa and Ajith B.S.*

Department of Mechanical Engineering, Sahyadri College of Engineering & Management, Mangalore, India

Abstract

This survey article portrays the fundamental ideas of an autonomous vehicle. The car business is one of the main ventures in the globe today. The ascent of driverless vehicles will significantly affect organizations and experts. Driverless vehicles could trade corporate task forces for transport, or moving delegates and employees would obtain valuable time in the day by functioning rather than driving during everyday drives. Advancements in the present field are additionally guaranteed to modify the vehicle insurance market by diminishing mishaps. The link and synchronization of radar and ultrasonic sensors and optical cameras permit driverless driving. Driverless vehicles having different degrees of automation are becoming famous in evolved nations because of their viability in diminishing the fatalities brought about by street mishaps. A nonindustrial nation like India, with the second biggest population on the planet, makes special street situations for a driverless vehicle which requires a great deal of testing and tweaking before execution. This prompts the significance of data sets giving data about different traffic circumstances in India. For preparation, driverless vehicles need to identify, order and gauge the profundity of deterrents they experience on the streets. Driverless vehicles can tackle traffic-related issues and ecological issues, and difficulties. AI-fueled driverless vehicles are broadly expected innovative progressions within recent memory. Individuals accept that driverless vehicles will give a superior future by expanding street wellbeing, bringing down framework expenses, and further developing portability for youngsters, the old, and the incapacitated. Sensor-independent vehicles use to figure out where they ought to travel and perceive things out and about to stay away from auto collisions. Information gathered by the sensors displays heterogeneous and multimodal qualities, which are combined to approach successful choice guidelines. Accordingly, sensors assume a significant part in the dynamic movement of Autonomous Vehicles (AVs).

**Corresponding author*: ajithbsuresh1991@gmail.com

Rathishchandra R. Gatti, Chandra Singh, Rajeev Agrawal and Felcy Jyothi Serrao (eds.) Self-Powered Cyber Physical Systems, (243–254) © 2023 Scrivener Publishing LLC

Keywords: Autonomous vehicle, innovation, the insurance industry, sensors, reducing accidents, artificial intelligence

12.1 Introduction

Driving is training that facilitates our routines by shipping us to and from work to social or exercises important to keep up with our everyday schedules. For more than a century, vehicles have impacted our versatile conduct. They have developed considerably about solace, wellbeing, and asset utilization; however, one thing has continued as before: the vehicle is driven by a completely centred individual. With expanding computerization and robotization, the vehicle turned out to be an increasingly more advanced buddy which upholds the driver in many regards; it cautions the driver before a crash or forestalls unintentional path changes. Regardless of this help, notwithstanding, the control and, accordingly, at last, the obligation stayed with the driver [1]. Lately, computerization has moved forward, and self-driving vehicles are being brought into the market to some extent. The Society of Automotive Engineers (SAE) has characterized a few degrees of mechanization in this manner. The initial three levels (0–2) require the consideration of the driver to screen the street and resume and assume control over control without warning. Levels three to five hand over an obligation to the vehicle to screen the street, with level five characterizing an independent vehicle. The word independent is taken from Greek roots.

$$\text{Independent} = \text{automobiles} + \text{nomos (SELF)} + \text{(GOVERN)}$$

In this way, an Autonomous vehicle is a vehicle that can drive itself starting at one point and then onto the next without help from a driver, at the end of the day, with an autopilot system. Another illustration expresses that an independent vehicle is a vehicle that is fit for detecting its current circumstance and exploring without human-being interaction. Independent driving innovation is fit for giving helpful and safe driving by staying away from crashes brought about by driver mistakes. Considering that we will probably be standing up to a few very long-term progress periods when independent vehicles share the street with human-driven vehicles, it is basic to guarantee that independent vehicles collaborate with human-driven vehicles securely. Since driving consistency is a huge variable in well-being, a huge test of independent driving is to copy, while remaining inside security limits, human driving styles, i.e., to accomplish human-like driving. Urban areas, these days, are continually confronted with quickly rising populaces, which have prompted changing examples of developments for shopping, recreation, and work. With the continuous development of the economy and an ascent in the metropolitan populace and vehicle proprietorship, there has been an expansion in rush hour gridlock interest, prompting delays, an expansion in movement time, contamination, and fuel utilization. The vehicles that work in blockage consume more fuel than those working under inconsistent state traffic conditions for a similar normal speed. Likewise, blockage prompts an expansion in movement time and street client costs. These issues are more unmistakable in the developing economies of the creating scene. Discussing agricultural nations, does the ascent generally impact the nations in ozone-harming substance outflows (GHC) because of their fast development in transport interest? Even though the impacts of GHG emanations are more articulated in agricultural nations, the continued expansion in GHG outflows influences all countries worldwide because the vehicle area is perhaps the biggest producer of GHG. Another tremendous outcome of the expansion in vehicle utilization and possession because of rapid monetary and populace development is traffic security and the lamentable ascent in fatalities from streetcar accidents [2]. Traffic fatalities and wounds influence the livelihoods of those impacted and harm the economy of nations with high paces of car accidents. The majority of these traffic fatalities are a result of human blunders. All the above incidental effects or results, for example, commotion and air contamination, blockage, security, fuel use, and transportation costs because of financial development, populace development, expansion in vehicle proprietorship, and traffic request, put an imperfection on the vehicle area. This has driven various countries and foundations, for example, the auto business and innovation industry, to attempt to find creative approaches to making the vehicle area more secure, less contaminated, less clogged, more eco-friendly, and less expensive. The outcome is the creation and utilization of autonomous vehicles.

12.2 History

The research and experiments on autonomous vehicles started in the 1920s, and trials started in the 1950s; since then, the work on autonomous cars has never taken a pause. The first-ever autonomous car was seen in the 1980s with Carnegie Mellon University's Navlab and ALV projects in 1984 and Mercedes-Benz and Bundeswehr University Munich's EUREKA Prometheus Project in 1987. Since then, many companies have conducted experiments and developed working prototypes of autonomous vehicles. The companies like Mercedes-Benz, Toyota, Renault, Nissan, Audi, Hyundai, Volvo, Tesla motors, and many others have been in the race to develop autonomous vehicles. At present, Tesla is near to achieving fully self-driven cars on the market with their full self-driving package. The cars in the market need to be self-driven, utilizing all possible estimation methods.

12.3 Degrees in Automation

The classification of autonomous vehicles is based on six different levels published in 2014 and updated again in 2016 by SAE International, an automotive standardization body. The six levels of automation are listed below:

1. Zero level (Without automation): This is the level under which most vehicles available now fall. By studying the surroundings through his eyes, the driver controls the steering, throttle, and braking, as well as navigating and determining when to utilize turn signals, change lanes, and turn. However, there may be some warning mechanisms in place. Here all the decisions are taken by the driver himself.
2. The first level (driver assistance): The steering, throttle, and braking can be handled by the vehicle but not in all conditions, and the driver must be ready to take over these situations when the vehicle cannot handle the condition. That means the driver must know what the vehicle is doing and step in when needed.
3. The second level (partial assistance): The vehicle handles steering, throttle and braking yet quickly allows the driver to take over, assuming that he identifies instances where the automobile is not answering. The driver is in charge of the first three levels for observing environmental factors, such as traffic, weather, and street condition.
4. The third level (conditional assistance): The car scans the environment and handles all steering, throttle, and braking in particular situations, such as motorways. Be that as it may, the driver should be prepared to mediate, assuming the vehicle demands it.
5. Fourth level (highly automated): The vehicle controls steering, throttle, and braking, as well as monitoring environmental parameters in a wider range of situations. However, not all of them can be tracked, such as inclement weather.

When it is safe, the driver activates the automatic driving mode. From that point onward, the driver is not needed.

6. Fifth level (fully automated): The driver only has to set the location and start the vehicle; the vehicle takes care of the rest. The vehicle can lead to any legal destination and go with its own choices. The above levels are significant because they are common principles for how technologically progressed a vehicle is. That is the biggest difference starting at level 3. The automated driving system is equipped to screen the driving conditions.

SAE level	Name	Narrative Definition	Execution of Steering and Acceleration/ Deceleration	*Monitoring* of Driving Environment	Fallback Performance of *Dynamic Driving Task*	System Capability *(Driving Modes)*
***Human driver* monitors the driving environment**						
0	**No Automation**	the full-time performance by the *human driver* of all aspects of the *dynamic driving task*, even when enhanced by warning or intervention systems	Human driver	Human driver	Human driver	n/a
1	**Driver Assistance**	the *driving mode*-specific execution by a driver assistance system of either steering or acceleration/deceleration using information about the driving environment and with the expectation that the *human driver* perform all remaining aspects of the *dynamic driving task*	Human driver and system	Human driver	Human driver	Some driving modes
2	**Partial Automation**	the *driving mode*-specific execution by one or more driver assistance systems of both steering and acceleration/ deceleration using information about the driving environemnt and with the expectation that the *human driver* perform all remaining aspects of the dynamic driving task	**System**	Human driver	Human driver	Some driving modes
***Automated driving system* ("system") monitors the driving environment**						
3	**Conditional Automation**	the *driving mode*-specific performance by an *automated driving* system of all aspects of the dynamic driving task with the expectation that the *human driver* will respond appropriately to a *request to intervene*	System	**System**	Human driver	Some driving modes
4	**High Automation**	the *driving mode*-specific performance by an automated driving system of all aspects of the *dynamic driving task*, even if a *human driver* does not repond appropriately to a *request to intervene*	System	System	**System**	Some driving modes
5	**Full Automation**	the full-time performance by an *automated driving system* of all aspects of the *dynamic driving task* under all roadway and environmental conditions that can be managed by a *human driver*	System	System	System	**All driving modes**

12.4 Benefits and Drawbacks

Benefits—an expansion in the utilization of autonomous vehicles would make possible such advantages as:

- Fewer car crashes and car accidents.
- Expanded street limit and decreased gridlock because of diminished need for wellbeing holes and the capacity to better oversee traffic stream.
- The higher speed limit for autonomous vehicles.
- Decrease in the actual area expected for automobile parking.

- Decrease in the requirement for traffic officers and expense on automobile insurance.
- Decrease in vehicle robbery because of the automobile's expanded consciousness.
- Elimination of the steering wheel.
- Efficiency.
- Ease automobile inhabitants from driving and route errands, permitting them to do different undertakings, in other words, to relax throughout their long traffic ventures.
- Speed differentials between cars are kept to a minimum.
- Throttle reduction and braking control are both automated.
- The number of genuine street signs is decreasing, and the journey is becoming easier.
- Because everyone could drive, there would be a purpose to pass a driving test or obtain a driver's license.
- Improved portability for children, the elderly, the disabled, and those in need.
- Commercial automobile sharing.

Drawbacks—despite numerous rewards of comprehensive automobile automation, a couple of unsurprising challenges persevere:

- Risk of life because of accidents.
- Loss of driving-related employment.
- Loss of security.
- Autonomous vehicles might be loaded with explosives utilized in the form of bombs.
- Vulnerability of the vehicle's navigational set-up to various kinds of climate.
- Provisional development regions which are not present on any maps or information databases.
- Assurance of the seriousness of roadway deterrents, as in securely riding a pothole or trash.
- Current cops and other people walking nearby motions and nonverbal prompts are not adjusted to self-driven driving.
- If there is a primary failure sensor and backup sensor, the automobile builds an opportunity for an accident.
- A breakdown or a virus in the system can prompt deadly collisions and deaths-programming dependent.

- Network security—the chances of car hacking through data distribution through the automobile-to-automobile and automobile Infrastructure.
- Execution of legal structure and foundation of unofficial laws for autonomous vehicles.
- Dependence on self-driven drive delivers unskilled drivers whenever the physical drive is required.
- Obstruction to people to let go of control of their vehicles.
- Reduce the utilization of general public automobiles.
- AI must still be ready to work appropriately in turbulent interior-city road conditions.
- While programming the product that determines what action the car takes in the event of an unavoidable accident, ethical and moral considerations come into play.

Challenges:

- The eventual fate of transportation and humanity.
- Vulnerability of vehicles' navigation system to various climates, for instance, snowy circumstances.
- Legal issues.
- Higher manufacturing expenses.
- Traffic signal detection.

12.5 Working Principle of Autonomous Vehicles

The driver selects a destination, and the car's software calculates the route and initiates the trip to the destination. A revolving, roof-mounted LIDAR sensor scans a 60-meter radius around the automobile, producing a detailed three-dimensional representation of the current weather. A sensor detects sideways movement on the left rear wheel, compared to the three-dimensional map, to identify the vehicle's location. Obstruction distances are calculated using radar sensors focused on the front and back bumpers. The artificial intelligence programming in the vehicle is linked to all the sensors and includes Google Street View, and camcorder data. By imitating human perceptual and dynamic cycles, AI governs driving systems such as steering and braking. The vehicle's product tells Google Maps to warn you about sites of interest, traffic signals, and lights. A repeal work allows a person to anticipate a vehicle order. Individual vehicles could benefit from data collected from nearby vehicles, particularly data relating to traffic and security threats. In a circulated association, the vehicular correspondence system uses automobiles and roadside units as conferring centres, outfitting data between them [3].

12.6 Mechanics Involved

Self-driving cars use various techniques to recognize their surroundings, including radar, laser beams, GPS, odometers, and computer vision. An advanced control system decodes sensory input to identify appropriate navigation paths, deterrents, and significant indicators. Self-driven vehicles have managing systems that are prepared for researching actual data to perceive different vehicles making the rounds. This is incredibly useful in organizing a way to reach the best objective. The systems used in the self-driven vehicle contain

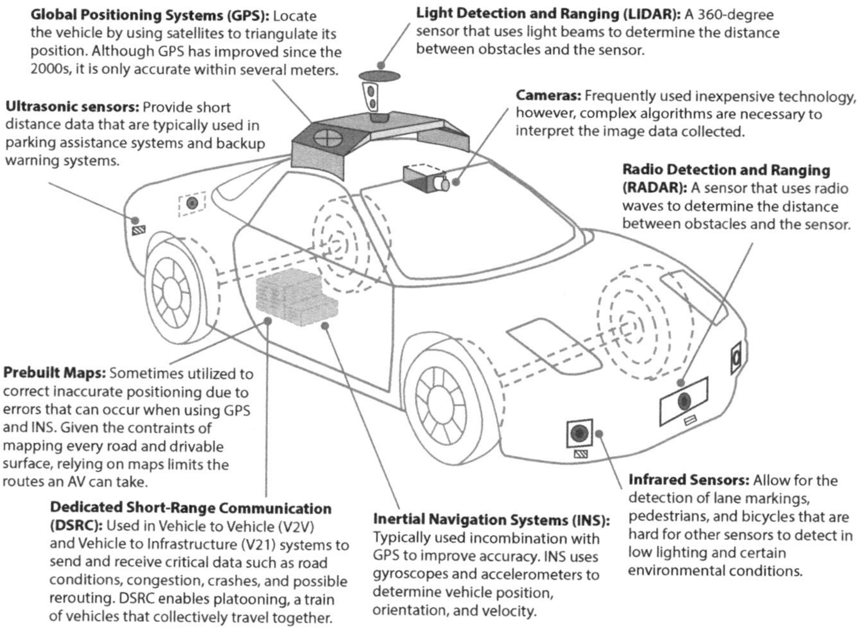

LIDAR: A distance identification automation calculates space by shining a ray of light on a target and separating the reflected light. The device is mounted atop the vehicle's highest point on a barrel-shaped fenced-in

section that rotates 360 degrees and is the primary device in a self-driving vehicle. It includes a collector, mirror, and emitter. It creates a three-dimensional map of the surrounding area that rotates 360 degrees. It visualizes objects using infrared, visible, ultraviolet, or laser. The emitter fires a laser beam that bounces off a ten-times-per-minute spinning mirror near the spherical, empty housing. After reflecting on the obstacles, the laser beam returns to the mirror and is rapidly returned towards the beneficiary, where it is unravelled into data. This data is taken care of by the computer, which creates high exactness three-dimensional maps of the region. The automobile can utilize maps to escape obstacles. The precision of the given map is in centimetres as the recurrence of light utilized is tiny and can bounce back the wide scope of the top layer as well as small objects.

RADAR: Using electromagnetic waves, this device can determine the common speed of the article and the vehicle. It transmits a signal during the evaluation and waits until it is returned. The repetition of the passed-on back message is predicated on a shared development event that is marginally different (Doppler's effect); hence, the speed of the forehead-driving vehicle can be processed fast. The radar, diverged from Lidar, uses a more prominent recurrence and lower signal energy. Portraying the condition of the checked space is not competent. It can likewise detest non-metal things or things that have a particular shape. A huge part of the radars works to around 77 GHz and is fairly directional to investigating bar—up to 200 metres in front of the car, the radar channels the street. Two radars with different scopes are used by several vehicles. The Radar system is installed on the vehicle's front and rear guards. Radar detects the weather, which is then fed into the computer, along with data from the Lidar system. The radar technology is used to identify moving vehicles, their speed, and other obstructions, as well as for automatic parking, blind-spot identification, route assistance, adaptive cruise control, optional impact cautioning, and cross-traffic alert [4].

ULTRASONIC SENSORS: These are fixed on different sides of the automobile to recognize objects astoundingly close to the automobile or calculate the spot of different automobiles while parking. These sensors give crash cautioning, parking aid, and lane crossing among different limits.

VIDEO CAMERAS: They are fixed above the front glass, close to the back viewing mirror, and foster ongoing three-dimensional photos of the street ahead. They are utilized to recognize traffic lights, traffic signals, and astonishing entities like creatures or other people walking. They similarly become conscious of street signals like "STOP" signals, zebra crossing points, and signboards. Camcorders, in like manner, help in seeing specific signs that various sensors cannot understand, like hand waving or traffic cones.

GPS: A satellite navigation system provides continuous real-time location and time information wherever in the world where there is an unhampered view of something like four satellites. It is the reason for the large number of maps that automobiles utilise inside a city. The two frequencies used by all satellites are 1.57542 GHz and 1.2276 GHz. GPS uses satellites to collect information about the vehicle's current location. With a precision of 30 centimetres, GPS leads a car on its pre-planned route. A map of the region is overlaid on the computer with the help of GPS. Various GPS methods are used to determine the all-out position [5].

IMU: The information obtained solely from GPS is only partially accurate. As a result, the information obtained is combined with the IMU's yields. Accelerometers, gyroscopes, and magnetometers are used in the IMU. The IMU is an electronic device that measures and reports the vehicle's speed, heading, gravitational forces, and other parameters. When signals are difficult to reach, such as in tunnels or bad weather, and when the electromagnetic impedance is present, the IMU aids GPS operation [6].

CPU: All of the data gained from every sensor system is taken care of by the computer, which processes this data at high speed. The computer is an extraordinarily powerful data-handling unit mounted inside the vehicle. With the help of significantly advanced programming seeks after the important decision and directs the outcome to electro-mechanical units like automatic steering, accelerator and braking mechanisms. The computer is similarly linked to the internet and GPS to give consistent, continuous real-time monitoring and updates [7].

12.7 Conclusion

Soon, our desire and lead will continuously be facilitated toward vehicles equipped with the front line and modern devices. These systems are planned to simplicity and work with the drive; however, they may be the justification for intricacy over the capable assessment of the accident. Self-driven vehicles are at this point a making advancement. Many associations and scientists have estimated future turns of events and the logical impacts of automobiles. The need to diminish accidents and biological pollution and make a viable transportation future that best obliges people's prerequisites have driven the movement in vehicle automation. This innovative driving innovation hopes to bring many advantages like street security, regular and financial advantages, rearrangement of time and resources for different exercises, and more eco-friendly metropolitan habitats. The change from self-driving vehicles with various degrees of automation to totally autonomous vehicles is yet to be made. Current AI and machine learning developments are rapidly moving in this direction, propelling the industry forward. Top automakers, such as GM, Ford, and Tesla, are in the last stages of testing their completely autonomous vehicles, implying that we are on the verge of witnessing a sea change in how we drive.

References

1. Faisal, A., Kamruzzaman, M., Yigitcanlar, T., Currie, G., Understanding autonomous vehicles. *J. Transp. Land Use*, 12, 1, 45–72, 2019.
2. Kato, S., Takeuchi, E., Ishiguro, Y., Ninomiya, Y., Takeda, K., Hamada, T., An open approach to autonomous vehicles. *IEEE Micro*, 35, 6, 60–68, 2015.
3. Sun, X., Cao, S., Tang, P., Shaping driver-vehicle interaction in autonomous vehicles: How the new in-vehicle systems match the human needs. *Appl. Ergon.*, 90, 103238, 2021.
4. Bilik, I., Longman, O., Villeval, S., Tabrikian, J., The rise of radar for autonomous vehicles: Signal processing solutions and future research directions. *IEEE Signal Process. Mag.*, 36, 5, 20–31, 2019.
5. Zein, Y., Darwiche, M., Mokhiamar, O., GPS tracking system for autonomous vehicles. *Alex. Eng. J.*, 57, 4, 3127–3137, 2018.
6. Ilci, V. and Toth, C., High definition 3D map creation using GNSS/IMU/LiDAR sensor integration to support autonomous vehicle navigation. *Sensors*, 20, 3, 899, 2020.
7. Iturbe, X., Venu, B., Jagst, J., Ozer, E., Harrod, P., Turner, C., Penton, J., Addressing functional safety challenges in autonomous vehicles with the arm TCL S architecture. *IEEE Des. Test*, 35, 3, 7–14, 2018.

13

Meeting Privacy Concerns in Intelligent Transportation Systems

Sharon D. John

Department of Mechanical Engineering, Sahyadri College of Engineering & Management, Mangalore, India

Abstract
This paper presents the basics of intelligent transportation systems and their protection worries throughout their applications. Intelligent Transportation Systems (ITS) focus on incorporating detecting, investigating, and executing correspondence into movement, well-being, and comfort. All vehicle producers continuously make more brilliant and futuristic vehicles, changing the feel of movement. Travelling is turning out to be more productive in all conditions. More secure vehicles are presented consistently with thought for walker well-being and travellers. However, the new innovation and expanding availability of ITSs are assault vectors for vindictive programmers. Metropolitan urban communities with great associated public transportation frameworks present new security worries with the information gathered from travellers. Moreover, we examine the difficulties in tending to security and protection issues in ITS. The revelations in this paper intend to give direction on the most effective way to address the current, and future challenges that the transportation business faces and make ITS more secure, safe, and security shielding.

Keywords: Intelligent transportation systems, ITS, framework, security shielding

13.1 Introduction

In present-day transportation systems, Intelligent Transport System (ITS) is significantly altering how our streets and urban communities look and work. The ITS has the ability and potential to make an experience superior and effective, more secure and agreeable experience for voyagers

Email: sharondeannajohn@gmail.com

Rathishchandra R. Gatti, Chandra Singh, Rajeev Agrawal and Felcy Jyothi Serrao (eds.) Self-Powered Cyber Physical Systems, (255–264) © 2023 Scrivener Publishing LLC

everywhere. However, when individuals expect protection inside their regular routines, worldwide worries for better being and utmost security of the voyaging public frequently are principle concerns. For instance, data about airline travellers are frequently circled among numerous organizations. Remember, the retail proprietors or security officials continually observe customers at a retail location. For example, phone and any movement route administrations take and store our information for some time later or any improvement programs. Various organizations and urban areas have begun a drive to help the advancement of such ITS. In any case, progresses made in ITS also develop extra difficulties. As the general population hopes to have the most security in their day-to-day lives, protection for privacy concerns plays a vital element that cannot be neglected at any cost in issues of current or future circumstances.

Expanding integration of protected devices and applications with transportation systems presents opportunities for assailants to take advantage of these systems. Security is an intricate issue vital to the public in light of the related emotions and the numerous settings where it is applied. For instance, there needs to be a standard implemented to measure the rate of privacy.

An ITS that depends on surveillance of the transportation system can be considered to compromise the protection of the travellers. Different ITS applications consolidate electronic ringing, photo execution, vehicle mile reading, and GPS tracking, recently known as Vehicle-Infrastructure Integration (VII) [1]. These gadgets work by observing, distinguishing, following and business vehicles. They all can disregard the security assumptions. Because of the chance of broad physical and individual harm, a risk check is done on the entirety of ITS advances, and many experimental models have been made. In any case, many of them had unanswered inquiries; hence further exploration on this point needs to be continually done and keep the data refreshed for future reference.

How much security does the public request, for the most part? What actions can be taken to safely monitor all data put away from being unveiled or utilized by information hackers? What steps can be taken if the trust is disregarded? What level of security insurance is expected for purchasers to acknowledge and even commitment the assortment of this information?

In this paper, we study the present status of security insurance connected with ITS innovation and address changes to the systems that have been advanced for resolving issues. They feature security and protection issues in current organizations. Also, inspecting it from a broad system view and examining long-awaited security issues. This study points in a different direction in an extensive examination of ITS courses inside security and protection and future exploration for heading inside the field.

13.2 Synopsis of ITS

Let us now see the run-down of all the elements in ITS. We will be covering all possible segments that would help us understand ITS better for greater understanding.

A. Components of ITS

i. Smart Vehicles:

ITS is most frequently connected with smart vehicles, whether they have driver-assist innovation, are semi-independent, or even completely independent. Smart vehicles are a significant part of ITS because of the sheer volume of individual vehicles on the street. As the transportation business keeps advancing, smart vehicle advancements largely affect how we travel and the hidden foundation of transportation. Smart vehicles can make "convoys" utilizing vehicle-to-vehicle (V2V) correspondence tech to develop travel effectiveness further. This correspondence and bunching of vehicles have been made conceivable because of examination in vehicular ad-hoc organizations (VANETs).

A VANET is a correspondence network coordinated by vehicles through specialized remote gadgets that empower vehicles to trade information, like crisis circumstances, the distance between vehicles, and so on, to upgrade the well-being and productivity of transportation frameworks.

Other than smart vehicles speaking with each other, the parts inside the vehicle, like electronic control units (Ecu), need to impart Intelligent transportation systems.

With one another to execute different conveyed control capacities. This in-vehicle correspondence is accomplished through in-vehicle associations, for instance, Controller Area Network (CAN) and CAN with Flexible Data-Rate (CAN FD). A past investigation has shown that in-vehicle networks are feeble to security strikes where an attacker can remotely control the vehicle while disregarding the driver's feedback.

ii. Public Transportation:

The public transportation system in numerous urban communities is a powerful method for exploring metropolitan regions. Rail and transport courses work almost nonstop to give residents effective and prudent travel assistance. ITS may fabricate the capability and the throughput of the public transportation system. Smart bus stations can furnish travellers with data regarding the transport plan (approach and take-off times). Headway of courses that considers persistent traffic conditions (e.g., obstruct achieved by various vehicles and incidents) can work on the public's comfort and lessen the development time. Moreover, transport and rail frameworks can likewise profit from communication with the Internet of Things (IoT) gadgets in ITS that hand off data, for example, travellers holding up at a stop or objective point [2].

iii. Controllers:
The regulators or controllers of ITS are those parts that direct, control, or change the components of the transportation framework or system. The ITS controllers display the framework's confirmation, response, and extra action to the saw data. Instances of controllers incorporate traffic signals, traffic declaration systems, computerized street signs, and railroad switches. Controllers go with choices because of the noticed information from the sensors in ITS. Controllers inside ITS are, in many cases, incredibly straightforward state machines that carry out the essential rationale, for example, an if-else structure. As concentrated, some regulator gadgets have natural weaknesses that might permit aggressors to alter traffic designs. Control contribution of controllers can take many structures, for example, a public transport bus acknowledging a wayfarer waiting up at the bus station or signals from the parking space sensors that empower controllers to change the leaving accessibility sign (e.g., from empty to full when it detects that the last accessible parking space has been involved by a vehicle).

B. Benefits of ITS

- Time savings
- Better emergency response times and services
- Reduced crashes and fatalities
- Cost avoidance
- Increased customer satisfaction.

C. Challenges Faced
Tending to security and protection issues in ITS presents various difficulties for designing and advancing common-sense arrangements. While a considerable lot of these difficulties crossover with comparative standards like IoT, which is very undermining [3].

i. Heterogeneity:
ITS involves various advancements and parts with various functionalities and targets. This innate heterogeneity compounds the errand of examining

and guaranteeing the protected adoption and mix of such parts or components inside the ITS. Other than the weaknesses that may now exist in every part, the cooperation of various parts can lead to new weaknesses and exploitable dangers. Besides, components from sellers intended for a similar errand might observe various plans and guidelines, which need to be clarified for in-depth security examination and guarded arrangements. For example, various executions of a similar V2X correspondence might introduce weaknesses that emerge from the connection of conveying gadgets, which can be taken advantage of and compromise the security of ITS components and devices.

ii. Complexity:
ITS, in general, is, involves various interfacing parts of heterogeneous sorts and nonlinear elements. The highlighted structure forms Complex Adaptive Systems (CAS) characterizing qualities. In CAS, the relationship of constituent parts produces higher-demand lead that is missing in individual parts, yet rather is created from the framework overall. Examination and control of such higher-request occasions are frequently muddled and recalcitrant. Consequently, assuring the dependability and security of such emanate conduct in ITS is also a difficult issue, especially under ill-disposed conditions. A case of such new weaknesses in ITS is introduced, which shows that consecutive messing with traffic stream sensors can bring about significant gridlocks or traffic jams in metropolitan regions, resulting in immense trouble for all netizens.

iii. Secure Storage and Sharing of Secret Keys
A combination of safety primitives, like privacy, trustworthiness, and confirmation, in ITS representatives, depends on secure mystery keys. Large numbers of the ITS gadgets will probably need more capacity or assets to safely store and oversee secret keys created for secure correspondence or information move with other ITS representatives, which endangers the security of ITS and makes the protection of gathered explorer information very risky. Other than the certain limit of secret keys, secure key assignment of secret keys between ITS agents related to a given application presents another test. Considering the resource objectives of ITS delegates, new lightweight techniques for secure key exchange are being made, which present new risks and shortcomings for ITS agents.

iv. Dynamic Security and Privacy Requirements
Parts or components and clients of ITS innovations might have different security and protection inclinations, which might change over the long run

without notice. For example, a metropolitan ITS arrangement's security requirements might shift if there is a danger to public safety or cataclysmic events. Concerning the general public's security, clients' inclinations might shift from requiring outright protection and insignificant data openness to deliberately picking into new administrations that require more client data. Dealing with these unique inclinations proficiently and straightforwardly is basic to giving versatile, strong, and compelling ITS solutions to all problems [4].

13.3 Future Research Direction

The previous area developed the need for additional innovative work in different parts of safety and security in ITS. Some are considered essential because of the wide scope of issues that require further innovative work. This segment presents an outline of such regions and promising examination settings for future researchers. Some of the ways that can be considered are elaborated on below.

A. Artificial Intelligence
The gathering of AI and AI methodologies inside ITS advancements is creating at a rapidly uplifting rate. At the same time, the advantages of utilizing such approaches are unquestionably best in class, and the security implications of their coordination with ITS stay largely unstudied. As examined in this paper, basically all AI procedures are inclined to inherent weaknesses that can be taken advantage of to compromise the security of ITS in many ways. While AI well-being and security research is moving forward, it is important to concentrate on the significant parts of this exploration of ITS innovations. Then again, late exploration recommends that AI strategies are critical in computerizing the revelation, moderation, and guarding against security dangers inside the profoundly intricate ITS. Further exploration of such strategies can work with more productive ways to deal with the plan and the board of secure ITS advances. Development in this field will solve half of the general public's problems and would be very useful.

B. Privacy by Design
Protection by plan idea or system design underlines the proactive job of information regulators and processors intending to the security parts of the related system during the full life pattern of every system and through the planning and arranging stages. While the worldview of security by

configuration is not new, it has, as of late, acquired expanding consideration from scientists and the business because of the adherence of the General Data Protection Regulation (GDPR) to this idea [4]. While certain investigations have applied the standards of security by the plan to the ITS area, there stays a wide hole between the cutting edge and fulfilling the protection necessities of GDPR in ITS innovations. Promising specialized settings to investigate this path incorporates the as-of-late proposed block-chain-based approaches and isomorphic encryption, which may give an appropriate means to sharing and examining information while safeguarding singleness, protection, and privacy [5].

C. Complex Adaptive System

One more set of exploration is the security parts of the ITS all in all inside the reflection of CAS. The imagined worldview of ITS structures a CAS, wherein the collaborations of numerous constituent parts make higher-request impacts that can be deciphered as emanate peculiarities. Seeing such complex elements and controlling developing ways of behaving ITS from a security viewpoint is of central significance since, in such a system, nearby disappointments might lead to fountains of dissatisfaction, heightening the issue to the whole order or structure.

13.4 Contributions to this Research

This paper adds three bits of information about security in ITS. To start with, discoveries from the overviews recommend current country patterns for tending to the protection of the voyaging public for a couple of normal ITS applications. The first contribution is that we have got to know better about an Intelligent Transportation System (ITS). We have a better-revised version of basic information that might be needed to refer to and understand ITS in the future. The second contribution is the connection or link between the best studies that have been previously done based on its literature. Almost all literature has been understood for a better quality of this research paper for protecting data and information of the general public. Ultimately, this paper proposes a meaning of designated observation of transportation to recognize it from relaxed observation. This prior commitment gives a foundation to the transportation strategy producers or generators that are thinking about indicating the job of their office in observing rush hour traffic jams. Since it is becoming more standard, the blend of these three commitments will give an asset to foster state or public protection strategies for ITS applications throughout time [6].

13.5 Conclusions

This study was conducted, analyzed and broken down to current situations and changing security protection rehearses for state transportation and policing utilizing insightful transportation systems.

The reactions from different offices upheld past review discoveries that ongoing individual and business ringing policies proceed to safeguard security satisfactorily. When associations got some data about future discoveries and vehicle following examples, they exhibited that public security concerns would likely be the fundamental challenges to executing both vehicle mile ringing and associated vehicles. To address these difficulties, respondents recommended that conglomerating and concealing information were best practices for safeguarding the security of drivers while gathering travel timings and pace. For catching video reconnaissance, maintaining legislation or office arrangements was the favored strategy for safeguarding security. Sadly, restricted government guideline exists in our country and states must rather depend on direction reports.

The outlines have found that a crucial test in specific insurance measures for transportation surveillance describes such basic terms as "loosened up insight" and "assigned perception." To fill this opening in our data, the makers included the discoveries to define both terms. These definitions can act as beginning stages for organizations making, refining, or reconsidering their security approaches.

References

1. Fries, R.N., Gahrooei, M.R., Chowdhury, M., Conway, A.J., Meeting privacy challenges while advancing intelligent transportation systems. *Transp. Res. Part C: Emerg. Technol.*, 34–45, 2012.
2. Hammoudeh, M., Epiphaniou, G., Belguith, S., Unal, D., Adebisi, B., Baker, T., Watters, P., A service-oriented approach for sensing in the Internet of Things: Intelligent transportation systems and privacy use cases. *IEEE Sens. J.*, 21, 14, 15753–15761, 2020.
3. Guevara, L. and Auat Cheein, F., The role of 5G technologies: Challenges in smart cities and intelligent transportation systems. *Sustainability*, 12, 16, 6469, 2020.
4. Hahn, D., Munir, A., Behzadan, V., Security and privacy issues in intelligent transportation systems: Classification and challenges. *IEEE Intell. Transp. Syst. Mag.*, 13, 1, 181–196, 2019.

5. Lederman, J., Taylor, B.D., Garrett, M., A private matter: The implications of privacy regulations for intelligent transportation systems. *Transp. Plann. Technol.*, 39, 2, 115–135, 2016.
6. Mokaddem, Y.E. and Jawab, F., Researches and applications of intelligent transportation systems in urban area: Systematic literature review. *ARPN J. Eng. Appl. Sci.*, 14, 3, 639–652, 2019.

14

Feasibility Study of Digital Twin in Automotive Industry—Trends and Challenges

Preethesh B., Hithesh Alen D'Costa, Sohan R. Alva, Vishal H. Kanchan and Rathishchandra R. Gatti*

Department of Mechanical Engineering, Sahyadri College of Engineering & Management, Mangalore, India

Abstract

A digital twin is a cyber-clone or a cyber-representation of a physically existing real product or process that serves as its real-time digital counterpart. It is a concept formed due to the continuous evolution of computer-aided product design and engineering. The creation of digital twins results from continuous improvement in product design and engineering activities. Over the years, new product development has progressed from manual drafting to CAD/CAE. Now it is evolving toward model-based-systems engineering. The basis of any digital twin depends on the digital thread, which forms the basic design of the product it represents in the cyber world. Whenever the product requires design changes, the change in the digital thread is initiated, and the design parameters being affected are studied. Then there will be subsequent changes in the real product through engineering change orders.

The physical vehicle, the virtual vehicle twin, and the connections between these entities comprise the digital twin concept in automotive vehicles. Data that flow from the physical vehicle to the digital vehicle and information available from the digital vehicle to the physical environment are the connections between the physical vehicle and its virtual vehicle twin. The proposed chapter discusses different types of digital twins, such as digital twin prototypes, product digital twin, production digital twins, and performance digital twins and their applications in the automotive domain. It also explains the different levels of digital twins, such as descriptive twin, informative twin, predictive twin, comprehensive twin, and

**Corresponding author*: gattirathish@gmail.com

Rathishchandra R. Gatti, Chandra Singh, Rajeev Agrawal and Felcy Jyothi Serrao (eds.) Self-Powered Cyber Physical Systems, (265–282) © 2023 Scrivener Publishing LLC

autonomous twin, which are essential during the product's lifecycle. State-of-the-art digital twin applications in the automotive domain are discussed, highlighting the benefits and challenges in adoption. The feasibility of adopting digital twin technology across the different functions and structures is analyzed considering the current and anticipated commercial and technological opportunities and issues of automotive OEMs.

Keywords: Digital twin, industry 4.0, digital thread, automotive, product life cycle, predictive maintenance

14.1 Introduction

In the era of Industry 4.0, where the industries are experiencing a major digital transformation, Digital Twin is gaining a lot of competitive and economic advantages over other competitors. The term "digital twin" refers to a cyber-representation of the finished product. To better understand what was created against what was designed, compare a Digital Twin to its engineering design, closing the loop between design and execution [1]. Through models and data, it encompasses the characteristics, settings, and behavior of the real-life object. A digital twin is a detailed digital representation of a single product. The digital twin is created concurrently with the physical twin and serves as the virtual equivalent throughout the product lifetime [2]. It can identify physical copy issues, from design through production, and create, compare, and discuss physical twin issues [3].

The Digital Twin concept is divided into three parts: a) physical items in physical space, b) virtual products in virtual space, and c) data and information links that connect virtual and physical products as in Figure 14.1. Michael Grieves [3] first proposed the concept of digital twins in 2002, and NASA had been utilizing elaborate simulations of spacecraft for decades before that. Digital twins are now more available than ever because of the rise of the Internet of things (IoT) and the resulting decreased costs of related technologies.

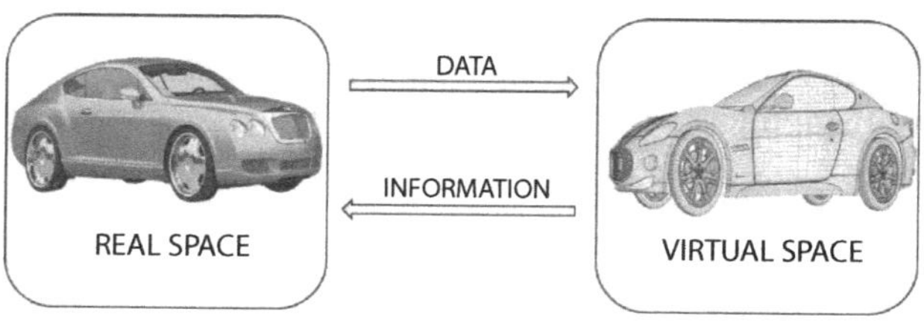

Figure 14.1 Digital twin concept.

14.2 Industrial Revolution

Industrial revolutions are periods when manufacturing methods and humankind are entirely transformed. The four industrial revolutions can be viewed as a journey through several stages of industrial manufacturing system development, from manual labor through automation principles.

14.2.1 Industry 1.0

It lasted from the 18th century to the beginning of the 19th century. This era saw the rise of mechanization, where agriculture was replaced with industry as the foundation occurred during this period. The mass extraction of coal and the development of steam engines and railways have accelerated monetary, human, and material trade. The Steam Revolution Era is another name for this period. This era mainly improved the quality of life.

14.2.2 Industry 2.0

In the nineteenth century, the inventions of electric current, mass production, and engines ushered in the second industrial revolution. With the advancement of the combustion engine, steel, chemical, and textile demand rose drastically in the second industrial revolution. Also, this revolution initiated the idea of mass production where the products carried the conveyer belt and performed further tasks. Henry Ford carried these principles to automotive production and made a drastic change in the process where the construction was made in a single station. After implementing the conveyer belt steps, the process became substantially faster and less expensive. This development resulted in developing and establishing large factories.

14.2.3 Industry 3.0

The Third Industrial Revolution began in the 1970s with the development of memory-programmable controllers and computers to partially automate processes. Technological advancements allow businesses to automate an entire production process without human participation. This is well-known in the form of robots that obey pre-programmed instructions without the need for human interaction. This aids flexible manufacturing in the assembly line, where a wide range of items is made on flexible production lines with programmable machinery.

14.2.4 Industry 4.0

The Fourth Industrial Revolution is currently in progress. This is sometimes referred to by alternative names, such as "Business 4.0" or "Industrial Internet of things." It refers to the use of emerging ICT in the manufacturing industry. It is founded on the achievements of the Third Industrial Revolution. Computer-based production systems are strengthened by a network link and have a digital counterpart on the Internet. These allow them to communicate with other systems and output data about themselves. This is the next step in the production automation process. All systems are linked, resulting in "cyber-physical production systems" and "smart factories," in which production systems, components, and people interact through a network, and production is almost autonomous. Industry 4.0 integrates emerging ICT domains, namely, Robotics and Automation, IoT, advanced wireless communication networks, Cloud computing, Data analytics, Big data, distributed computing, virtual simulation technologies, augmented and virtual reality-based data visualization, and cyber-physical systems, cyber security with the existing physical industrial infrastructure.

14.3 Influence of IoT on Digital Twin

It is evident from the definition that digital twins must depend on IoT technologies. IoT is the foundation for digital twins, as it uses precise data about physical assets to help companies make better decisions. It is powered by sensors, artificial intelligence, machine learning, data, and analytics. As a result of its capacity to expand IoT installations, the digital twin is likely to expand in popularity. In industrial plants, IoT can be integrated into a variety of applications. It aids in the understanding and simplifying of the process flow, product tracking, work-in-process management, line balance, and material flow movement in the warehouse, as well as inventory level optimization. Based on the number of sensors and devices at reach, it can capture a massive amount of real-time data from the ambient environment to where they are deployed.

IoT data can be integrated with the digital twin tool to evaluate the system's efficiency and accuracy without changing the real-world system. Digital twin can collect data from IoT sensors and run a series of trial-and-error experiments to either comprehend the system's behavior or evaluate feasible alternative strategies to operate the system.

14.4 Digital Twin in CPS Applications

14.4.1 Health Care CPS

A digital twin can create a more accurate simulation of hospital CPS management and patient body condition monitoring. A digital twin can solve logistic issues in hospital management through simulation such as medical pathway planning, health resource scheduling, and medical activity prediction. It can also replicate the physical M2M communication of medical devices within the hospital CPS. At the patient monitoring level, the patient's physiological conditions can be replicated as a digital twin to conduct pre-dosage trials virtually before actually using the therapy with the patient. The concept of IoT-enabled wearables, IoT-enabled implants, Body area networks, patient health and community health data, personalized and targeted treatments, and the convergence of engineering disciplines with bio-medical fields are all critical driving forces in the smart and connected health realm.

14.4.2 Manufacturing CPS

The key to smart manufacturing is digitizing all possible manufacturing processes and systems. This digitization is possible by applying the digital twin concept in the existing manufacturing environment. Several frameworks have been suggested, finally converging to the standardized framework such as ISO 23247. Some applications where digital twins can be employed in smart manufacturing are in-process product tracking [4], production process simulation for planning and scheduling, machinery twins for health condition monitoring of employed machines, M2M communication simulation of CNCs and other automats in the shop floor.

14.4.3 Retail CPS

Retailers can utilize the digital twin to replicate the performance of a supply chain and discover areas for improvement. Retailers can use digital twins' real-time visibility to develop innovative non-linear supply chain fulfilling models, including curbside pickup or micro-fulfilment.

14.4.4 Smart Cities and Infrastructure CPS

Cities can also benefit from the digital twin to become more economically, ecologically, and socially sustainable. Virtual models can help planners make better judgments and solve modern cities' numerous complicated problems. Real-time solutions to problems, for example, can be informed by real-time data from digital twins, allowing assets, such as hospitals to respond to a crisis in real time.

14.4.5 Intelligent Transportation CPS

The majority of the new automotive development takes place in a virtual environment. In the automotive sector, digital twins generate a virtual replica of a linked car. Even before production begins, car businesses employ technology to design the ultimate product. They replicate and assess the manufacturing process and any issues that may arise once the car is on the road. Companies can ensure that unforeseen damage and injuries are minimized by building a digital twin of an automobile and evaluating every part of the vehicle. Testing and vehicle maintenance are two examples of digital twin applications in the intelligent transportation sector.

14.5 Digital Twin Types

14.5.1 Product Digital Twin—Using Digital Twins to Create More Efficient New Product Designs

Digital twins may be used to evaluate the product's performance virtually and assess how it works in real-world scenarios. The word "product digital twin" bridges the virtual and physical space, allowing one to evaluate and understand how the product will perform in different circumstances and make virtual adjustments to ensure that the new product released works precisely as expected in the market. It makes no real difference how complicated your system is, and the specification used in the product digital twin can help in managing the difficulty and crucial decision making possible in a better way. This process results in reduced production time, increased product quality, and eliminates the repeated need for the prototypes, decreasing the time to get back the customer feedback.

14.5.2 Production Digital Twins—Manufacturing and Production Planning Using Digital Twins

The production digital twins can help the manufacturing process virtually before it hits the market. With the help of the digital twin, the company replicates the circumstances and effective production technique and understands the changes to be done using digital thread.

The output of the industrial equipment enhances even further by optimizing the manufacturing process. The data obtained by replicating the events can be beneficial in reducing the timely maintenance of the machinery, and the company will be able to predict the exact time for the maintenance. Manufacturing becomes even faster and more reliable by using this process with the help of the continuous flow of data throughout.

14.5.3 Performance Digital Twins—Operational Data are Captured, Analyzed, and Acted on Using Digital Twins

With the continuous use of smart technologies, much data about how they have been used and how effective they are. The performance digital twin collects and analyses data from operational products and plants to give helpful information for better judgment. Companies can use performance digital twins for

- (i) business expansion.
- (ii) enhance virtual mockups,
- (iii) efficient data acquisition,
- (iv) improve product performance.

14.6 Levels of Digital Twin

There are five levels of sophistication in digital twins. The most basic models combine data from multiple sources; the most advanced models are capable of acting independently.

14.6.1 Level 1: Descriptive Twin

The descriptive twin is a visual replica of a built asset that is an alive, editable version of design and construction data. Users can specify what kind of data they want to extract and what information they want to include.

14.6.2 Level 2: Informative Twin

A layer of operational and sensory data relevant to the real-world object has been added to this level. The twin collects and aggregates defined data, such as historical, maintenance, and real-time data, and verifies it to ensure that systems are compatible.

14.6.3 Level 3: Predictive Twin

This twin can gain insights from operational data. With the help of the sensor data, the digital twin system understands when the downtime of the process may occur and can be predicted effectively to avoid further problems that might occur in the future (think of a car that alerts you when it is time to change the oil).

14.6.4 Level 4: Comprehensive Twin

This twin considers "what-if" scenarios and simulates future scenarios. This enables the experiment on the product virtually and process to find the ultimate limitations, leading to optimal product functioning.

14.6.5 Level 5: Autonomous Twin

The trend of IoT, AI, and machine learning is opening up many opportunities in areas like maintenance and manufacturing, allowing the systems to have autonomy where the digital twin infrastructure facilitates self-repair features where the systems can learn and act on behalf of other users, where the system algorithms are capable of immediately detecting issues and recommending or initiating corrective action.

14.7 Digital Thread

The goal of digital threads is to achieve data uniformity and easy universal access. In digital thread, the flow of information occurs right from the beginning of the product concept, manufacture, and sales until the disposal and recycling. It also gives the pertinent details on the product at each life cycle stage and can trace the product information in any stage based on the customer usage.

Digital thread can be developed for various stages of the product like life cycle, manufacturing process, production management, supply chain, and customer data, which goes on until the end of the lifecycle. Digital thread can be benefited in predicting the business growth and having good up and down communication with the customers to get their valuable feedback where an immediate response can be initiated for the upcoming development of the product [5].

14.8 State-of-the-Art Digital Twin Deployment

One successful state-of-the-art proof of concept of digital twin implementation was the bending beam test conducted at the Dik research lab by Haag and Anderl [1]. The test setup comprised a typical digital twin setup with a physical twin, a cyber twin and the ICT-enabled interconnections between the twins. The physical twin was the bending beam with linear actuators for displacement and load cells for force measurement. The digital twin was the exact CAD replica of the beam. The ICT interconnection was designed using publish-subscribe architecture using the MQTT protocol. To start the test, the displacement of the resulting force was input in the GUI, which is messaged through the MQTT protocol to set the parameters in the physical twin, and the physical output was measured. It was then sent to the digital twin, which performs automated FEM analysis using the real value of the input parameter. Then the results of both automated FEM of digital twins and the physical measurements of physical twins were compared. It was found that the deviations were within the margin of error.

O'Sullivan *et al.* [2] suggested a DT framework based on a closed loop method that allows the team to learn as they use the framework and iterative improvements in different portions as work on different areas is done. This framework was created using action research and a participatory approach in their large-scale manufacturing CPS. Their proposed framework was designed to help a large-scale manufacturing facility install a maintenance digital twin. Issues that inhibit such effort are discovered and categorized to emphasize the difficulty of putting this concept into practice. A digital twin framework is offered to aid in the installation process, which breaks down the digital twin development process into parts that may be accomplished separately.

Schleich *et al.* [4] presented a digital twin implementation reference template for various applications, considering essential model criteria such as model features, scalability, interoperability, expansibility, and fidelity. The notion of Skin Model Shapes, which serves as a digital twin of the real product in design and manufacturing, was used to develop their suggested comprehensive reference model. The steps of the product life cycle were discussed, including conceptualization, representation, and implementation.

Son *et al.* [6] created a digital twin (DT)–based cyber-physical system (CPS) that can forecast if an automotive product can be built according to a customer's schedule on an automotive body manufacturing line when anomalous scenarios occur. The digital model was built using the product, process, plan, plant, and resource information for automobile body manufacturing lines. According to the trial results, the proposed system achieved an average prediction performance of 96.83% for the actual production plan.

One of the essential steps in advancing digitalization and Industry 4.0 in this area is to create a link between the virtual and physical worlds by establishing both simulation and virtual commissioning (VC). Hence, a case study was done by Royo *et al.* [7] with the modelling and emulation of an Automotive Volkswagen (VW) Paint Workshop located in Navarra, taking into account the importance of applying these digital-twin principles to existing manufacturing plants. The digital twin technique resulted in a 66% decrease in commissioning (emulation) and a 16% reduction in cycle time losses (simulation).

da Silva *et al.* [8] introduce the cyber digital twin (CDT), a novel concept that applies the DT concept to automobile software for security research. The ECU programme (i.e., firmware) is converted into a CDT, which contains automatically extracted security-relevant information from the firmware. This allows us to assess vehicle security requirements utilizing automated security requirements verification, policy enforcement checks, and security vulnerability detection. The evaluation may be done in real time using newly integrated checks and publicly available security vulnerabilities.

The above are some of the mentions of the current developments in deployments of DT in automotive CPS. Further integrations, especially the interconnections between the DT and Physical twin, the accuracy of the representation of the physical twin by its digital twin, security and privacy issues, and the complex engineering systems intertwined in the automotive CPS and automotive manufacturing CPS pose several challenges for deployment of DT.

14.9 Benefits of Digital Twin

Risk assessment and manufacturing time are both accelerated: Industries that use digital twin technologies can test and evaluate before it emerges into the physical world. This technology enables the creators of the product to anticipate the process has to put up for starting the different manufacturing processes of the product. By creating the digital mockups, one can optimize the model anytime and anyhow and simulate the product according to the real-life situation where failures are more common and come up with a solution. This helps improve the product quality and brings better and sustainable changes to the product. A massive amount of time can be saved in the virtual process.

Predictive maintenance: During the extensive manufacturing process, it is necessary to ensure the machines or any product operates as smoothly as possible. Instead of checking on the machine every time, there will be a massive loss of time in the process to complete. Instead of facing the immediate rush, the digital twin helps in the predictive maintenance of the machines or products where the intelligent systems inform us where exactly the problem lies, enabling more excellent product lines and reducing a lot of time and money in production.

Real-time remote monitoring: Managing products and increasing productivity is crucial in any product life cycle. Implementing a real-time monitoring system accurately enables a watch at any stage of the product. On the other hand, the data can be accessed anytime, allowing the users to work remotely and control the system.

Better team collaboration: A better team collaboration helps to make a vast improvement in the workflow wherein accessing data remotely enables advanced collaboration and communication, which results in customer satisfaction and, eventually, team members.

Better financial decision-making: Due to the availability of a large amount of data and advanced financial methods, by gathering the financial data and integrating them into a virtual representation of the physical object, we can predict the labor cost, material cost and other components costs. By the changing trends in the market by integrating data, one can make better financial decisions based on the industry value and can decide on making the necessary changes.

14.10 Digital Twin Life Cycle

Creating a digital twin is not limited to product developments. It can be done using existing data or scanning and modelling the facility. The life cycle for existing facilities is fundamentally the same, beginning with a

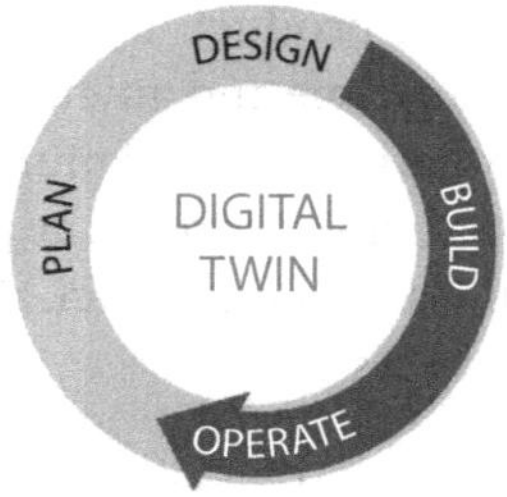

Figure 14.2 Digital twin life cycle.

thorough understanding of the desired operational outcomes and the data required to achieve them as in Figure 14.2.

The beginning of digital twin of any new product begins with planning. Owners and firms work together to understand the desired operational outcomes and the data needed to achieve them. As the planning phase progresses, the respective technologies capture and map the digital twin data, which precisely reflects the idea proposed. Later the digital twin enters the subsequent operational phase. The digital twin can be linked to other systems to collect operational performance data, and system models for simulation can be created. The customers can review the data and check the room for improvement and evolve to support new needs like facility utilization and improvement in the future.

14.11 Digital Twin in Automotive Industry

Mass production, lean adoption, and globalization have all been critical enablers for the automotive industry's growth and stability in recent decades. However, with data becoming the new oil and Industry 4.0 gaining popularity, data-driven manufacturing is expected to drive the industry's future growth. This model allows businesses to use data throughout the product life cycle to create faster, more cost-effective products of higher quality.

By leveraging different technology capabilities made possible by IoT, Big Data analytics, and simulation techniques, this concept can be extended across the automotive value chain to perform operations efficiently.

The role of digital twins in addressing the automotive industry's current challenges, particularly in terms of vehicle product design, manufacturing, sales, and service.

14.12 Applications of Digital Twinning Technology in the Automotive Industry

The automotive industry is one of the most revenue-generating sectors in the world. Implementing the digital twin helps the automotive product development cycle in different stages and helps enhance and optimize the product's efficacy. It also will help in better decision-making starting from concept, manufacturing, sales, and maintenance. Further digital twin helps engineers and customer service communicate effectively with the customer.

14.12.1 Vehicle Development

The digital twin helps track the whole development cycle of the automotive systems. Based on the data collected with the help of digital twin can help the manufacturers make significant performance improvements that can be made for the systems, which improve the efficiency and reliability of the products. The team behind the development can make better decisions and communicate effectively in the whole process.

14.12.2 Vehicle Manufacturing

Digital twins can help you with a variety of production processes. Machine downtime can be predicted and avoided by manufacturers. Similarly, Digital twins can help create a safe and ergonomic work environment which will eventually reduce the cause of workplace accidents. Furthermore, it can control modern manufacturing methods, such as automated robots. It is capable of controlling autonomous robots and equipment without the need for human intervention. Even the complexity of the vehicle production process could be reduced with the help of technology.

14.12.3 Vehicle Sales

The automotive industry sales have transformed the whole sector through good connectivity and communication among the customers. As new players enter the market, the vehicle sales have to be more creative, and customer friendly, where convincing the customers willing to pay for the new feature plays a key role. A virtual vehicle twin will help manufacturers access the customer data and provide the necessary updates. Nowadays, customers visit dealers less often and choose to do it virtually. This digital twin concept can be beneficial in understanding customers better. Usage of online portals and websites has increased very rapidly. A digital twin can help you manage your customers better. Manufacturers have rarely been involved in customer service, understanding customer behavior and preferences.

The data gathered by the digital twin can be beneficial in evaluating vehicle conditions and assist the customer by informing them to have the scheduled maintenance and other servicing. Manufacturers can inform the updates to the product for better performance, assist in gathering the ownership data, and the list goes on.

14.12.4 Vehicle Maintenance and Servicing

The data gathered by the digital twin can be beneficial in evaluating vehicle conditions and assist the customer by informing them to have the scheduled maintenance and other servicing. Manufacturers can inform the updates to the product for better performance, assist in gathering the ownership data, and the list goes on. Manufacturers can save money on warranties and learn more about their products' performance. If necessary, they can even be anticipating breakdowns. This way, the manufacturers can safeguard their brand image and reputation.

14.12.5 Product Life Cycle of the Automotive Sector

The product life cycle of a vehicle in the automotive industry includes several stages: conceptualize, design, procure, build, stock, sell, service, and recycle as

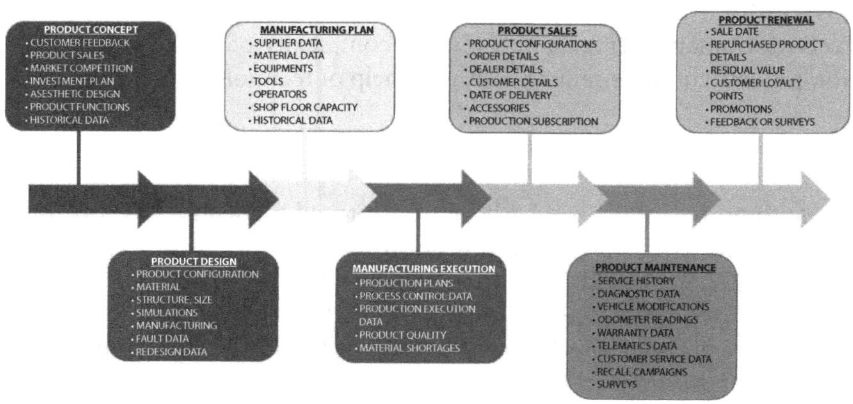

Figure 14.3 Product life cycle data (automotive).

in Figure 14.3. As part of routine activities, an enormous amount of data is generated at each stage. The goal of every organization is to acquire data that can be built faster and more cost-effective. Also, the automotive industry uses the data in each stage effectively [9].

14.13 Role of Digital Twins in Addressing Current Automotive Challenges

The use of the digital twin in the automotive industry can tackle many real-world challenges faced by the industry. The challenges like data loss, downtime, and lack of helpful information to understand customer feedback can be eased away with the help of using the digital twin concept.

14.13.1 Unifying Data

Manufacturing in the automotive industry necessarily involves using diverse data like historical data, performance data, and user habits. The digital twin solves this problem quickly by combining data from multiple sources. Analyzing and examining the data will give an accurate picture of the manufacturer.

14.13.2 Easy Verification

Automotive manufacturers dedicate a significant amount of time to testing the new product. Manufacturers must wait until their designs are put into production to see if they are viable. Naturally, the procedure takes time and extends the time to market. A digital twin can be used to verify designs quickly and reliably. It can gather all the information needed to simulate and produce reliable outcomes.

14.13.3 Minimization of Failures

Machines will eventually experience downtime or malfunction. Your production will come to a halt, and your profits will suffer. Data can be used by digital twin technology for the prediction of machine downtime. By understanding the data, such circumstances in the future can be avoided. Your production continues without interruption and without incurring any financial loss.

14.13.4 Predict Customer Demands

Manufacturers rarely learn what their customers think because information never reaches them. Suppose data from the car can be obtained even on the road. Manufacturers, for example, will be able to see which features customers are bound to use the most. So, the producers can predict the customer demand more accurately. They will understand what products customers use and are willing to pay for.

14.14 Challenges for Implementing Digital Twin in Automotive Industry

Not many firms have started using the digital twin concept for production despite its many benefits. This is because many firms are yet to thoroughly understand the process involved in the implementation of digital twin systems as it requires a large gathering of the information and infrastructure, which mainly relies on the IoT where the requirement of many computing devices, digital machine's ability to store a considerable amount of data is a tedious task for a company where the idea has not yet been introduced. The cost of implementation associated with bringing the digital twin may cost very high. A more complex process will require a significant amount of investment. With the proper deployment of digital twin and careful planning, the challenges mentioned above can be decreased and will not affect expected returns.

14.15 Bridging the Gap

The digital twin market is expanding with the expansion of IoT technology. As per Markets and Research, by 2026, up to 91% of IoT platforms will support digital twinning, and by 2028, digital twinning will be a standard feature in IoT application enablement. To have a better automotive product life cycle, that process involves the collection of many data and managing that data efficiently will result in a better end-to-end process. Every bulk data obtained will be utilized at each development stage step and implemented for future phases. To bridge the gap between the virtual and physical world digital twin is most important. With the use of the digital twin, challenges in the automotive industry can be addressed. Finally, the early adopters of the digital twin will be able to stay strong in the upcoming technological advancement trends and bring all the departments together to outperform the current industry trend and challenges.

References

1. Haag, S. and Anderl, R., Digital twin – proof of concept. *Manuf. Lett.*, 15, 64–66, 2018, DOI: 10.1016/j.mfglet.2018.02.006.
2. O'Sullivan, J., O'Sullivan, D., Bruton, K., A case-study in the introduction of a digital twin in a large-scale smart manufacturing facility. *Proc. Manuf.*, 51, 2019, 1523–1530, 2020, DOI: 10.1016/j.promfg.2020.10.212.
3. Grieves, M., Digital twin: Manufacturing excellence through virtual factory replication digital twin: Manufacturing excellence through virtual factory replication this paper introduces the concept of an a whitepaper by Dr. Michael Grieves, April 2015.
4. Schleich, B., Anwer, N., Mathieu, L., Wartzack, S., Shaping the digital twin for design and production engineering. *CIRP Ann. - Manuf. Technol.*, 66, 1, 141–144, 2017, DOI: 10.1016/j.cirp.2017.04.040.
5. www.ptc.com/en/blogs/corporate/what-is-a-digital-thread.
6. Son, Y.H., Park, K.T., Lee, D., Jeon, S.W., Do Noh, S., Digital twin–based cyber-physical system for automotive body production lines. *Int. J. Adv. Manuf. Technol.*, 115, 1, 291–310, Jul. 2021, DOI: 10.1007/s00170-021-07183-3.
7. Royo, L., Martinez, A., Bacaicoa, L., Rodriguez, F., Lopez, B., A digital twin based approach for simulation and emulation of an automotive paint workshop. SAE Technical Paper, 0148–7191, 2021.
8. da Silva, A.C.F., Wagner, S., Lazebnik, E., Trail, E., Using a cyber digital twin for continuous automotive security requirements verification, ArXiv Prepr. ArXiv210200790, *IEEE Software* 40, no. 1, 2022, 69–76, 2021.
9. Paper, W., Digital twin in the automotive industry: Driving physical-digital convergence. Tata Consultancy Services White Paper, 2018.

15

State-of-the-Art and Future Applications of Farming Robotics

Badrinath A.R., Abhishek Kamath, Veerishetty Arun Kumar, Nishan Rai and Rathishchandra R. Gatti*

Department of Mechanical Engineering, Sahyadri College of Engineering & Management, Mangalore, India

Abstract

The current state of farming robots is presented in this paper. The extensive use of sensors, microelectronics, and computers has resulted in farming robots' most significant recent advancements. Compared to well-known industrial robots, some unique characteristics of these machines emerge. Agriculture is quickly becoming an advanced field, bringing in new companies. The present paper deals with challenges faced by agricultural robotics, such as urban sprawl, rapid urbanization, increasing demand for good fruits and vegetables, and a lack of skilled labor are all difficulties that agricultural robotics is working to tackle. The application of robots in farming enables farmers to get a high yield in less time with reduced human errors. Sowing the seeds and cutting down the crops, environmental management, and caring for crops are all processes that can be automated by robotics. The present paper provides an overview of agricultural robots' technologies, types, and applications. It also discusses how agricultural robotics can be employed in different farming processes.

Keywords: Robotics, agriculture, farming, engineering, technology

15.1 Introduction

The study of the designing, making, uses, functions, fabrication, and application of robots is known as robotics. Robotics brings together various engineering branches and expertise for traditional farming needs. Farming machine manufacturers and researchers have created a wide range of small

**Corresponding author*: gattirathish@gmail.com

Rathishchandra R. Gatti, Chandra Singh, Rajeev Agrawal and Felcy Jyothi Serrao (eds.) Self-Powered Cyber Physical Systems, (283–296) © 2023 Scrivener Publishing LLC

and large agricultural machines, but precision agriculture requires employing robotic and pneumatic mechanisms [1]. The worldwide need for agricultural goods is expanding very fast [2]. Several groups throughout the world have used sensors, actuators, control systems, and other advanced inventions like drones to automate different farming processes, including sowing seeds and cutting down final crops, managing growth conditions, water and fertilizer delivery, and plague and disease detection and treating the crops, in recent years [3]. According to Market Research Consultancy, robotics applications and markets in agricultural and associated industries are rapidly developing, with the industry predicted to reach $11.58 billion by 2025 [4]. Agricultural robots can be fully autonomous systems that can go through multiple stages of a process to solve a problem. They have been utilized to reduce the workforce, growth time, and money in repetitive operations such as soil preparation, irrigation, plant protection, trimming, harvesting, surveillance and control, and mapping [5]. Fortunately, due to advanced technologies and inventions, we can maintain and get profit and high-quality yields in agriculture and mitigate the challenges posed due to increasing population, high rate of urbanization, and high competition for good-quality goods. We cannot find a qualified workforce, and human labor is vulnerable to health hazards [6]. Future agriculture will include several tasks, such as replacing the workforce, reducing cost and enhancing grade, and performing activities more efficiently than larger machines currently in use. Smooth handling tools are used to pick mushrooms as part of an investigation. Other uses, such as broccoli, can be accomplished using crop-cutting tools, but the gathered crop must be cared for and kept in reasonable condition [7]. We will also learn about the procedures involved in farming and how robots have assumed the role of people. Specifically, we will learn about the technologies and innovations of scientists that have resulted in a significant shift in agriculture. Farming robots are mostly highly advanced and capable of performing jobs much faster than people. We will also learn about robots and their duties, including how they can execute various tasks several times without tiring [8].

Agriculture relies heavily on robots. Many artificial robots and devices are currently accessible on the market, and by employing such robots, we may finish work very simply. Use automated machines, such as robots, to save personnel while improving yield quality and precision. The workforce was primarily required in harvesting, but robots have since taken their place. While harvesting fruits, due care must be taken, such as handling them cautiously and tenderly and preserving them in the needed condition.

We also discovered that there are several hazardous agricultural techniques. Moreover, they impact human health, for example, when spraying fertilizers and pesticides, many farmers have health difficulties; we also discovered that they cause breathing problems, which leads to lung cancer and other respiratory-related health problems. To overcome all of these challenges, they constructed machines to replace humans, which has resulted in a significant shift in the agricultural area; also, robots are more advanced than humans.

Tractors that can carry huge loads are also examples of robots because a man cannot carry weight on his back. Thus a tractor was created to replace that. Robots may be employed for various agricultural tasks such as spraying fertilizer, watering plants, monitoring weather conditions, and carefully maintaining crops.

When a set of instructions is given to the robot, it automatically takes them. It does the work according to the instructions given by the instructor. We can also automatically code, such as what task must be done at what time, so all of this is possible because of technology and scientists who research and who have given us these beautiful machines that are robots. Essentially, a robot is a machine that can execute specified tasks when necessary. For example, consider a pick-and-place robot. In this robot, the object is picked up and placed in its proper location, which may be repeated without interruption. The robot does not weary as we do, and robots are utilized in many different industries such as horticulture, pisciculture, military activities, hospitals, and hotels.

In the agricultural area, robots may assist farmers from sowing the seeds to harvesting the produce, sorting the good and poor goods, and packing and processing. Regarding agriculture, we all know that proper care must be taken while growing crops, and a workforce is required. However, by using robots, we can code so that the robot does all the tasks, such as watering the plants in the morning and afternoon, spraying fertilizers and pesticides when needed, and removing weeds. They may also replace labor with robots for weed removal and another process.

15.2 Components of Agricultural Robots

The Components of a typical agricultural robot are shown in Figure 15.1. It consists of a robotic arm system mounted on a mobile wheeled base. It consists of a control system, robotic arms with a specific arm configuration, end effectors specific to work, sensors, a battery and a power management system.

15.2.1 Control System

A robot's control system is similar to its human brain. It does all the processing work, such as commanding the robot, and is accountable for its

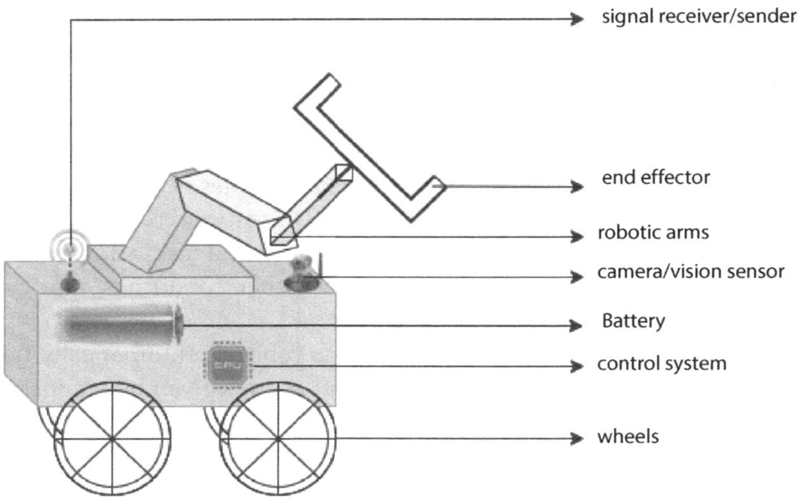

Figure 15.1 Schematic of a typical agricultural robot.

activities. It functions similarly to a human brain in that it displays the response to a series of instructions. Moreover, the control system decides on movement and functioning actions.

15.2.2 Sensor and Actuators

Sensors act as input devices for the robot. It is mainly used for input, such as an IR sensor, which is utilized for object detection. The IR sensor, for example, sends rays that strike an object and are reflected; the reflected rays are analyzed, and the object's distance is determined. Numerous sensors are utilized in the construction of robots, and sensors work as feedback devices, assisting the robot in directing movement and also assisting the robot in functioning without mistakes. Vision sensors are used to detect what objects are there, investigate their attributes, and determine what they are.

Actuators are in charge of robot movements; if a robot does not move, it has no utility. Because all of a robot's tasks are performed by movement of the robot's body components, and every little movement is performed by actuators, of which there are several sorts. However, we use electric actuators powered by electricity and can perform precise movements. They are made up of motors that produce linear and angular motion when the motor's rotation occurs.

15.2.3 Power Supply

We, humans, consume food and water. Robots, like humans, require electricity to function correctly. Stationary robots do not want to move and require alternating electricity supplied to a socket through wires to execute their activities. However, when it comes to non-stationary robots, they cannot hang cables around them and move around since it is difficult and dangerous, so they came up with the notion of employing portable electric current, which is batteries. Furthermore, the usage of batteries was safe and had a long life. All of these will be more important when building a robot.

15.2.4 End-Effectors

The grippers and other end tools do most of the robot's physical labor. Grippers and other tools are adjusted in the arms and, in particular, at the ends of the arms to function freely with a degree of freedom; it plays a significant role when the grippers are used to grasp and put things. They have various purposes, and most grippers are constructed of silica and rubber to handle items softly and protect the object being carried. Sensors, such as force, displacement, and IR, are added to operate effectively.

15.2.5 Artificial Intelligence

Artificial intelligence (AI) is an advanced technology. Where the machine or robot can function like a regular human being without making mistakes, where the robot can think and react to actions, and where the reactions are voluntary. Many software and applications use AI as the backbone and have succeeded and are available in the market. It automatically recommends what you need without searching. It acts like a human brain and reacts to instructions or conditions. Moreover, AI is blooming very fast in the present times, and everywhere AI is used for everything, not only in robots but also in all other fields like automotive and other fields it is highly used.

In farming robots, they play an essential role that i.e., an apple-picking robot is a mechanism that autonomously selects apples from a tree. It selects the good apples, scans them, and then only picks the ready apples with the necessary color, shape, and internal features. The waste apples are thrown away by picking the good ones, and the excellent apples are carefully moved through a conveyer belt and stored in the required conditions.

Weed removal robots automatically recognise the weed and pick the weeds, after which the weed is chopped. The robot distinguishes between weed and crop plant, and after scanning, the robot picks the weed, and the weed is plucked. This is all done by AI and is beneficial since it selectively eliminates weeds from plants without consuming breaks.

15.2.6 Robotic Arm

Industrial robotic arms are one of the most prevalent forms of robots today. Robotic arms, also known as articulated robotic arms, are quick, dependable, and precise, and they can be trained to do a permanent job in various conditions. Previously, teaching a robotic arm to do tightly specified tasks, such as collecting a particular type of object from a specific place with a given position, was necessary. Robots could not distinguish one sort of object from another, establish an object's position with some degree of accuracy, or alter their approach depending on visual basic. With the help of artificial intelligence and network technology, robotic arms can now perceive, evaluate, and respond to their surroundings while passing vital data and insights back to operations and methodologies.

15.2.7 Driving System

It may be driven in various methods, including hydraulic, pneumatic, or electrically driven. Hydraulics is a fundamental approach in which the internal flow of fluid causes the movement of arms and other parts. There is also a reservoir where the hydraulic fluid is kept and a pump to move the fluid from the reservoir to other parts. Compressed air is utilized to create movement in pneumatics, and a compressor and air tank are present. The usage of an electric motor for linear and rotational motion is referred to as electrically powered.

15.3 Types of Agricultural Robots

15.3.1 Weed Removing Robots

This is a robot that assists in the removal of weeds in farming fields. Initially, it was removed by hand, but now it is replaced by a robot, which is more advanced than humans and has excellent precision. Autonomous weeder and Z440 refer to a robot that can eradicate weeds without the intervention of humans.

15.3.2 Pest and Infection-Spotting Robots

It indeed detects the sickness and cause of the infection in the plants, as well as the reason and source of the infection, and it aids in managing pests and diseases. It benefits farmers by doing so and also serves as a caretaker. ROBHORTIC is a robot that assists in the detection of pests and crop illnesses.

15.3.3 Seed Sowing Robots

The seeds were being sown in the traditional method, which usually takes more time and workforce, and there were many errors, such as sowing in different places or not even spacing between two successive plants. They were also unable to complete the process in less time. To replace humans, robots were created, which can work constantly and finish tasks in less time without making mistakes. They created a seed-planting robot using Arduino.

15.3.4 Robots for Scouting Crops

Crops are thoroughly examined to determine whether a disease or illness is harming the crop and identify insects and pests. The farmer should inspect every corner to see if there is an issue, but this is a time-consuming operation that is tough for the farmer to do, but we can employ many robots to complete this work quickly. TERRA-MEPP is a robot used for scouting.

15.3.5 Robots for Spraying Fertilizers and Pesticides

They also influence human health, while spraying fertilizers and pesticides, many farmers have health problems; we also learned that they cause breathing problems, which lead to lung cancer and other respiratory-related health problems. To address all of these obstacles, spraying robots are used. The R150 is a spraying robot that is used to spray fertilizer.

15.3.6 Robots for Harvesting

It is the last operation in farming; crops are commonly chopped once the grains or pulses are mature, and this procedure is time demanding and must be done with caution. This generally necessitates an additional workforce, and to address this issue, machines have been developed that can do all the jobs that a man usually does manually. Virgo is a harvesting robot that can recognise and harvest various crops.

15.4 Implementation of Robotics in the Agricultural Process

15.4.1 Ploughing/Tilling

Robots play a critical part in planning. It effectively substitutes man labor, and the ploughing robots generally do the work of ploughing the soil for planting crops. It replaces castles, and these robots can complete the work without taking a break, and they can also be controlled by humans sitting in someplace. They work more accurately than cattle and other machines, and their performance is excellent compared to other machines and humans. Their input is just codes or controls that can be done by sitting far from where it is working, and this robot can precisely do the ploughing job. Essentially, ploughing involves soil loss to sow the seeds [4].

15.4.2 Sowing Seeds

Seeding is an essential duty in agriculture, and in this process, seeds are sown in the ground using various techniques, and there are several methods of sowing seeds, such as conventional and non-traditional methods and traditional methods. Traditional methods demand more workforce, so the labor for them is expensive, which means that more people are needed, and daily pay must be paid to them. In contrast, non-traditional methods are cost-efficient, and planting seeds play a significant part since it is directly related to the profit and loss of the entire crop. The critical challenge is where and how many seeds must be distributed in a particular area, and it must be done precisely at the specified rate of seeds per square foot or metrical foot. Moreover, many kinds of literature tell us the same thing, some material tells us the number at which it must be placed, and we also know that crop density is essential. Therefore, it is vital to sow the seeds with adequate spacing between them.

For planting seeds, seeding robots have been developed. Moreover, this robot assists in spreading seeds at the proper distance and also in placing the seeds equidistant from one another. We can modify the distance between two sowing seeds, and the seeing robot is primarily used for rice and wheat, and it is more sustainable for the wetland. There is research being conducted on planting robots. The seed's depth and place will be sown as essential for planting. In summary, seeding robots have been created in recent years for seeding at both the scientific and industrial levels, although to a limited number due to the intricacy of the operations. Even though existing sensor advances are critical in dealing with the challenges associated with these robots, they should be enhanced further by eliminating mistakes and improving precision. Following seeding, the sowing efficiency should be checked, particularly in fields with rough ground [8].

15.4.3 Manures and Fertilizers

Manure and fertilizer are critical components of farming. Previously, the farmer would spray or apply fertilizer in powder form. However, sprayers were introduced, in which he had to mix the pesticides with water and put the solution into the spraying machine, which was a hand-driven pump with a sprayer that he had to pump with one hand while spraying with the other. During this unit, he had to carry on his back for a long day, which was tiring and had a high impact on his back. He had to roam the entire field, which was very difficult, and later on, they introduced labor, but they demanded high wages because it had a high health risk. They also thought it was harder to do the whole workday, and many health hazards resulted from spraying pesticides and insecticides. Many patients had lung problems, and some had respiratory issues due to the spraying of pesticides. As nothing more than a consequence, the farmer is exposed to high levels of hazardous active components [8].

Later, they produced spraying robots, which truly benefit and prevent farmers from suffering from various health problems and replace much labor and their daily wages. Then they built spraying robots based on artificial intelligence concepts and another computer programming. This protected farmers from pesticides and other hazardous chemicals, as well as saved them from a variety of diseases. The robots must have modern sensors that recognise and spray accurately and uniformly throughout the agricultural field. Moreover, the robot's pace of movement must be predetermined so fertilizers are distributed evenly [6].

15.4.4 Weeding

Weeding is the most critical aspect of farming; it is performed multiple times and cannot be omitted. Moreover, the weed eradication operation consumes around 40/100 of the crop's investment. Weed removal is time-consuming and requires more labor for this process, and nowadays, there is a massive demand for labor, and they demand higher wages, which will become a burden for formers, and farmers. To avoid all of this, they use pesticides and weedicides, which significantly impact the soil and the crop. More pesticides and weedicides are used, which significantly influences the health of those who consume fruits and vegetables. Moreover, when it comes to soil, it loses fertility and lust for a shorter amount of time, but weeds grow in a shorter period as well, so we should use weedicides and pesticides less in farming.

Traditionally, they would pick the weeds off their hands and had specific equipment to assist them in eradicating the weeds. Farmers must locate the weed in the crop and carefully clip the weed; subsequently, they built a robot that assisted them in weed eradication. It recognizes the weed and cuts it off, and it was one of the most excellent options since it could work instead of the farmer and afterward eliminate the weed without tiring. These robots have sensors and cameras for weed detection, and they must carefully identify the weed and accurately chop it off without harming the produce. This sort of robot has a lower negative influence on farmers and assists farmers in solving their problems through robots, which replace people and save farmers money on labor costs.

Another robot was based on pesticides, weedicides, and other chemicals, and it sprayed the chemicals on the weed to destroy it. However significantly influences the crop and the end product, directly and indirectly affecting farmers and customers. It also affects the soil of the field, which loses fertility and impairs the feature growth of crops in the field, resulting in a significant loss for the farmer later. To prevent all of this, the robot had to identify the weed using artificial intelligence and other computer technologies with the assistance of sensors. Once the weed is recognised, just the weed is sprayed with the chemical so that it does not influence the rest of the field and does not harm the soil [8].

15.4.5 Protection of Crops

Crops must be disease-protected and due care must be given. Some crops can be ruined by diseases, causing the farmer to lose a large amount of money. To avoid this, the farmer must carefully identify and care for the plants. We must visit individual plants, and then he must identify what type of disease or pest affects the plant. If he fails to identify it, it will be a massive problem for him, and he must know what chemicals or pesticides are to be sprayed. Farmers must go to each plant and spray individually because spraying on the entire crop may affect yield and farmland. So to protect it from the image of the disease selectively is a very time-consuming process, and you may get tired so replacing all of these robots have come into existence where the robots can selectively select the plants which are affected by the diseases and they can spread the fertilizers or pesticides individually by suction.

The plant growth condition also plays a vital function in demonstrating to farmers that farmers must take care of sufficient nutrition for the plants and correct irrigation, including watering plants. The water must be applied, the plant should not be allowed to dry, and the farmer must apply fertilizers and manure at regular intervals to receive proper nutrition throughout its growth. Later, all of these processes can be automated using a robot without the involvement of a farmer by using technology such as artificial intelligence and software by coding the robot. The robots use sensors such as moisture sensors to determine the moisture content of the soil and to water the plants regularly. The robots also have sensors to identify the nutrition and nutritional requirements of the plants so that, when necessary, the robot automatically provides the plant with the proper amount of nutrition, which aids in health. Proper plant development would result in a profit for the farmer. As a result, robots can better care for plants than farmers [6].

15.4.6 Harvesting, Threshing, and Winnowing

Harvesting is a time-consuming procedure that must be conducted regularly on an agricultural property. Physical harvesting is the removal of plant end products. Consider a fruit plant, for example. When the fruit is matured correctly, it is ready for harvesting, so we must remove it from the plant. This process must be done carefully and with care, and it is easier for farmers to harvest if there are fewer plants. However, farmers can only individually collect some fruits in significant quantities by selecting the best ones. So, they hire laborers, who demand exorbitant rates, and any mistakes made by laborers harm the plant growth. To avoid all of this, robots for harvesting various types of crops were designed.

Moreover, the robot invented for the harvesting operation plays a vital role because it replaces the workers and is accurate and precise. Labors cannot identify whether all the fruits and vegetables are suitable for harvesting. However, robots are coded in such a way that they can quickly identify the good and bad ones and can easily separate the bad ones. The robot can work continuously without tiring, and the entire job can be completed by a single robot rather than 10 laborers. The time required is also reduced compared to laborers. Thus, robots are the best option. Harvesting crops such as rice, maise, and other things is a little challenging because it must be done correctly and precisely if any error is made; otherwise, it is not feasible for a single farmer to do it by himself since he will have to walk All Around the Farm to finish this procedure. Machines have been invented in which the equipment goes about the farm, quickly chops down the plants, and removes the produce [2].

15.5 Challenges

Operating a robot is not a simple chore for farmers because if something goes wrong, the farmer will lose the entire harvest, and a faulty robot might cause him a significant loss. Moreover, commercial robots are costly, and the farmer cannot afford to invest much when he is not making enough money from just one robot. Furthermore, if he buys, he will have to spend money on upkeep because the sensors and other elements of the robot are expensive. Spending more with less income is worthless. Moreover, the robots must be handled with extreme care so that they do not need to be fixed or the parts do not break.

Moreover, the network must be robust because when two or three robots work together, they will have a link and work more, making the task easier and faster. In addition, if the network fails, it cannot communicate, and the function may be disrupted. Moreover, as a result of the problem, a malfunction may occur.

There are vision sensors in certain robots, such as weed eradication robots, and they must be highly exact because if the robots recognise the wrong plant as a weed, it will cause a significant problem. In other words, the entire harvest will be ruined. In pest and disease detection, robots must observe the issue closely. The camera must have a high-definition camera so the robot can capture and identify it precisely. When we go to a high-definition camera, it usually costs more than expected, and it may be challenging to maintain and protect the sensors and other parts.

Moreover, because replacements for the robot are highly uncommon and must be purchased directly from the manufacturer, we do not have access to them in local markets. As a result of the delay in the delivery of spares, agricultural work will be disrupted. Furthermore, the majority of the robots are controlled and powered by electricity. Farmers would find it difficult to work if the power went out.

15.6 Conclusions

The farming functions that were done manually are now entirely automated by robots, and they can also be replaced by new technology that upcoming technologies can further improve. It would take a long time for the farmer to do it manually, so using robots has become easier for them and takes less time, and the work done by the robots is precise and error-free.

Furthermore, we can see that there is a specific robot for each operation that can replace human labor and solve the problem. When a farmer uses robots, he gets excellent harvests and high earnings. Instead of spending much on labor, he might invest in a robot to do the work for him. Moreover, the money spent on labor is just for one time, after which the farmer must pay the laborers again, but the investment in the robot is a one-time investment that will last until the robot is damaged or destroyed.

Health issues usually cause some jobs like spraying fertilizers and other things for the farmers that the use of robots for such activities can skip all. It is perfect to use a robot because it protects the farmer from other health issues, and work is done faster and better. Robots are replacing humans, and it is preferable to have a single robot rather than numerous human beings.

The farmer's old conventional procedures that he had followed were successful. From ancient times, new technologies and machinery that are more advanced than farming methods have been substituted. The farmer spent much time preparing the land, a time-consuming operation for which he had to pay for labor and animals. Later, this operation was mechanized or finished by Tractors, which took less time and accomplished all of the labor effortlessly without the farmer having to make more effort. Sowing the seeds was done manually before being replaced by robots with precision and accuracy of identical spacing between seeds. The farmer made money by using robots.

Moreover, crop management is essential in agriculture. That is, the plants must be carefully observed, and care must be taken; the farmer must individually visit each plant and identify the situation of the plant; if he fails to visit the plant individually, the affected plant with the disease will spread to the entire crop, and the entire crop will be lost, causing the farmer a considerable loss, and robots were invented to overcome all of these problems. Robots have sensors and powerful cameras, and I will choose a plant experiencing a problem and resolve it. Certain robots assist farmers with everyday tasks such as watering so that if the farmer is not there, the robots may continue the process, and the farmers can remain pleased. Harvesting machines are also available, which have replaced laborers and allowed the work to be done accurately.

References

1. Mohiuddin, S.M., Agricultural robotics and its scope in India. *Int. J. Eng. Res.*, 4, 7, 1215–1218, 2015.
2. Kootstra, G., Wang, X., Blok, P.M., Hemming, J., Van Henten, E., Selective harvesting robotics: Current research, trends, and future directions. *Curr. Rob. Rep.*, 2, 95–104, 2021.
3. Roldán, J.J., del Cerro, J., Garzón-Ramos, D., Garcia-Aunon, P., Garzón, M., De León, J., Barrientos, A., Robots in agriculture: State of art and practical experiences, in: *Service Robots*, pp. 67–90, 2018.
4. Paul, P., Sinha, R.R., Aithal, P.S., Ricardo Saavedra, M., Aremu, B., Mewada, S., Agricultural robots: The applications of robotics in smart agriculture—Towards more advanced agro informatics practice. *Asian Rev. Mech. Eng.*, 9, 1, 38–44, 2020.
5. Zimmer, D., Plaščak, I., Barač, Ž., Jurišić, M., Radočaj, D., Application of robots and robotic systems in agriculture. *Teh. Glas.*, 15, 3, 435–442, 2021.
6. Oliveira, L.F.P., Moreira, A.P., Silva, M.F., Advances in agriculture robotics: A state-of-the-art review and challenges ahead. *Robotics*, 10, 2, 52, 2021.
7. Mahmud, M.S.A., Abidin, M.S.Z., Emmanuel, A.A., Hasan, H.S., Robotics and automation in agriculture: Present and future applications. *Appl. Model. Simul.*, 4, 130–140, 2020.
8. Fountas, S., Mylonas, N., Malounas, I., Rodias, E., Santos, C.H., Pekkeriet, E., Agricultural robotics for field operations. *Sensors*, 20, 9, 2672, 2020.

16

Review on Robot Operating System

G. Vijeth and Rathishchandra R. Gatti*

Department of Mechanical Engineering, Sahyadri College of Engineering & Management, Mangalore, India

Abstract
In the present and future era of automation being implemented into industrial technologies, the capabilities of the Robot Operating System (ROS) are advancing swiftly. Even though this technology is relatively new to the robotics world, it is available openly to any early adopters. The fundamentals of ROS are presented in this paper. The significance of ROS will be discussed in this study. In the realm of autonomous vehicles and swarm robotics, the ROS is critical. The several frameworks of the present ROS are discussed in this article. We will discuss the benefits and drawbacks of ROS in this article. The scope of ROS in the field of robotics and automation will be discussed. We will go through how ROS is employed in various research fields. Different areas where ROS is making an impact in increasing productivity and efficiency are presented in this paper.

Keywords: Autonomous vehicle, computer vision, SLAM, automation, ROS

16.1 Introduction

16.1.1 What is ROS?

ROS is not an operating system. ROS operates on a broad range of operating systems, most often Linux. ROS is a collection of tools for designing, setting, constructing, troubleshooting, and coordinating complicated robotic systems. ROS is fundamentally a publish/subscribe framework that could be used to communicate data and signals amongst tiny computer applications which comprise an enhanced robotic system [1]. Consider an

**Correspooonding author*: gattirathish@gmail.com

Rathishchandra R. Gatti, Chandra Singh, Rajeev Agrawal and Felcy Jyothi Serrao (eds.) Self-Powered Cyber Physical Systems, (297–308) © 2023 Scrivener Publishing LLC

instance there is a camera setup, and the User needs to link it to a vast industrial arm. ROS enables the User to integrate these components, transmit and acquire information to and from them, and create code to connect. ROS is a versatile framework with various tools and libraries for developing robotic applications. It has various sophisticated features that aid developers in tasks like message forwarding, spreading computation, code reuse, and the development of cutting-edge algorithms for robotic systems. Morgan Quigley founded ROS in 2007 as part of the Stanford STAIR project incorporating robots. Willow Garage was the primary site for the development of ROS. The community of ROS developers is rapidly expanding. There are quite a lot of ROS users, and ROS developers present all around the world. Most sophisticated robotics firms are adapting their program to run on the ROS platform. This pattern is also seen in industrial robotics, as corporations shift away from proprietary robotics software and towards ROS. The ROS industrial concept has grown in popularity recently due to many studies on that subject. ROS Industrial can use ROS's enhanced capabilities to manufacture. The growing number of ROS applications may result in a plethora of job options in this industry. Some reasons why people choose ROS over other robotic systems, YARP, Orocos, MRPT, player, and others are listed below.

High-end features: There are pre-installed features like the Adaptive Monte Carlo Localization (AMCL) [2] and Simultaneous Localization and Mapping (SLAM) [3] packages in ROS which can be used in robotic systems for autonomous navigation for robot manipulator motion planning. The MoveIt package can be utilized. These characteristics can be utilised immediately in any robot program without any difficulty. Because these abilities are the best way to accomplish them, creating new code for existing capacity is akin to reinventing the wheel. Furthermore, these features are incredibly customizable; the User may fine-tune any feature by adjusting several settings.

Tons of tools: ROS includes many tools for troubleshooting, visualizing, and running simulations. Some powerful open-source debugging tools, visualization and computation include rqt GUI, RViz, and Gazebo. A software framework with this number of tools is quite rare.

Sensors and actuators of the highest quality are supported: ROS contains device drivers and interface packages for a wide range of robotics sensors and actuators. Velodyne-LIDAR, Laser scanners, Kinect, and other high-end sensors, as well as actuators such as DYNAMIXEL servos, are examples of high-end sensors. These components can be easily connected to ROS by users.

Interoperability between platforms: ROS message-passing middleware allows nodes to communicate with one another.

The nodes could be written in any programming language that supports ROS-client libraries. Other nodes could be implemented in Python or Java, while high-performance nodes can be written in C++ or C. Other frameworks do not provide this level of adaptability.

Adaptability: One of the challenges that might arise in most independent robotic frameworks is that if any of the primary code strings crash, the entire robot application can stop. In ROS, the scenario is different; users write separate nodes for every procedure, and if any node fails, the program remains functional. Furthermore, ROS offers robust techniques for resuming activities even if any sensors or motors fail.

Can handle several resources simultaneously: Managing a hardware resource through two or more procedures is usually painful. If a user wants to utilise a camera to recognize faces and movements, they must first analyze the image, User may write a program as if it were a single unit that can accomplish both or as a single set of code for concurrency. When a user inserts more than two features into a thread, the application's behaviour becomes tricky and complicated to troubleshoot. On the other hand, ROS users can connect to the devices through the ROS drivers' ROS topics. Any number of ROS nodes can subscribe to the image message from the ROS camera driver, and each node can perform different functions. It can lower computing complexities while simultaneously improving the whole system's debugging ability.

Thriving community: When customers select a software framework or library, particularly if it comes from the open-source community, one of the essential elements to consider is its software assistance and development community. An open-source tool will be supported. Some tools enable excellent support, while others still need to. ROS has an active support community. There is also a web portal to address customer assistance inquiries. The ROS community is growing steadily in terms of developers worldwide. Other than the points mentioned above, there are several reasons to select ROS.

16.1.2 Characteristics of ROS

Plumbing: Plumbing refers to the middleware mechanism by which nodes communicate with each other. ROS uses a publish-subscribe messaging model, where nodes can publish data to specific topics, and other nodes can subscribe to those topics to receive the published data. This messaging system allows nodes to exchange information and coordinate their actions. The plumbing infrastructure in ROS consists of nodes, topics and messages. Nodes are the individual software modules or processes that perform specific tasks in a robotic system. They can communicate with each other by publishing and subscribing to topics. Topics are named buses or channels through which nodes exchange data. Nodes can publish messages, which are data structures, to a specific topic, and other nodes can subscribe to that topic to receive the messages. Topics are usually organized around specific types of data, such as sensor data, control commands, or status updates. Messages define the structure and type of data being transmitted over topics. ROS provides a message description language (based on the Interface Description Language, or IDL) that allows developers to define custom message types. Messages can contain various data types, including primitive types (integers, floats, etc.) and complex types (arrays, custom objects, etc.). Nodes use messages to communicate with each other by publishing and subscribing to topics. Nodes can act as publishers or subscribers (or both) to exchange data. A publisher is a node that publishes messages to a specific topic, while a subscriber is a node that receives those messages by subscribing to the corresponding topic. Multiple nodes can publish or subscribe to the same topic, enabling distributed communication.

Tools: ROS tools aid in the debugging and monitoring many types of data sent and received by ROS nodes. In ROS, there are both graphical and command-line tools. A camera, a laser scanner, and other sensors can be found in a robot. ROS tools allow users to visualize and troubleshoot numerous types of sensor data. Rviz and Rqt are two examples of ROS tools. ROS has numerous tools for setting up, initiating, introspecting, troubleshooting, monitoring, documenting, validating, and stopping distributed computing systems.

Capabilities: Many robot software blocks are constructed using the 'plumbing' function of ROS for performing various capabilities in robots, such as navigation, vision, manipulation, and so on. If someone wants to include these features in their robot, they may reuse the ROS software and save time. ROS includes an extensive library collection that implements practical robot capabilities focusing on mobility, manipulation, and perception.

Ecosystem: The ROS open-source robotics platform is powered by thousands of developers worldwide who contribute and maintain thousands of ROS packages, tutorials, Q&A, etc. ROS has a dynamic community of developers and users worldwide, making it a one-of-a-kind platform in robotics. A big community heavily supports and improves ROS, emphasizing integration and documentation. ros.org is a one-stop shop for discovering and learning about the thousands of ROS packages from developers worldwide.

16.2 Nomenclature

Packages: Packages make up the ROS software. The ROS package is a logical grouping of files that fulfils a specified purpose and typically includes executables and supporting files. It is not an exaggeration to claim that all ROS software is part of one or more packages. This includes any new programs that you develop. ROS comes with several commands for dealing with installed packages. Packages are listed and located. A manifest, which is a file called package.xml, defines each package. This file contains details about the package, like its name, version, maintainer, and dependencies. The package directory is the directory that contains package.xml. This is how a ROS package is defined: A package directory is any directory found by ROS that contains the file package.xml. This directory contains the majority of the package's files.

The Master: One of the most important goals of ROS is to allow roboticists to develop software as a collection of small, usually independent programs called nodes that operate concurrently. These nodes must be able to communicate with one another for this to work. The ROS master is the component of ROS that permits this communication. To start the master, run the following command: roscore. roscore does not accept any arguments and does not require configuration. It would be best to leave the master running for your ROS session. Start roscore on one terminal, then open other terminals for your "actual" job. Except when you have stopped working with ROS, there are a few reasons to cease roscore. When you get to that stage, you can terminate the master by pressing Ctrl-C in its terminal.

Messages: Nodes communicate with one another by sending messages to topics. A message is a straightforward structure of data made up of fields that have been typed. Some conventional types (such as integer, floating-point, and Boolean) are available, as are primitive-type arrays.

Messages can contain structures and arrays that are arbitrarily nested (much like C structs). Nodes can also send and receive requests and responses as part of a ROS service call. Srv files define these request and response messages. The msg files are simple text files that specify a message's data structure. These files are kept in a package's message subdirectory.

There are different communication protocols in ROS: topics, services, and actions. Although the topic publishes messages to many subscribers, this type of communication is unidirectional. A request is sent to a service server by a service client and it receives a reply but information about the status is not provided. Action clients, like services, send a request to an action server to achieve a goal and receive a response. Unlike services, an action server delivers progress data to the client as the action is being done.

Topics: Some nodes offer data to other nodes, such as a video feed. A node like this is said to disseminate information other nodes can receive. The information in ROS is referred to as a subject. The topic defines the sorts of communications sent about a topic. Buses are called by the topics about which nodes communicate directly. Topics feature anonymous publish-subscribe semantics, which separate the creation and utilization of data. Nodes are only sometimes aware of whom they communicate. Instead, data-hungry nodes subscribe to the appropriate topic, while data-generating nodes publish to the same topic. A subject can have numerous publishers and subscribers. Topics are designed for one-way, real-time conversation. Some nodes, such as a camera feed, transmit data to other nodes. This type of node sends out information that other nodes can receive. ROS data are referred to as a subject. A subject means the kind of messages to be delivered with that topic. Topics are referred to as buses, and nodes exchange messages across them. The unnamed publish–subscribe syntax of topics distinguishes data creation and consumption. Generally speaking, nodes need to be made aware of whom they are connected to. Those engaged in data should register to the right topic, while those who create data should publish on the same subject. There may be multiple publishers and subscribers for a given topic. Topics are essential for one-way, asynchronous conversation.

Services: Though the publish-subscribe architecture is a reasonably adaptable network protocol, Its one-way, many-to-many transportation is inadequate for communication between RPC requests and responses, frequently required in a distributed system. Request-publish is accomplished through a Service. This is characterized by two messages: one is for the inquiry and the other for the response. A supplying ROS node provides a service with a string name, and a client invokes the service by publishing a request message and waiting for a response. Client libraries typically expose this conversation to the user as a remote protocol call. The ROS client library then compiles the Srv files to specify services into program code. A client can create a link to a service, providing improved performance at the cost of less resilience to service provider changes.

Actions: ROS actions interact using a specified protocol on a client-to-server basis. ROS topics are used in the actions to communicate goal messages from the client to the server. Using the action client, you can cancel goals. When a goal is received, the server processes it and returns information to the client. This information contains the server's status, the current goal's state, feedback on that goal during operation, and a result message after the goal is completed.

16.3 ROS Implementation

There are lots of projects and research done based on ROS and are being implemented in the development of existing technology. Few of those works are explained below.

16.3.1 Smart SEAL: A Building Automation Framework for Smart Buildings Based on ROS

SEAL is a research project to develop Home Automation (HA) solutions for energy conservation in buildings, consumer customization, and better occupant safety [4]. One of the primary issues with home automation is the variety of connectivity standards used by commercial products. Typically, this limits the constructor to select a few manufacturers and equipment, restricting the equipment approach and functionality. Smart SEAL is a system that aims to link a range of sensors and controllers from various manufacturers by offering networking features, protocols, and interfaces that are simple to install and continuously adjustable. ROS, a robotics middleware, lies at the heart of this system. Developers adapted ROS capabilities to the home automation problem, building network and protocol architectures to meet these specific requirements. This software framework enables complicated home automation functions that could not otherwise be performed by using the services supplied by various equipment. The research center has examined the technology and is installed in real-world settings. There currently needs to be a universal standard for the interoperability of home automation systems. Various suppliers manufacture the gadgets in an automated house, which creates a barrier to their efficient use. Current home automation protocols include Zig-Bee, ZWave, and KNX. A standardized approach is a significant impediment to actual smart building incorporation. To address these issues, engineers advocated the usage of ROS, a middleware that enables the abstraction and connection of diverse technology. A central server manages the primary system, and every device, from individual sensors to vendor-specific regulators, is a network node capable of communicating with other nodes over a conventional TCP-IP network. The protocol for interaction has a topic publisher–subscriber architecture, which makes it possible makes it possible for complete device decentralization if required. A collection of standardized topics and commands has been created to provide a universal language for all devices to communicate with each other. Services that are more complicated and can function independently can be added to this communication architecture. When the system is up and running in place, surrounding data can be collected and delivered quickly and easily. This enables, for example, the collecting of temperature, moisture, quality of air and pollution levels over time, the development of prediction techniques, and the implementation of approaches based on user habits in the usage of the infrastructure, thereby improving convenience and security. The SEAL project necessitates adapting a home automation system for diverse consumers, and applying machine learning strategies to simulate user behaviors and habits is critical.

16.3.2 The Development of an Intelligent Drilling Robot System Based on ROS

An industrial robot system has been used in this project to determine the position of a randomly positioned part for a drilling procedure. The system's software is constructed in two steps and steers the robot to the target. In the first stage, the robot examines the working space for the part to drill by taking a continuous image from the camera while in the home position. When the robot arrives at the working space of the determining section, it determines the circle's radius using the Circle Hough Transform (CHT) algorithm [5]. The drilling tool is directed toward the circle's center if the radius exceeds a predetermined threshold value. It performs the drilling operation and returns to the home position to restart the process. Drilling can thus be accomplished without needing a fixed point and location, as is required in typical manufacturing procedures that require fixtures and fittings. For safety, a fixed tool is utilized instead of the actual drill. The KUKA Agilus KR-6-r900-Sixx industrial manipulator is employed and controlled using open-source licensed software ROS with OpenCV running on Linux. This project used ROS to process photos and sensor data from the external world. With the software built, it has been proved that the robot can complete the drilling process. This project used ROS to process photos and sensor data from the external world. With the software built, it has been proved that the robot can complete the drilling process. This project used ROS to process photos and sensor data from the external world. With the software built, it has been proved that the robot can complete the drilling process.

16.3.3 AgROS: A ROS-Based Computing Tool for Agricultural Robotics

AgROS's goal is to establish a farm management system that supports agri-food producers to effectively use cutting-edge digital technology, such as intelligent and autonomous unmanned ground vehicles (UGVs) [6], for field operations. Analysis shows that computer modelling and simulation have been widely utilized to examine sophisticated agricultural machinery even though many existing endeavors in academia focus on operations such as harvesting, picking, mowing, fertilizing, and spraying. The majority of studies take a generic agricultural structure into account. As a result, Agros was created, an emulation tool based on ROS that could be used to analyze the performance of real-world robot systems in specialised segments. Farm owners may use the AgROS to choose their actual life crop layout from a map layout, as well as import the landscape of the field, add features of the real-life agricultural layout, select agricultural machinery from a predefined list of commercial systems, import the selected unmanned ground vehicles into the simulation environment, and evaluate the robot's effectiveness in an atmosphere that is similar to that of the real world. AgROS helps in the former study and effectiveness evaluation of automatized precision farming activities while laying the groundwork for agriculture.

16.4 Conclusion

ROS is a software framework utilised in many universities for robotics research. ROS is utilised in robotics companies to prototype software in addition to academic study. Enthusiasts also use ROS to develop various robot applications. ROS will be an excellent tool for programming robots if you are a roboticist or an aspiring roboticist. Knowing ROS does not imply that you are an expert in robotics; it will be an excellent tool for prototyping your robotics algorithms. There are several prospects for ROS programmers in academic research and industry. Thus mastering ROS will be an excellent addition to your abilities if you want to follow a robotics career. Maintaining the equilibrium of the interests and requirements of a worldwide and heterogeneous community of participants is one of the most challenging issues. ROS and Gazebo users include hobbyists, scholars, academics, innovators, and veteran programmers who, among other things, produce code for vehicles, ships, aircraft, robots, and gadgets. To date, ROS frameworks have shown to be generic and adaptable enough to meet the needs of many, if not the majority, of these crowds. However, users must consistently need help to maintain equilibrium and prioritize the application they hope to enable. This document describes the ROS's basic structure and terminology in brief. ROS's transdisciplinary nature opens up a world of possibilities for customization. Furthermore, some platforms developed using ROS, and their implementation in real life are discussed.

References

1. ROS, ROS introduction, [Online]. Available: http://wiki.ros.org/ROS/Introduction.
2. dos Reis, W.P.N., da Silva, G.J., Junior, O.M., Vivaldini, K.C.T., An extended analysis on tuning the parameters of Adaptive Monte Carlo Localization ROS package in an automated guided vehicle. *Int. J. Adv. Manuf. Technol.*, 117, 1975–1995, 2021.
3. Thale, S.P., Prabhu, M.M., Thakur, P.V., Kadam, P., ROS based SLAM implementation for autonomous navigation using turtlebot, in: *ITM Web of Conferences*, vol. 32, p. 1011, 2020.
4. Bellocchio, E., Costante, G., Cascianelli, S., Valigi, P., Ciarfuglia, T.A., SmartSEAL: A ROS based home automation framework for heterogeneous devices interconnection in smart buildings, in: *2016 IEEE International Smart Cities Conference (ISC2)*, pp. 1–6, 2016, doi: 10.1109/ISC2.2016.7580798.

5. Bingol, M.C. and Aydogmus, O., The development of an intelligent drilling robot system based on Robot Operating System, in: *International Eurasian Conference on Science, Engineering and Technology EurasianSciEnTech*, Ankara-Turkey, pp. 808–813, 2018.
6. Tsolakis, N., Bechtsis, D., Bochtis, D., Agros: A robot operating system based emulation tool for agricultural robotics. *Agronomy*, 9, 7, 403, 2019.

17

An Overview of Collaborative Robots and Their Applications

Rao S. Krishna and Lawrence J. Fernandes*

Department of Mechanical Engineering, Sahyadri College of Engineering & Management, Mangalore, India

Abstract

Modest program design and the capacity of robots to complete tasks alongside human workforces devoid of the need for bulky, space-consuming barriers were thought to offer the missing link toward flexible manufacturing. However, before these expectations can be realized, we must first determine the present hurdles and development needs for collaborative robots from both a technical and a human perspective. Academics manufacturers and distributors of robots, system integrators, consumers, and educators are all involved and responded to an online survey about the deployment of collaborative robots. There is lack of awareness of prospective The most critical factors were applications, source cases, security legislation, and easiness of use—an essential hurdle to using collaborative robots in all categories. New techniques of dividing work between human workers and collaborative robots, as well as safety technology, were the most pressing development needs. There was a mismatch in the responses from representatives from various industries. The current needs of the industry are described in this paper, demonstrating that slightly different challenges have been raised in academics. Aside from discovering new ways to collaborate with collaborative robots, safety is still a problem, and there is a general lack of understanding of collaborative robots in industries.

***Keywords*:** Collaborative robot, technology adaption, hazard, analysis, safety assurance

17.1 Introduction

In current years, there is a surge in interest in collaborative robots or cobots. With collaborative robots, the concept of combining the endurance

**Corresponding author*: lawfds85@gmail.com

Rathishchandra R. Gatti, Chandra Singh, Rajeev Agrawal and Felcy Jyothi Serrao (eds.) Self-Powered Cyber Physical Systems, (309–320) © 2023 Scrivener Publishing LLC

of robots with humans' adaptability and problem-solving abilities is elevated to a different extreme. On the manufacturing floor, human-robot cooperation can keep many procedures, fluctuating from cooperation in a collective office to teamwork. When people and collaborative robots work on the same shared object at the same time. However, as evidenced by a recent survey in the German sector, there appears to be a need for more collaboration on the production floor. On the other hand, the reported applications were functioning normally, with the bulk of them being used in production lines.

Increased impact on organizational performance, advancement, cognitive and behavioral buildability improvements, modern production processes, as well as reduced assembly time, lowered repetitiveness, enhanced speed and reliability, and demographic change are all factors to consider all conveyed reasons for indicating collaborative robot uses in the German industry. The research found that organizations remained more tolerant with reimbursement periods than when it came to traditional manufacturing and that factors had influenced the acceptance of practically every collaborative robot use they looked at. Certification, the arrangement process, picking a tool, picking a robot, and participatory scheduling with workers were among the things that organizations said they needed help with. Collaborative robots were primarily used to grasp and connect components, transfer parts (machine loading, packaging, and palletizing), and employ tools.

Trendy contrast to the factory's hopes for co-robots, workers' attitudes toward collaborative robots have changed. Workers expected collaborative robots to relieve physical and mental strain, and they raised concerns about the robot's functionality and payload, program design, and object picking. Research also suggests that some collaborative robots could handle the objects at a suitable elevation for the employee and adjust to the employee's language and working speed. Workers in some other studies predicted that such physical tasks would lower, while the level of task work and safety would increase (due to robots delivering more customers and thus more work).

17.2 Art of Study

Researched automated systems and automobiles; inventory monitoring and decision-making systems; intelligent technologies and Internet networks for enhancing the flow of information; genuine big data analytics of vehicles; merchandise and facility locations for finding materials and components' optimal scheduling transportation are among the significant uses of Industrial revolution 4.0 in logistical support. Generally, the Internet of Things (IoT) is the most applicable area of logistical issues 4.0, which is aided by various technologies. As a result, productivity and quantity have increased. Collaborative robots engaged in selecting, stacking pallets, or manufacturing tasks, for example, are spreading rapidly, as indicated by the rise in the number of publications published in the literature. Consequently, this article presents the findings of a bibliometric of 64 science articles in the logistics field that deal with this problem. Two software applications, Microsoft Excel™ and VOSviewer, have been used to analyze the data. The findings are encouraging since they show that there are a lot of opportunities [1].

The research article provides a novel method for collaborative robotics that intends to make computing easier and more efficient for non-technical consumers. It is based on the following principles: (1) a primary level and diligent robot programming framework known as the simple programming path; (2) a blended communication configuration in which robot activities could be coded utilizing human language conversation, and, if essential, accomplished and finalized throughout a block-based functionality; and (3) a robust intellectual contest between both the user's conceptual model and the coding user experience. The recommended technique has been developed and evaluated by creating a model examined to evaluate that it can be customized to various application domains by successfully listing destinations and activities. The model has been tested with two experiments conducted to verify the suggested new technique and examine its influence on terminal coding. The findings suggest that a hybrid strategy utilizing human language chat and a block-based interface may assist non-technical customers in making lines of code simple and fast [2].

Over the last few generations, the function of mechanization in contemporary manufacturing has been increasingly important. Factories, on the other hand, have yet to be able to properly automate numerous processes, forcing them to rely on personal employees with limited physical capabilities. In the past few years, autonomous robots have shown to be a low-cost means of completing manual tasks. As a result, this unique technique may constitute a cost-effective option for (partial) manufacturing automating manuals. To establish cost-effective system designs, we investigate a production line problem with the system in which collaborating robotics and people employees may be allocated to the line's units. We create a computational formula and use case studies to demonstrate its functioning. Following that, we determine the elements that influence the use of industrial robotics [3].

The study's proposed solution is a minimal generic robotic technology with unique security mechanisms to boost operational flexibility in the automobile metallurgy industry. This example is inspired by an actual situation in a metallurgical production firm for automobiles. The industry research aims to show the possibility of using collaborative robotics in automobile metallurgical operations. The research focused on the Hungarian medium-sized corporate sector. To illustrate how types of technology are employed in a Hungarian midsize firm to optimize their operations, AI-powered and specific mobile robotic security systems—adjusted by social cognition—are being used. The fundamental issue in the automobile metallurgy factory is versatility, such as factory line cutover and computation. As a result, the process involves determining the cost of the product, and the only option to improve profit is to lower development and manufacturing expenses. This necessitates effective management and operation of the organization as well as production. Several of the keys to success is having a good team [4].

Simple coding and the capacity of robots to operate alongside human employees without any need for bulky and space-consuming barriers were thought to offer the gap beyond manufacturing technologies. Nevertheless, before such assumptions can be satisfied, we must first determine the existing impediments to cobot advancement from academics and industry perspectives. Robotics manufacturers and suppliers, technology providers, terminals, and universities responded to an online survey about implementing collaborative robots. The lack of awareness of prospective applications, precedent cases, safety regulations, and convenience was the most critical hurdle to using cobots in all categories. The most pressing development demands were new means of assigning labour [5].

Automation of specific operations, for example, can only be successful if considerable changes in organization, design, and manufacturing are undertaken. Integrating the latest machinery with the current system and human employees can successfully automate other processes. According to the author, the adoption of automation into a system fails to produce the promised performance gain and also leads to catastrophic system breakdowns. As a result, increased automation only sometimes implies increased productivity, safety, or operator well-being, as it might lead to other issues such as misunderstanding, complacency, or the loss of specific abilities. Such circumstances can arise, as per the author. (1980), since the just one element The issues experienced by factory professionals throughout the deployment of advanced technologies over the last three decades are also a concern for today's executives. However, in contrast to the rapid technological developments that have led to AMTs, the implementation environment presents additional obstacles. As a result, the introduction of collaborative robots (cobots) presents a new number of co-management issues. Technical difficulties begin with the rising sophistication of industrial robotics, notwithstanding the ease of programming [6].

Collaboration robotics may be a way to address the issue of musculoskeletal conditions (MSDs) in the workplace; however, developing similar robots effectively responds due to the need for ecological evaluation methods for such devices. This work describes a general technique for extensive ecological evaluations of co-manipulation tasks and their relevance to the cooperative robotic system. Many ecological markers are established to quantify the various biomechanics demands that occur during manual interventions. Those metrics are monitored for every particular task using dynamic digital human scenarios with variable human and robot attributes. Sensitive indexes are then calculated to measure the impact of every robot variable and suggest those that should be improved most to improve ergonomic effectiveness. Scenario analysis also enables you to Identify the criteria that effectively characterize the entity's general ecological effectiveness. In order to develop an effective robot architecture, an iterative approach is utilized to optimize the robot's influencing parameters with some of the most relevant ergonomic indications. The entire process is used to improve the morphology of a robot that assists in drilling. The robots' outcomes demonstrate the suggested method's applicability [7].

For cooperative robotic devices, close contacts with a human operator are a significant safety risk. Robot integrators must indicate that they have made efforts to detect possible risks, which may be incorporated into collaborative tasks or inside the collaborative workspace, to ensure safety. As a result, design protections are created to prevent unintentional touch, which might be dangerous to a humans colleague when doing cooperative learning duties. When incorporating direct and indirect protections on robot systems or developing collaborative workplaces, the ISO 15066 standard gives rules that designers, integrators, and users should consider. The standard premises, for example, include protections on organized hazard analysis and risk assessment methods. However, based on research, it is not always evident how architects, programmers, or other professionals work. as well as risk analysis When robot systems and social coworkers undertake participatory jobs in manufacturing, it can often be unclear which measures designers should take or which techniques they should use to investigate dangers and estimate risks. As a result, determining whether design protections are sufficient to mitigate dangerous interactions in a production cell is difficult. This report analyzes the criteria for collaborating robot network safety assurance addressed in the latest ISO 15066 standard for collaborative robotics, as well as how these precautions are implemented in the research described in the literature [8].

The focus of this research was to see what collaborative strategies had a good impact on science process skills, collaborative problem-solving, and learning motivation among people studying collaboration robots assignments. The study will examine how past robotics experiences affected students' science process skills, collaborative problem-solving abilities, and effective learning. The findings revealed that students' levels of experience and cooperation initiatives could have an influence. Students' motivation and collaborative problem-solving improved once they were assigned group roles. Student learning motivation and collaborative problem-solving were similarly affected by the experience level, with novice status being connected with more significant levels than students with more expertise. A collaborative strategy has been found that has the opportunity to benefit students participating in collaborative robotics projects, as well as aid teachers in the classroom instructors in the intentional creation of collaborative robotics projects using scientifically supported methodologies to increase behavioural results for students of all ages—expertise with robots [9].

Collaborative robots have made news recently due to recent industrial purposes. However, they are less common in the industry than they appear. According to the authors of this article, one reason for the sluggish adoption is the high safety standards and the need for more technical tools for assessing cooperative robotics systems. Software architecture is a suitable foundation for developing engineering tools for quicker and more reliable deployment, but it has recently been applied to robotics difficulties. This study discusses state of the art for creating robotics applications with human-robot cooperation (HRC), followed by a review of existing systems engineering methodologies that might help. Our assessment seeks to aid the robotics community in the development of design tools in the future. That might be of assistance. This analysis seeks to aid the robotics community in the future creation of design methods that will help in understanding, designing and deploying collaborating robot systems [10].

17.3 Implementation of Collaborative Robots

1) Electronics and technology

Manufacturers of lights, mobile phones, speakers, computers, and other products employ collaborative robot automation. Collaborative robots may be programmed by non-experts and are chosen for quick installation since they are simpler and faster than traditional industrial robots. Agility is the key to competitiveness in this fast-paced market. Flexible, collaborative robots can increase productivity, quality, and consistency, whether you are creating sub-assemblies, loading and unloading test rigs, or kitting and packaging products. Cobots can do special activities, such as insertion, dispensing, screw-driving, or labelling with excellent repeatability while working alongside humans. You may redeploy talented people to more valuable activities as you automate these monotonous operations. Collaborative robots are also great for separating workers on assembly lines by filling in upstream or downstream operations. Because cobot arms are light and straightforward to program, they may quickly be shifted from one manufacturing line to another or from production to packaging or palletizing as needed. Cobots can even pay for themselves in competitive electronics production.

2) Food and agriculture
Food and farming production occurs 24 hours a day, seven days a week. Compared to human labor, collaborative robots can operate 24 hours a day, resulting in less waste and faster production times. Furthermore, cobots can work in extremes of heat, cold, and discomfort. As a result, many food manufacturing enterprises are turning to cobots to supplement human labor. When it comes to food processing, the first steps are to clean, sort, and package the items. Before being packed, fresh vegetables go through primary processing, while the rest of the food goes through secondary processing. Cobots with soft grasping and sensing are now being utilized to handle raw foods, thanks to recent advancements. It is difficult to pick and place fruits and vegetables, but cobots do so delicately and without causing any damage.

With automation in the food business, cutting and slicing jobs are also automated. While a food processor can do simple tasks, cobots can handle more complicated slicing and cutting. The human workforce previously undertook these duties prior to the development of robotics in robotic cake decoration. It entails applying icing to cakes with the help of a robotic arm. Even cake cutting is automated and done robotically. It is used to cut complicated figures in cakes using high-pressure water. Robotic food preparation can eliminate the need for human personnel. Manufacturers of robots have worked hard to make robot shells smoother so they may be thoroughly cleaned. This permits them to be properly cleaned at the end of the processing activity. One of the essential applications of robotics is in food packing. Packaging becomes more standard and straightforward as the collaborative robots market becomes more involved. The employment of collaborative robots to pack things is a popular option. They can use the pick and put action to identify, align, and place products. Raw and processed food products will be wrapped throughout the primary packaging process. The size and style of the packaging will vary depending on the food item that is being wrapped. It is done correctly with robots, coupled with quality analysis.

3) Pharmaceuticals and chemicals

Pharmaceuticals, chemicals, and medical equipment are all made with extreme accuracy. Mixing, counting, and dispensing a range of ingredients is a common task for UR cobots. Collaborating robots can be used to automate a wide variety of tasks. They simplify the automation of inspection, mixing, dosing, filling, and packaging. Cobots assist your staff by relieving them of arduous and repetitive tasks, resulting in a significant gain in inefficiency. They assist with laboratory analysis, testing, injection moulding, and pick and place applications. Collaborative robots can also be used to operate machines.

These robotic arms are a source of relief for your employees. A robotic arm does not bother with duties detrimental to one's health, unclean, or complex. At the same time, they complete their tasks with extreme precision and speed. This way, the highest quality standards can be maintained without jeopardizing competitiveness. The cobots are used in the hospital's laboratory rather than inpatient care. The UR5 robots from Universal Robots relieve lab personnel of time-consuming and tedious chores. They sort about 7-8 samples each minute and about 3,000 tubes per day. This is made more accessible by the use of a camera system. The robots then sorted the samples by colour and placed them in the proper container. The tubes are subsequently removed by the robots and placed in a centrifuge for analysis and centrifugation. As we observe the chart revenue from the various industries.

Furthermore, various sectors automotive sector gains 16% revenue. In the automotive sector, there is more need for collaborative robots because human work is very complicated and dangerous to employees. That is why there is a need for cobots in the automotive sector before there is less revenue compared to collaborative robots implementation increase of 16% revenu from the sector. Similarly, in the chemical and pharmaceuticals sector, there is an increase of 5.5% in the implementation of cobots in industries. In chemical industries, due to the implementation of cobots, there is an increase in revenue and accuracy is also increased cobots.

17.3.1 Collaborative Robot Revenue Split by Industries

In pharmaceuticals, there is a need for more concentration on medicine preparation, and collaborative robots can do it precisely. There will be an increase in revenue by 5.5%. By observing this, we can understand the need for collaborative robots in this industry. In the food and beverage industries, there is a 7.6% growth in revenue. In the food industries, there is a need for packing and placing food items and preparation of bottles for beverages. Preparation of soft drinks in the same amount and concentration of all material for everything collaborative robots have used that increase in growth of revenue. There is a need for cobots in this sector. In the field of electronics industries, there is a need for collaborative robots to assemble and prepare electronic items. Due to the implementation of collaborative robots, 34.1% of the revenue from cobots is grown. As in the chart from the plastic and rubber industries, there is revenue growth of 7.8% compared to traditional manufacturing as it increases the work lead time, and cobots can work efficiently for more time. In logistics also, we can use collaborative robots for picking and placing the goods and also they can take over or can pick up heavy loads. Also, that load that humans cannot pick, and it can quickly move from one place to another within a limited period. Similarly, in other industries, there is a constant increment in revenue after implementing collaborative robots in different industries, and they can gain production yield. So observing all this, we can conclude that there is a hike in revenue of the industries after the implementation of the collaborative robots in the industries and can high profits from the implementation.

17.3.2 Drawbacks of the Collaborative Robots

i) Regarding heavier loads, there are better options than collaborative robots.
ii) The speed of collaborative robots is restricted.
iii) Collaborative robots may not be as efficient as six-axis robots.
iv) Collaborative robot safety approval might be difficult.
v) Collaborative robots are only partially autonomous.

17.4 Conclusion

The more stressed research requirement among the respondents was a new method of dividing labour between employees and robots. While one of the reasons for adopting cobots has been described as new manufacturing techniques, the degree of cooperation in present industrial applications is still relatively low, so creating new collaborative activities might be one of the future priority areas. Task assignment is already being studied, which will aid in creating future collaborative projects. One significant challenge that must be addressed is safety technology, which has been mentioned in previous publications. However, safety was not identified as one of the most critical impediments in this study. Instead, The most major impediment was identified as a lack of understanding, as stated earlier. Although the respondents considered themselves relatively aware of cobots and robots, they needed more expertise on several topics, including the technology in general, possible applications, safety-related difficulties, and actual implementation. The survey participants also obtained relevant cases. Hands-on examples of cobots in novel collaborative activities and demonstrating the associated safety and risk work methods are thus desperately needed.

References

1. S.T.S. and Weckenborg, C., Assembly line balancing with collaborative robots under consideration of ergonomics: A cost-oriented approach. *IFAC-PapersOnLine*, 52, 13, 1860–1865, 2019.
2. D. A. L. H. L. F. V. P. J. and Tokody, H.P., Collaborative robotics research: Subito project. *Proc. Manuf.*, 46, 467–474, 2020.
3. A. C. S. A. L. and Simões, B.A.C., Factors influencing the intention of managers to adopt collaborative robots (cobots) in manufacturing organizations. *J. Eng. Tech. Manage.*, 57, 101–574, 2020.
4. A.P.J. and Quenehen, K.N., Process optimization using collaborative robots-comparative case study. *IFAC-PapersOnLine*, 52, 60–65, 2019.
5. P. P. V. M. Y. and Maurice, B.P., Human-oriented design of collaborative robots. *Int. J. Ind. Ergon.*, 57, 88–102, 2017.
6. P. and Francesco, P.G.G., AURA: An example of collaborative robot for automotive and general industry applications. *Proc. Manuf.*, 338-345, 11, 2017.
7. D. G. L. G. G. and Fogli, T.F., A hybrid approach to user-oriented programming of collaborative robots. *Rob. Comput.-Integr. Manuf.*, 73, 102–234, 2022.

8. P.P.L. and Chemweno, D.W., Orienting safety assurance with outcomes of hazard analysis and risk assessment: A review of the ISO 15066 standard for collaborative robot systems. *Saf. Sci.*, 129, 104–832, 2020.
9. G. V. G. T. L. and Atzeni, B.E., A bibliometric analysis on collaborative robots in Logistics 4.0 environments. *Proc. Comput. Sci.*, 180, 686–695, 2021.
10. I. and Aaltonen, S.T., Experiences and expectations of collaborative robots in industry and academia: Barriers and development needs. *Proc. Manuf.*, 38, 1151–1158, 2019.

18

State-of-the-Art and Future Applications of Powered Exoskeleton

C.P. Dheeshith, K. Abhijith, A. Shahaas, Rithin B. Nambiar and Rathishchandra R. Gatti*

Department of Mechanical Engineering, Sahyadri College of Engineering & Management, Mangalore, India

Abstract
Over the last decade, research on robotic exoskeletons has skyrocketed due to their importance in medicine and the military. Prototypes for testing on humans are now possible, thanks to advancements in the energy and hardware industries. This paper intends to report the most recent stages in the development of powered exoskeletons, their applications in the current world and their possible future improvements. This paper first describes a powered exoskeleton and how it impacts our current technologies and associated applications across different industries, such as the military, health care, and consumer electronics. Several state-of-the-art developed exoskeletons for critical applications such as the rehabilitation of a patient or boosting the physical attributes of a soldier are discussed. The different types of designs and some considerations to be concerned about in the design of these exoskeletons are also discussed here, along with the influencing design parameters such as safety, cost, flexibility, and adaptability. The future challenges that would be faced by this industry, along with some viable innovations in the current industry, are also reviewed here.

***Keywords*:** Exoskeleton, robotic devices, military, rehabilitation, prosthetics

18.1 Introduction

Robotics has taken over almost every aspect of our daily life, from manufacturing industries to the service industry. Robots can operate indefinitely and

**Corresponding author*: gattirathish@gmail.com

Rathishchandra R. Gatti, Chandra Singh, Rajeev Agrawal and Felcy Jyothi Serrao (eds.) Self-Powered Cyber Physical Systems, (321–332) © 2023 Scrivener Publishing LLC

precisely without becoming tired or distracted. While robotics is the use of engineering to replace humans with robots for small or risky tasks, exoskeletons are robotic devices that support humans in the execution of various tasks [1]. It can be pretty tricky to define a concise definition of a "powered exoskeleton." The Merriam-Webster dictionary describes an exoskeleton as "a wearable, external mechanical supporting framework." In an active or powered exoskeleton, the function of an exoskeleton is assisted by the use of actuators, electric motors, pneumatics, levers, hydraulics, and microcontrollers.

In contrast, a passive exoskeleton works without the help of an external power source. The powered skeleton can be further classified based on the types of actuators used in it and the control system used for its functions. Based on the type of actuators, powered exoskeletons are classified as hydraulic actuated, pneumatic actuated and electric motor actuated as shown in Figure 18.1.

Based on the control system, the powered exoskeleton can be classified as Admittance controlled, and impedance controlled and feedback-controlled exoskeletons.

The commercial market for technologies that improve human performance in the industrial environment has been rising for quite a long time. There are currently a variety of whole-body, lower-limb, and upper-extremity prosthetic devices available. Industrial exoskeletons are attracting much attention to increase human performance and reduce the effects of Musculo-Skeletal Disorders (MSDs). A powered exoskeleton can boost a person's strength, endurance, and other physical abilities, whether healthy or not. Putting on a robotic suit that can increase one's strength or even act as a prosthetic limb is intriguing.

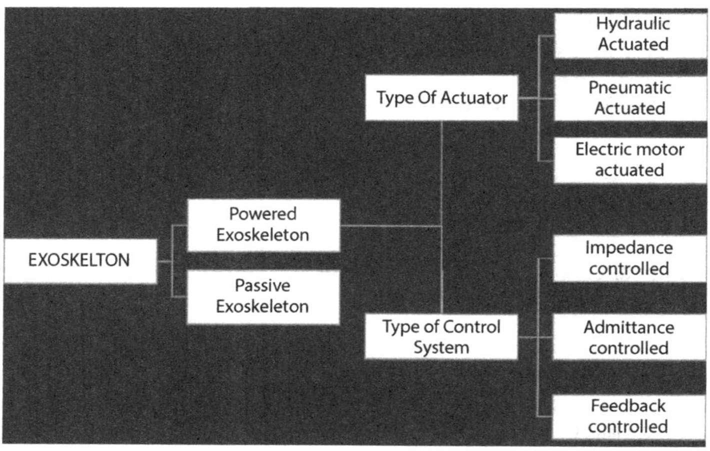

Figure 18.1 Classification of exoskeleton.

Contrary to popular belief, exoskeletons intended to provide superhuman strength to the user have been around since the 1960s [2]. Today, the powered exoskeleton industry is continuously evolving and re-inventing itself across every corner. The complexity of humans wearing exoskeletons has received very little attention in studies focused on industrial implementations. Various studies have been conducted on the ergonomics of exoskeletons to be worn. A wide range of issues is to be considered for its practical application. For example, even though wearing a powered exoskeleton is found to be effective in reducing the load on one part of the human body, in doing so, there can be an increase in the load experienced on another part of the body [3]. This can be a significant issue in military applications where soldiers have to carry approximately a load of 60 kg on them already [4]. As a result, when considering implementing any exoskeleton, the entire human body, both inside and outside, must be considered, even if the exoskeleton is just one part of the body [5]. Exoskeletons can be placed on the user's body to amplify, reinforce or restore human performance. For example, it can be used to help in the lifting and carrying of heavy objects.

According to the changing needs in its application, powered exoskeletons can be made to cover the entire body or even a specific body part exclusively. Powered exoskeletons will act parallel to the body structures and support impaired human walking [6]. This study aims to observe the changes powered exoskeleton has brought into this world and to discuss the challenges it may face in the coming future.

18.2 Powered Exoskeleton

Currently, there is a wide variety of exoskeletons in existence, ranging from the differences in their structure, the functioning action, the power technology used, the body part being focused on, the purpose, and the area of application [7]. Powered exoskeletons help perform the task at hand without the user exerting any energy to complete the task. Here, the energy required for the task will be provided by the attached external power source, unlike the passive exoskeleton, where the user needs to exert energy to perform the work since they do not have any other power sources as shown in Figure 18.2. Based on the energy conservation law, it can lower human metabolic costs [4]. Depending on the task, different kinds of actuators like hydraulic, pneumatic or simple electric motors can be used to impart the motion accordingly.

The task to be performed also influences the control system on the powered exoskeleton. The admittance control system defines the posture that

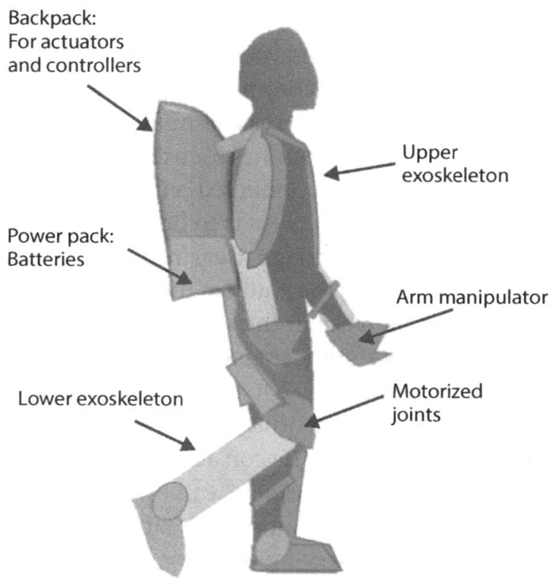

Figure 18.2 Components of a powered exoskeleton.

must be assumed by the exoskeleton, while the impedance control system corrects the changes caused by interactions with the external environment [6]. In the case of feedback-controlled exoskeletons, they can either be electromyography (EMG) driven, where electrodes placed on the user's muscles are used to generate a feedback signal in response to muscle activation or using electroencephalography (EEG), where the electrical activity from the brain is collected via sensors placed on the scalp. The powered exoskeleton can also be categorized into two based on its purpose. They are Recovery and performance. Recovery-oriented powered exoskeletons are designed to improve the user's physical health, whereas performance-oriented powered exoskeletons are developed to maximize the user's physical characteristics.

18.3 State of the Art

The powered exoskeleton has a long history dating back to the 1960s. Back then, exoskeletons were developed by the US military to improve and enhance a soldier's ability for military purposes. Apart from military purposes, research was also being carried out on other possible applications of powered exoskeletons [3]. Back The General Electric Company created a two-armed master-slave manipulator around the same period to aid in handling radioactive equipment. The research on powered exoskeletons, which started from scratch, has now entered a phase of optimization and innovation. There have been significant advancements in this industry in the last few years that may address some of technology's most pressing issues. Exoskeleton designers have recently shifted their focus away from stiff, steel exoskeletons to Soft exoskeletons, which are made of fabric, include flexible artificial muscles and thus provide several benefits over traditional metal exoskeletons [8]. As these exoskeletons resemble clothes rather than an external structure, they are more comfortable, allowing for more extended periods of use.

There is also a general search for solutions to energy issues faced, with some research pointing to the use of built-in solar panels as an alternative power source for the exoskeleton. Exoskeleton designers have also begun incorporating artificial intelligence (AI) and connected sensor technology into their designs. Here, Sensors embedded throughout the exoskeleton collect data on their active use and the users' physiological condition. This data is then interpreted by AI, which adjusts the exoskeleton as needed to achieve optimal performance. Some researchers have also developed exoskeletons that will alter torque to maximize users' running and walking economy depending on metabolic data. These technological advancements could prevent repetitive strain injuries in the workplace or improve physical therapy results. Guardian XO Max, a full-body, powered industrial exoskeleton robotic system produced by Serc's Robotics, is a notable result of these advancements [4]. With this exoskeleton, the user can securely lift and manoeuvre weights up to 200 pounds. The power required for the functioning of the exoskeleton has been reduced by more than 90% compared to the traditional humanoid robots, all the while enabling work sessions of up to eight hours on a single charge.

Even though exoskeletons are already stunning, these new developments illustrate how far the technology may advance [9]. This could become a multi-billion dollar industry in the short-coming future as the research funding for exoskeletons will expand as more organizations use these technologies and result in further advancements.

18.4 Design Parameters to be Considered

A. Safety of the user

In exoskeleton design, safety is a critical design consideration. These devices enclose the user, who may suffer from disorders that cause reduced motor control and muscle weakness. Safety should be addressed at all design phases, and measures should be implemented at all the hardware, software, and operator levels. Some of the potential risks associated with the use of exoskeletons may include pushing the patient's range of motion beyond its safe limit, Moving the patient at a high rate—abrupt movement from the device which can cause muscle strain and injuries and excessive torque being applied to the user or the user having to apply excessive torque to the exoskeleton.

B. Adjustability
A robotic exoskeleton must be able to accommodate individuals of any size, gender, joint range of motion, and other factors. As a result, careful consideration of human ergonomic and anthropomorphic data is essential. Adjustable frames are used in exoskeletons to adapt to users' various anatomical characteristics.

C. Biomechanics of humans
Lately, exoskeletons have passive safety features like the addition of mechanical end stops that would prevent the joints of the exoskeleton from exceeding the human's anatomical range of motion and having no sharp edges. Exoskeletons are mechanically operated wearable devices that may interfere with the normal limb movement of the user. This can result in low compatibility between the robotic and human joints, resulting in undesirable human-exoskeleton contact pressures. A thorough understanding of joint biomechanics and human walking biomechanics is required to create a viable powered exoskeleton. Although robots are often designed to mirror human kinematics in terms of structure, range of motion, and even the degrees of freedom (DOFs), due to the complexity of human joints, absolute clarity in robotic joints is often not possible as increasing the joint's complexity may result in higher expenses and decreased system dependability.

D. Affordability
Presently, robotic gadgets are costly, and only a few consumers can afford them. When consumers have more sophisticated or specific requirements not addressed by more generic equipment, the high device prices become a much more prominent issue. Some commonly recommended strategies for cost-cutting include – making a trade-off between accuracy and strength, reducing the complexity of the mechanical design of the exoskeleton

system and designing for cost by involving readily available cost-effective hardware.

E. Mechanisms needed

Ideally speaking, a robotic exoskeleton should be able to produce natural movements with the wearer not being subjected to any external vibration or sudden jerks. The actuator chosen for the exoskeleton will significantly affect the performance characteristics of the exoskeleton in terms of the generated output torque, efficiency and portability. Actuators like electric motors are widely utilized because of their comparatively larger power output. They are straightforward to handle with portable rechargeable batteries and may be controlled by digital or analogue input signals. Pneumatic actuators are lighter than electric actuators and have a lower intrinsic impedance, but their nonlinear nature makes them more challenging to regulate. As they demand pressurized air, the compressor's size increases the system's total size, and hence they are better suited for applications where the system is idle. Hydraulic actuators produce a large amount of torque and are extremely sensitive and responsive. However, they are less suitable for rehabilitative applications because of their weight, susceptibility to fluid leakage, and significantly larger size. Hydraulic and pneumatic actuators are often selected for exoskeletons for performance enhancement of the user because of their superior power output-to-weight ratio.

F. User interface

The physical interface design of an exoskeleton is critical due to its influence on the mechanical power transfer of the powered exoskeleton to the operator. Padding and straps are commonly used to secure an exoskeleton to a person. To avoid injuries, the user interface should be comfortable and give adequate support to the user. In the design, the contact manner, contact intensity, and contact locations on the body should all be considered. Ideally, the exoskeleton's interface should tap into the user's latent abilities.

G. Accessibility

Users' needs may vary depending on their needs and abilities. The accessibility of a powered exoskeleton is a critical factor in its design. The equipment should be adaptable and simple to use. This includes how simple the exoskeleton is to set up and how long it takes to do the same. The interface for the operation of the exoskeleton should be as simple as possible.

18.5 Challenges to Tackle

A. Power supply Challenges
Batteries are the obvious choice for a power source when it comes to portable devices, but they must be replaced or recharged regularly, and sometimes they may even explode due to thermal runaway. Serco claims to have overcome some of these concerns with their battery technology, especially consumption, by lowering the power consumption rate to operate their Guardian XO robotic exoskeleton to under 500 Watts while allowing batteries to be hot-swapped without having to turn the unit [4].

B. Material Selection issues
Steel and aluminium were employed in the early exoskeletons as they were affordable and easy to shape. Steel is a heavy metal, which means that the powered exoskeleton will have to exert energy to overcome its weight, reducing the efficiency of the exoskeleton. Even though aluminium alloys are lightweight, they often quickly break due to fatigue. Materials like fibreglass and carbon fibre offer significantly better strength-to-weight ratios than other materials.

C. Standardization issues
Exoskeletons need to be adaptable or [1]They are easily tailored to each user individually due to the broad range of size disparities in skeletal bone lengths and limb-torso girth. Individual differences in range, type of motion and speed of the motion make it challenging to create a uniform device to deliver the proper amount of help at the right moment. Algorithms are being developed to alter control functions to optimize energy expenditure in tasks like walking. It may be possible to overcome this in military applications by requiring users to be of a specific physical size before being given an exoskeleton. Body size restrictions are already placed in military jobs, making this problem less of an issue.

18.6 Applications of Powered Exoskeleton

A. Healthcare Applications
The powered exoskeleton can play an essential role in aiding the rehabilitation of injured patients and improving the mobility and dexterity of the world's elderly population. A powered exoskeleton can help patients with paralysis limbs regain the function of their limbs, making this technology

perfect for medical usage. Senior citizens and people with disabilities, in addition to people with paraplegia, can receive assistance from a lightweight powered exoskeleton, mainly helping patients recovering from an injury or stroke [2]. Today, over 20% of the population is 65 years or older, and this is expected to surpass 35% by 2050 [2]. As a result, these robotic devices are seen as a critical technology that will allow the elderly to maintain their independence while remaining active.

B. Defence Applications

The United States military began envisioning the usage of powered exoskeletons half a century ago, with prototypes completed within two decades. More concepts are becoming a reality due to recent breakthroughs and technological advancements. Sarcos, an American robotics company, has been building a robotic suit for the "army of tomorrow" at its research laboratory in Salt Lake City, Utah, since 2000, with financing from the American Defense Advanced Research Projects Agency (DARPA) [1]. The exoskeleton allows the user to lift 200-pound loads repeatedly without exhaustion. There is also no delay between human operation and the response of the exoskeleton, which reflects the user's operation in a direct and real-time manner. Sarcos also says that the exoskeleton may improve a soldier's agility, allowing them to navigate rugged terrain and perform more commonplace duties like kicking a soccer ball or climbing stairs [4]. The number of US soldiers dismissed from active duty due to musculoskeletal disorders has increased tenfold within the last decade. The US army spends more than 500 million dollars annually on treating soldiers with spinal injuries and other orthopaedic issues [4]. Soldiers would be able to run faster, extend endurance, minimize fatigue, carry large weapons and artillery, and avoid any barriers on the battlefield due to this technology. A powered exoskeleton would also shield soldiers from gunfire and explosions simultaneously.

C. Consumer Applications

Civilians can also use these exoskeletons in their day-to-day operations to aid them in performing their duties more effectively. Firefighters, emergency medical technicians, search and rescue professionals, and others fall into this category, especially in potentially dangerous and life-threatening situations. Exoskeletons are also being designed to aid firefighters and help in rescue missions by carrying heavy equipment to assist in their work. Powered exoskeletons are likely years away from widespread commercial use, with cost being the big obstacle.

D. Industrial Applications

Many industrial activities entail complex movements and functions that require human employees to move and adjust to the operation's needs. Exoskeletons enable workers to do jobs more efficiently and effortlessly by utilizing powered exoskeleton suits in industrial settings where automation is impossible and for the assembly of pieces with more accuracy and precision [3]. Workers using robotic exoskeletons are much less likely to overexert or harm themselves, which has safety and health implications.

18.7 Conclusion

The evolution of exoskeletons has taken a long time and is still ongoing. New technologies have evolved over the years that have the potential to be utilized in the design of powerful exoskeletons. Exoskeletons may become a regular part of life as they evolve and are integrated into civilization. According to studies, exoskeletons may, in the coming future, allow humans to work in unstructured areas such as the deep sea and outer space for building mining and other purposes once the current drawbacks in power supply and user interface are tackled. The research funding for this industry will expand as more firms adopt these technologies, resulting in further advancements in this field. Shortly, the exoskeleton industry may become a multibillion-dollar industry.

References

1. Sirlantzis, K., Larsen, L.B., Kanumuru, L.K., Oprea, P., Robotics, in: *Handb. Electron. Assist. Technol.*, pp. 311–345, Jan. 2018, DOI: 10.1016/B978-0-12-812487-1.00011-9.
2. Bogue, R., Exoskeletons and robotic prosthetics: A review of recent developments. *Ind. Rob.*, 36, 5, 421–427, 2009, DOI: 10.1108/01439910910980141/FULL/XML.
3. Fox, S., Arango, O., Heilala, J., Vahala, P., Exoskeletons: Comprehensive, comparative and critical analyses of their potential to improve manufacturing performance. *J. Manuf. Technol. Manage.*, 31, 6, 1261–1280, Nov. 2020, DOI: 10.1108/JMTM-01-2019-0023/FULL/PDF.
4. Zhou, J.-Y., Ye, L. I. U., Xin-Min, M. O., Chong-Wei, H. A. N., Xiao-Jing, M. E. N. G., Qiang, L. I., Yue-Jin, W. A. N. G., Ang, Z. H. A. N. G., A preliminary study of the military applications and future of individual exoskeletons. In *Journal of Physics: Conference Series*, vol. 1507, no. 10, p. 102044, IOP Publishing,

https://iopscience.iop.org/article/10.1088/1742-6596/1507/10/102044/meta, 2020.
5. Huysamen, K., Power, V., O'Sullivan, L., Elongation of the surface of the spine during lifting and lowering, and implications for design of an upper body industrial exoskeleton. *Appl. Ergon.*, 72, 10–16, Oct. 2018, doi: 10.1016/J.APERGO.2018.04.011.
6. Moreno, J.C., Mohammed, S., Sharma, N., Del-Ama, A.J., Hybrid wearable robotic exoskeletons for human walking, in: *Wearable Robot. Syst. Appl.*, pp. 347–364, Jan. 2019, DOI: 10.1016/B978-0-12-814659-0.00018-7.
7. de la Tejera, J.A., Bustamante-Bello, R., Ramirez-Mendoza, R.A., Izquierdo-Reyes, J., Systematic review of exoskeletons towards a general categorization model proposal. *Appl. Sci.*, 11, 1, 76, 2021, Dec. 2020, DOI: 10.3390/APP11010076.
8. Strickland, E., Iron man' suits are coming to factory floors. *IEEE Spectr.*, 56, 1, 27–29, Jan. 2019, DOI: 10.1109/MSPEC.2019.8594789.
9. Malcolm, P., Derave, W., Galle, S., De Clercq, D., A simple exoskeleton that assists plantarflexion can reduce the metabolic cost of human walking. *PLoS One*, 8, 2, e56137, 2013, DOI: 10.1371/JOURNAL.PONE.0056137.

19

An Overview of Recent Trends in Consumer Robotics

Pramod Rao M., Shrihari P.C., Manoj, Shankar Gouda S. and Rathishchandra R. Gatti*

Department of Mechanical Engineering, Sahyadri College of Engineering & Management, Mangalore, India

Abstract
Robots have widespread use in homes, health care, traffic management, law enforcement, and military applications, to name a few. We utilize robots to perform daily repetitive, monotonous tasks too dangerous tasks. However, consumer robotics plays a significant role in everyday life and also becoming a flourishing business. From utility robots, such as cleaning and lawn mowing robots, to entertainment robots, such as pet and toy robots, consumer robotics will be the next wave of the robotic business. The present paper discusses the state-of-the-art development of consumer robots, different types of consumer robots that are commercialised and the future trends of consumer robotics.

Keywords: Consumer robots, robotics, cleaning robots, humanoid robots

19.1 Introduction

Consumer robots and robotic gadgets are designed for everyday use, usually in a household or personal setting. Many of them are made for mundane tasks like cleaning or carrying bags, lawn mowing, and entertainment.

Wearable robots, robotic gadgets, and robots used for gaming or entertainment also come into this category. Household robots, vacuum robots, lawn mowing robots, pool cleaning robots, and other robotic devices are also among them. Consumer robots are devices you may purchase

**Corresponding author*: gattirathish@gmail.com

Rathishchandra R. Gatti, Chandra Singh, Rajeev Agrawal and Felcy Jyothi Serrao (eds.) Self-Powered Cyber Physical Systems, (333–340) © 2023 Scrivener Publishing LLC

and use for entertainment or domestic chores. Aibo, the robot dog, the Roomba vacuum cleaner, AI-powered robot assistants, and an increasing number of robotic toys and kits are examples. Consumer robotics is now transitioning from being dominated mainly by cleaning robots to robotic personal assistants or family companions. Improved product quality, one of the primary reasons a business would want to automate with industrial robots is to increase product quality, make affordable products, have faster delivery times, and improve customer service. These robots are frequently created to assist persons with specific needs, such as the elderly, autistic children, and disabled adults. They generally try to assist in a specific setting, such as a home, school, nursing home, or hospital. The capacity of robotics to construct an ever-changing assortment of commodities and kinds of items is driving their growing importance in consumer product production. These household robots are making a difference in people's lives. Even though we are still a long way from the science-fiction future, numerous robots are taking over our duties. Because they handle our jobs independently, these robots effectively free up our time. We may spend more time on things that matter to us, such as hobbies and family, if we spend less time cleaning up our homes. Some domestic tasks such as caring for the aged and children, front desk reception, driving a car, entertaining people, making love and answering queries or telephonic calls require higher cognitive intelligence and different levels of robot kinematics. Therefore, consumer robots must be built very differently from industrial robots.

19.2 Entertainment Robot

19.2.1 Actroid

The primary purpose of the Entertainment robot is to entertain humans with less or no physical work. Some of the entertainment robots are actroids. Actroids creators claim that their (mainly female-appearing) artistic performances will aid all androids in assimilating into civilisation. Their presence in theatre, music, and film/TV is expected to increase. Interaction with robots that closely resemble humans will be used to acquaint and entertain the public. The actroid robot can imitate noises and other basic human activities such as blinking, speaking, and even breathing in its chest [1]. Its features are internal sensors, silicone skin, speech recognition software, actuators for air, artificial intelligence (AI), microphones and speakers, Servo motors and the ability to move.

19.2.2 Driving Partner Robot

Driving partner robot is a vehicle dashboard-based robot penguin. The Robot is part of Pioneer's continuous robotic research to elevate automobile driving safety to new heights. The driving partner robot detects unexpected changes in automobile direction and senses risks along the roadside.

As you drive, this penguin-shaped robot swings its wings and blinks its blue eyes. Any deviation from safe driving, such as driving in the wrong lane or running a red light, causes the Pioneer's Driving Partner Robot to admonish the driver with its bird chirps and orange blinks [2]. Its features, detection of integrated images road photos may be processed quickly with this camera, voice synthesis capabilities with colour-changing LEDs on the eye, and simple dashboard installation.

19.2.3 Manus

Manus is a biped robot that stands 46cm tall and weighs 2 kg. This Robot is highly advanced, with 17 degrees of freedom (DOF) ascribed to 17 joints. Motors power these joints, which must all be controlled at the same time in order for the robot to move. The joints are human-like, and each leg has six degrees of freedom, allowing it to mimic how people walk.

Manus can avoid any obstructions in its route because of a vision sensor. Eight extra sensors have been added to this soccer-playing humanoid. When coordinating its sensory information and walking, a two-layered control system must keep all sensors in sync [3].

19.3 Educational Robot

19.3.1 Robokind

RoboKind is the outcome of this collaboration, and it is an educational robot platform that can be utilised for study in psychology, cognitive science, and neuroscience. It can also be utilized in therapeutic research for autism.

RoboKind can walk, talk, and display facial emotions. With the support of its walking control technology, it can walk with grace while speech and expression are performed through voice recognition and face tracking. RoboKind is equipped with high-definition cameras, inertial sensors, and Maxon motors [4]. It Features 60-63.5 cm in height, 6.5 kilogrammes, 25 degrees of freedom, two lithium-ion rechargeable batteries, Windows, Mac OS, and Unix all supported, high-definition cameras, sensors of inertia, Motors by Maxon.

19.3.2 TERRI

The National Oceanic and Atmospheric Administration, or NOAA, funded the TERRI robot from Concept Visions Corporation. NOAA hopes to promote and support scientific education using this Robot. TERRI, which stands for The Educational Resource Robot Initiative, is a fully integrated interactive humanoid robot that can be turned into a fully automated mode to engage with people without needing any external intervention or monitoring. This humanoid Robot has full artificial intelligence and access to an extensive database stored remotely. This Robot delivers data to its database and then reacts remotely since it travels often [5]. It features blue LED eyes, a helium voice, a robot on wheels, speakers, a microphone, a camera, a robot with full integration, and an LCD display.

19.4 Social Robot

19.4.1 BHR Series: BHR 3

The Beijing Institute of Technology's Institute of Intelligent Robotics collaborated with the China Aerospace Science & Industry Corporation to create BHR 3, a humanoid robot. This Robot is one of China's few full-sized humanoid robots, and its exoskeleton differs from other Chinese robots. The BHR 3 is more similar to its predecessor, the BHR 2. It has a height of 170 cm and a weight of 60 kilos. This robot is also capable of performing Tai-Chi moves. This humanoid Robot, like the BHR 2, can execute a variety of duties without the use of a cable. It can perform pre-programmed routines and fundamental movements such as walking forward, backward, and sideways. BHR 3 also has 28 degrees of freedom (DOF) and operates on RT-Linux software, much like its predecessor. This Chinese humanoid robot is available for hire or sale as a high-tech advertising item for use at museums, exhibitions, tech shows, and other robot-related events. However, this robot or its technology needs to catch up to ASIMO, BIT and other Chinese humanoid robot businesses that are making strenuous efforts to catch up with the finest in the world. It is no longer a pipe dream for China to become a world leader in robotics, thanks to the progress of the BHR 3 humanoid robot [6]. It Features a Tai Chi exoskeleton robot with 28 degrees of freedom, a robotic biped, Sensors for force and torque, Gyro sensors, and Recognition of sight and sound.

19.4.2 Mertz

MERTZ is a humanoid robot created by the MIT Media Lab to investigate scalable learning in social environments. It has 12 degrees of freedom thanks to the actuators (five for the active-vision head and eyes, two on its force-controlled neck, one on its coupled eyelids, and four on its facial features, such as eyebrows and lips). Two digital firewire cameras, a GN Netcomm microphone, and encoders and potentiometer sensors for proprioceptive feedback are also included.

It was recently tested and discovered that it has learnt to associate items and people by distinguishing people's routines and commonly stated phoneme sequences [7].

19.5 Toy Robot

19.5.1 AIBO

This robot dog comes with a pink ball with which to play. When the AIBO ERS-110 was initially presented in the United States, it sold 3000 goods online in only twenty minutes and another 2000 units in less than a week. Sony ERS employs a top-notch sound controller, even though it does not listen to voice instructions. Owners must provide long-term care and patience for this adorable little guy. To grow from a puppy to an adult, the dog robot needs interaction over some time. It comes with a heat sensor that may be used to reward or punish the dog. The memory stick, which comes in 8 MB and 16 MB sizes, can assign a gender.

It can display happy and angry emotions and impulses for friendship and connection. It has 18 degrees of freedom, with three for each leg, three for the head, two for the tail, and one for the mouth, allowing it to walk flawlessly on four legs and do intricate manoeuvres. The color CCD camera on the AIBO ERS-110 also predicts an object's form and colour. It features a Range Finder, which allows it to determine how far away an item is. Its LED lamps allow owners to recognize its mood, with green representing happiness and red representing anger. Thanks to the mini stereo microphone, it can hear the sound commander remote control and determine the sound source [8].

19.5.2 DragonFly

The DragonFly is a flying robot built by WowWee, the company that created the Robosapiens. It is made to fly with the help of a controller. The DragonFly has four flapping wings that allow it to fly. The Dragonfly robot, which has a tail rudder, can fly like a real dragonfly—a gigantic one—at 18 mph in a specific direction. Though flying this robot is simple, it takes time to perfect every turn and change of height without colliding with a wall or a human face [9].

19.5.3 Robopet

WowWee of China introduced the Robopet to robot lovers on November 16, 2006. It is an interactive pet robot capable of numerous exciting activities. Robopet can walk, run, sit, lie down, and roll over in a realistic biomechanical manner. It can speak with its owner and another WowWee robot, the Robosapiens V2, as befitting an interactive pet robot. Robopet's IR interaction and motion sensors enable this. It can avoid obstacles while moving thanks to the same process and some programming. It is an acoustic and motion sensor that responds to noises and motions. The Robopet, which looks like a guard dog, also makes digital dog noises like barking, snarling, and panting. It can also do a variety of free-roam manoeuvres. Robopet, like a natural dog, can be trained to do tricks. It features a shut-off and sleep mode that activates when it is not in use, and the remote control is susceptible [10]. It Features 9x6x4 Dimensions (HWD), Battery Requirements: 7 "AAA" alkaline batteries (weight: 370 g), sensors include infrared vision, acoustic, and motion sensors, dogs that move realistically, sounds of cloned dogs, and trainable.

19.6 Conclusion

These consumer robots are frequently created to assist persons with specific needs, such as the elderly, autistic children, and disabled adults. They generally seek to assist in a specific setting, such as a home, a school, a nursing home, or a hospital. An entertainment robot is built for the pure subjective enjoyment of the human rather than for practical purposes such as production or domestic services. It generally serves the owner, his housemates, and visitors or clients. Educational robotics is a valuable learning tool for children and should be included in every school's curriculum. An entertainment robot is built for the pure subjective enjoyment of the human rather than for practical purposes such as production or domestic services. It generally serves the owner, his housemates, and visitors or clients. Before being packed and marketed, the entertainment robot is programmed. The robot is usually controlled via a remote during programming. It may be configured to do a different job for each button pressed. Robots are already being used in places like entertainment venues and interactive point-of-purchase systems, and their use in the film industry is rising. A camera-wielding robot can film thousands of identical movie scenes or control precise, repeated special effects actions. Educational robotics is a valuable learning tool for children and should be included in every school's curriculum. Computational thinking is one of the core cognitive abilities of mathematical thinking that educational robots assist children in acquiring at a young age. That is, they aid in developing the mental process we use to solve issues of all types by following a set of steps. Social robots may engage with humans in collaborative and intimate situations, such as shopping malls and completing household and healthcare jobs. Some social robots serve as helpers, assisting with duties such as lifting, while others serve as companions and even imitate emotions. Children will be the people most affected by robots on a social level. Robots will infiltrate classrooms and aid students with their studies. However, robots can also do chores. Robots are incredibly adept at producing schoolwork, as children will rapidly discover.

References

1. Knox, E. and Watanabe, K., Ethical considerations of gendering very human-like androids from an interdisciplinary perspective. *2017 26th IEEE International Symposium on Robot and Human Interactive Communication (RO-MAN)*, pp. 539–546, 2017, DOI: 10.1109/ROMAN.2017.8172355.
2. Liu, Z., Koike, Y., Takeda, T., Hirata, Y., Chen, K., Kosuge, K., Development of a passive type dance partner robot. *2008 IEEE/ASME International Conference on Advanced Intelligent Mechatronics*, pp. 1070–1075, 2008, DOI: 10.1109/AIM.2008.4601810.
3. Stanger, C.A., Anglin, C., Harwin, W.S., Romilly, D.P., Devices for assisting manipulation: A summary of user task priorities. *IEEE Trans. Rehabil. Eng.*, 2, 4, 256–265, Dec. 1994, DOI: 10.1109/86.340872.
4. Silva, V., Soares, F., Sena Esteves, J., Mirroring and recognising emotions through facial expressions for a RoboKind platform. *2017 IEEE 5th Portuguese Meeting on Bioengineering (ENBENG)*, pp. 1–4, 2017, DOI: 10.1109/ENBENG.2017.7889480.
5. Newell, Z., Creative pedagogy and information literacy in higher education, in: *Information Literacy in a Post-Truth Era. April 2021. Communications in*

Computer and Information Science, vol. 1533, S. Kurbanoğlu, S. Špiranec, Y. Ünal, J. Boustany, D. Kos (Eds.), Springer, Cham, 2022, https://doi.org/10.1007/978-3-030-99885-1_29.

6. Zhang, X., Rong, K., Chen, G. *et al.*, Effect of dual-tasks walking on human gait patterns. *J. Bionic Eng.*, 19, 4, 991–1002, 2022, https://doi.org/10.1007/s42235-022-00196-y.
7. Penčić, M., Čavić, M., Oros, D., Vrgović, P., Babković, K., Orošnjak, M., Čavić, D., Anthropomorphic robotic eyes: Structural design and non-verbal communication effectiveness. *Sensors*, *22*, 3060, 2022, https://doi.org/10.3390/s22083060.
8. Tanaka, K., Makino, H., Nakamura, K. *et al.*, The pilot study of group robot intervention on pediatric inpatients and their caregivers, using 'new aibo'. *Eur. J. Pediatr.*, 181, 1055–1061, 2022, https://doi.org/10.1007/s00431-021-04285-8.
9. Penčić, M., Čavić, M., Oros, D., Vrgović, P., Babković, K., Orošnjak, M., Čavić, D., Anthropomorphic robotic eyes: Structural design and non-verbal communication effectiveness. *Sensors*, 22, 3060, 2022, https://doi.org/10.3390/s22083060.
10. Aho, K., "We are protecting them to death"—A Heideggerian interpretation of loneliness among older adults in long-term care facilities during COVID-19. *Phenomenol. Cogn. Sci.*, 1–14, 2022, https://doi.org/10.1007/s11097-022-09803-z.

20

Soft Robotics in Waste Management

S. Rithvik, Vijith Rai, Surya Dornal, Deepak J. and B.C. Pramod*

Department of Mechanical Engineering, Sahyadri College of Engineering & Management, Mangalore, India

Abstract
The advancement of technology has had a life-changing impact on society, and it is now smarting all aspects of existence. With rising energy needs, benefactions to global warming, harmful gas emissions, and contamination from manufacturing to incorrect waste scrapping, the aforementioned advancements frequently come at a cost to our ecosystem. Getting rid of these negative consequences is one of society's key issues. Soft, biologically inspired automation and robots are now gaining traction, and they have the potential to become not only as good as their natural counterparts in terms of execution and capacity but also relation to environmental impact. This criticism examines the emergence of imperishable materials in bioinspired robotics, focusing on actuators, energy storage, and electronics. In the subject of robotics, soft robotics is a novel technique. Soft robots are made of natural polymers known for their flexibility and durability, allowing for new and innovative implementations. Soft robotics is now at the forefront of its field. Soft robotics' progress in waste management might point the way to a more sustainable future.

***Keywords*:** Ecosystem, soft robots, recycling, life-changing influence, multidisciplinary, bibliometric

20.1 Introduction

Environmental effects as a criterion for technical breakthroughs are required in a society plagued by climate change and frequent waste generation. Technology disposables, in particular, are a fast-rising segment of our waste, with over 100 000 tonnes produced daily. Consumer electronics

**Corresponding author*: pramod.mech@sahyadri.edu.in

Rathishchandra R. Gatti, Chandra Singh, Rajeev Agrawal and Felcy Jyothi Serrao (eds.) Self-Powered Cyber Physical Systems, (341–348) © 2023 Scrivener Publishing LLC

and other end-of-life appliances are frequently discarded because the different product designs and material anatomies are difficult to reprocess while being inexpensively manufactured. Soft robots are employed to assist in the sorting of waste goods in order to recycle them.

Soft robotics is an emerging area. Researching novel mediums, drive capabilities, vision technologies, and servo closed-loop systems in soft robotics are challenging since the concepts of stiff robotics, such as industrial and space exploration robots, are fundamentally different. The environmental factor is pushed to the background or ignored entirely in these many disciplines of study. However, with the advancement of climate change and the well-known fact of our finite resources, there is a pressing need for sustainability [1].

It has the power to transfigure reprocessing through automatic material sorting. Although environmental agitations have made it critical to improve recycling processes, entity sorting remains a substantial hurdle. Although some recycling centres utilize automated sorting techniques for metals such as eddy currents and magnets and optical inspection considering polymers, most still rely on manual labour to grab and sort items that are not mechanised. This can create dangerous working conditions, particularly where rubbish and recyclables are mixed.

For automated recycling, soft robotic grippers are a logical choice. It would take much work to estimate and devise a strategy for a typical stiff gripper to grasp the things that escape on a conveyor belt in the recycling business [2]. The need to hold an object is greatly reduced with a compliant gripper that complies with the body's surface. The soft robotic gripper employs sloppy hardware and basic software to grab the fleeing items. Similarly, there is a fluid-driven soft robotic gripper that is employed for a variety of objects of complexity. Finally, this study discusses how soft robots might help with trash management and sustainability.

20.2 Soft Robotics Insights

Soft robots feature continuous body kinetics, large-scale deformation, peak submissiveness, and adjustable electrical resistance, unlike traditional robots with peak resistance. Because of these characteristics, this type of robot is well suited to highly dexterous tasks and interactions that need compliance with environmental ambiguity.

Soft robotics has a turbulent half-century history, although it has only lately become a major topic of discussion among scientists and the general public. As the robotics community becomes more aware of these breakthroughs, more scientists and engineers are eager to participate. This is shown in a rise in the number of laboratories, global partnerships, rising publications, soft robot-related organizations, and special sessions at international conferences, professional events, and activities. Even though soft robotics is still in its early stages, many review article papers have been prepared to summarize accomplishments, assess methodology, and investigate barriers and prospects. These evaluations were written to provide a new viewpoint on the soft robotics research area by using a bibliometric survey to emphasize the historical map and overall image of the subject.

20.2.1 Materials and Actuators

Traditional robots can be designed using a mechanised system comprising actuators and joints, powering units, and control systems. Each module is well separated and has a distinct purpose within the robot's body. In nature, movement, energy storage, sensing, and processing are all ravelled and constructed around the same soft material architecture. The same dissolving divisions affect soft robotics, with the outcome that most of a soft robot's functioning is decided by body design rather than a mix of electromechanical components. Similarly, environmental challenges begin as questions of material science. The function of soft or hard (lightweight) robots is determined by their mechanical qualities. Young's modulus (E) of human building blocks ranges from 1 to 10 kPa for brain tissue, 0.1 to 1 MPa for skin, 0.5 to 1 GPa for tendons, and 10 GPa for bone. Because these stresses are frequent for pneumatic or dielectric elastomer actuators, elastomers for soft robotic actuators must-have Young's modulus of 0.1 to 10 MPa and a stretchability of more than 200% of ultimate strain. The less the material's (repeatable) stretchability, the less probable it is to fail. The elastomer's form recovery determines the accuracy of reversible actuation under cyclic stretching and component fatigue. Plastics and other structural support mediums, like biological tendons, have Young's modulus of 1 to 5 GPa and a stretchability of a few percent. Thermosets and thermoplastic materials are attractive for mechanical components due to their various manufacturing procedures and processing capabilities [4].

20.3 Soft Robots in Waste Management

Soft automation can revolutionize reprocessing through automated entity sorting. Although environmental and perseverance considerations have made it imperative to enhance reprocessing procedures, entity sorting remains a critical tailback for scalability. Failure to identify recyclable entities accurately gives on to waste; in the United States, 25% of all reprocessed materials are contaminated to the point that they must be disposed of in landfills. Single-stream recycling has a higher contamination rate than pre-sorted dual-stream recycling, even though it is more user-friendly [5].

Although some recycling centres utilize automated sorting techniques for metals, such as eddy currents and magnets, and optical inspection for polymers, the majority still rely on manual labor to grab and sort items that are not mechanised. This can create dangerous working conditions, particularly where rubbish and recyclables are mixed.

Soft robotic grippers are an obvious choice for automated recycling. It would be hard to devise a method for a normal, stiff gripper to snare the twisted and deformed articles that pass by on a recycling conveyor belt. With a compliant gripper that conforms to the item's surface, the computational requirements for capturing an entity are considerably decreased. A soft gripper allows faulty hardware and simple software to pick up a wide range of bodies instead of scanning an entity, calculating directions and forces, and flawlessly executing a method.

20.3.1 Operation of Soft Robots in Recycling Separation

This study describes the design of an autonomous motion planning algorithm for a soft planar grasping manipulator that can grasp and place operations despite unpredictability in the location and form of the item. Instead of the typically used multilayer lamination procedure, the soft manipulator's end effector is manufactured in one piece without weakening seams using lost-wax casting. The soft manipulation system can grab and move things that are randomly placed inside its reach envelope without the need for human involvement. The autonomous planning system uses the soft grasping manipulator's compliance and continuum bending to produce reproducible grasps in the face of uncertainty. We use the gripper to simulate an MRF recycling process to demonstrate its full capabilities. The objective is to remove products from a conveyor belt and put them in the allocated container using a Baxter robot with a gripper. The conveyor belt brings items to Baxter robot's workstation, coming to a halt when the object passes over an IR break beam, signalling to Baxter that the object is ready to be gripped. The object will next be grabbed by Baxter, who will categorize it according to the classifier's instructions.

Paper [2] depicts a handful of the items that were put to the test. Across three tests, with one selection every six seconds, they were 63 percent correct in identifying the bodies of our thirteen test entities. They correctly identified all-metal bodies, the paper cup, and the plastic cup in all three different trials but had the most burdensome with paper things. This is likely due to the closeness of paper and plastic items in our average object's characteristic space. The reprocessing job's lower accuracy rate than the motionless task is most likely due to sensors shifting when the reprocessed entity travels. As a result, our calibration would remain consistent with our expectations, resulting in more erroneous estimations. The existence of the conveyor belt's engine and a huge metallic object, which might be a source of electromagnetic interference, could potentially result in false readings.

In [2], the soft robotic gripper reveals the table's internal layers. They remove the glove on the index finger. A set of chirally opposite-handed shearing auxetics makes up each finger (HSAs). The rear of the fingers has a strain sensor, while the interior has a pressure sensor. The pressure sensor rubs on a paper backing, while silicone pads and gloves help to improve touch.

20.3.2 Recycling of Soft Robotics After Its Shelf Life

Products, like nature, should be designed to be as recyclable as feasible.

The three cycles that may be classified as post-consumer recycling are the primary cycle (pre-consumer reprocessing), the secondary cycle (recycling during product use), and the tertiary cycle (reprocessing after product use).

The primary cycle refers to the manufacturing operation, which includes reprocessing production waste, which can be recycled either directly or indirectly (pre-treatment).

The second cycle happens when the product is in use. The goal is to keep the product as it is now, in its most useful condition, for as long as possible. The product can be reused (for example, a canister) or reconditioned for future use after its original usage.

When a product is used for an occasion other than its primary use (for example, tyres even as port buffers), it loses value.

After goods have been utilized, the tertiary cycle is concerned with reconditioning them. Different materials must be removed for additional processing if a product cannot be reused or is not cost-effective. This might lead to the production of the same product (if the materials used are 100% reusable) or, if that is not possible, the discovery of a new use for the resources (e.g., shredded tyres as playground surfaces).

Silicone rubber has proven to be the most efficient and frequently used material for fluidic elastomer actuators (FEAs). They are hard to reprocess, however, because they cannot be liquefied and must be crushed into powder to be used. They can only be used as fillers when blended with other materials. Appropriate businesses must be engaged in material recycling. This necessitates classifying the materials.

After examining FEAs for their reprocessing-aligned design, it can be determined that, although the silicone-based construction of soft actuators, reprocessing may be relevant in every phase of the product life cycle (production, product use, and disposal).

Particular consideration must be given to the materials utilised regarding disposal. The number of different materials used should be kept to a minimum. FEAs allow for creating a soft robot made completely of one material. A surge in stiffness can also be obtained by a surge in the thickness of the strain-limiting layer rather than selecting an additional material.

20.4 Are Soft Robots the First Step for a Sustainable Future?

When the quantity of robots in society expands, and they are used in a broader range of applications, the difficulty of resilience arises. Traditional robotics aims to create robots that are as long-lasting as feasible [1]. This is in contrast to the robots' negative impact on the environment as they near the end of their useful lives. Post-life impacts and resilience will become increasingly important as the number of soft actuators rises. Biodegradability is essential for resolving this problem. Soft biodegradable robots that degrade to nothing are necessary, precisely as they are in nature, where all organic materials are reprocessed when an organism dies. Biodegradable bodies, including biodegradable actuators and sensors, are required to achieve this capability.

Soft actuators made of biodegradable biopolymers, comprising pneumatically and electrically driven structures, have recently been shown. As several polymers have demonstrated, it is vital to remember that decomposition is different from safe breakdown. Even when the plastic has decomposed in many circumstances, the individual components might still have a severe environmental impact. On the other hand, environmental soft robotics must demonstrate a net unbiased or positive environmental aftermath over the course of their whole life cycle and long-term effects.

Robots must be sustainable to survive in potentially dangerous, fragile natural and artificial surroundings. This encompasses their environmental credentials (i.e., effect, cost, and persistence) and their capacity to deal with adverse conditions [3]. When a rigid robot is injured, it has limited capacity to mend; repairing a broken metal or stiff composite material is difficult. On the other hand, soft robotics can be made of damage-resistant and self-healing materials. Natural creatures employ self-healing as a means of extending their lives and ensuring that they reach adulthood safely. Heal-all in Diels–Alder materials is one example, while self-healing in biodegradable polymers such as konjac glucomannan gels is another. While these use molecular-level self-healing, they can also use a structural approach, such as capsule-based and vascular healing systems. These mechanisms are similar to those observed in nature, where local clotting and fibroplasia work together to seal and heal tissue damage. Heal-all is a crucial feature for both practical and long-lived soft robots.

20.5 Conclusions

Soft robotic organisms (SROs) are a new robotics prototype. They can be built in large numbers and collaborate in ways traditional robots cannot. They can decline safely to nothing in susceptible circumstances and remain for an unlimited time/period by scavenging energy. The next step in this subject is to bring together soft actuators' many parts and capacities to create a prototype. Ensure that excessive complex biological behaviors, such as self-replication and reproduction, will evolve. This will usher in a new era of soft robotic creatures that are "alive."

A soft actuator gripper that can sort reusable has been shown. To accomplish this, we developed a puncture-resistant gripper with tactile sensing. They achieved the design objectives and developed a classifier that could recognize different materialistic items by coupling electrically driven and durable HSA actuators with proprioceptive strain and pressure sensors.

More effort is needed to improve the apparatus used in MRFs and other recycling facilities. Additional sensors, particularly throughout the touch surface, might provide additional data to allow more complicated classification algorithms to be applied, enhancing classifier accuracy. A larger contact resolution, for example, enables the classifier to better know the object's entire form rather than its size. This gripper technology will need to be combined with vision-based technologies on a system level to improve accuracy and allow picking in crowded conditions. Future vision-based systems can recognize a broader range of entities, including glass and non-reusable materials.

Even as the research is still in its early stages, this soft robotic gripper has much potential for increasing recycling efficiency. Although the classifier could be better, any increase in upstream sorting can result in large quality gains downstream. Furthermore, because of our system's small design, it can be installed further demanding than MRFs, allowing big businesses and residential buildings to pre-sort their single-surge trash and minimize pollution rates. In any case, soft robotics can aid in improving recycling and developing a more sustainable economy.

References

1. Hartmann, F., Baumgartner, M., Kaltenbrunner, M., Becoming sustainable, the new frontier in soft robotics. *Adv. Mater.*, 33, 19, 2004413, 2021.
2. Chin, L., Lipton, J., Yuen, M.C., Kramer-Bottiglio, R., Rus, D., Automated recycling separation enabled by soft robotic material classification, in: *2019 2nd IEEE International Conference on Soft Robotics (RoboSoft)*, IEEE, pp. 102–107, 2019.
3. Tan, Y.J., Harnessing the circular economy to develop sustainable soft robots. *Sci. Robot.*, 7, 63, eabn8147, 2022.
4. Schmitt, F., Piccin, O., Barbé, L., Bayle, B., Soft robots manufacturing: A review. *Front. Robot.* AI, 5, 84, 2018.
5. Nguyen, A.M. and Seibel, A., Recycling-oriented design in soft robotics. *Actuators*, 8, 3, 62, 2019.

21

State-of-the-Art Review of Robotics in Crop Agriculture

A. Shahaas, Rithin, B. Nambiar, C.P. Dheeshith, K. Abhijith and Rathishchandra R. Gatti*

Department of Mechanical Engineering, Sahyadri College of Engineering & Management, Mangalore, India

Abstract

There has been a significant increase in the use of agricultural robots in recent years. Population growth, rapid urbanization, great demand for high-quality goods, scarcity of skilled workers, and preserving the environment are some major challenges conquered by the continual advances in agricultural robotics. In agriculture, robots help quickly complete repetitive tasks, improving crop yields. Robots can be effectively employed to lessen the work of farmers and relieve them from the tedious tasks, such as sowing, harvesting, picking, and disposal of Agri-waste farming. Farming was and still is a labor-intensive human sector in many parts of the world. Globally, agriculture employs almost 870 million people. It is one of the most susceptible industries to automation. This paper mainly focuses on the scope of the use of robots in the agricultural industry. We also look into the current methods for preparing the ground before growing, seeding, reaping, and measuring yield.

Keywords: Robotics in agriculture, automation, precision agriculture

21.1 Introduction

The world's population has grown dramatically from one billion in 1800 to 7.9 billion people. This rapid increase in the human population has forced farmers to change their farm management strategies to fulfil the rising demand for high-quality food as more people choose healthier foods free of herbicides and pesticides. The application of automation is

**Corresponding author*: gattirathish@gmail.com

Rathishchandra R. Gatti, Chandra Singh, Rajeev Agrawal and Felcy Jyothi Serrao (eds.) Self-Powered Cyber Physical Systems, (349–358) © 2023 Scrivener Publishing LLC

very favourable when it comes to agriculture. However, there are diverse challenges since agricultural environments need to be structured and controlled. This paper addresses various agricultural tasks which can be done with robots. Robotics is set to transform the agricultural industry. For increasing productivity and reducing waste, developing new Robotics and Autonomous Systems (RAS) will have crucial environmental and societal gains. Agricultural activities rely primarily on human labour, which is prone to health difficulties and errors [1].

Precision agriculture is a farm management technique that takes advantage of technology to ensure that crops and soil get exactly what they need to thrive and yield, which is being introduced. Agricultural robotics technologies, which include mobile sensing, computing, and actuators, are important to reaching this aim on the cultivation side by allowing precision farming even down to the basic level of farming.

The scale of a single plant is ever-increasing spatial and temporal resolutions. Such individualized plant care approaches are referred to as phytotechnology. Moreover, they promise to improve yield while lowering water, chemicals, and energy use. On the subject of the breeding front, the capacity to functionally connect the plant phenotype as a result of genotype, crop management, and field environment will be important for the quick production of dramatically enhanced crop varieties that are highly productive and drought and disease tolerant.

21.2 Scope

Our farm equipment firms and researchers have produced a wide range of small and heavy farm equipment for conventional farming demands, but precision farming necessitates using robotics and automation. Because robots have joined the fields indicated above, it is vital to consider why they have yet to invade the farming field. Robots may be used for weed management, which will help to minimize the use of pesticides and make the food organic. Robots can also be used to transplant seedlings, saving time and effort. Electric motors can be controlled remotely via mobile phones, which is especially useful for farmers during the summer when the power supply is inconsistent. Considering enhanced intelligent robots in farming, sensors, readers, and hand-held PDAs will be useful in calculating and accuracy. However, several obstacles exist in multiple nations for automation in the agricultural sector. The topic of this article is confined to agricultural robots used in crop yields, especially surface robots that perform open-field tasks, including plant breeding, crop planting, maintenance, and collecting. Hard automation technologies have historically serviced post-harvest processing, but mechatronics and robots have rapidly become part of post-harvest and food production systems. To solve labour shortages in the sector, agriculture automation has become a trend in which agricultural tasks, including cultivation, inspection, spraying, pruning, and picking, are carried out autonomously. The creation of an effective farm automation system has been a focus of contemporary agricultural research to assure food security's long-term sustainability. Examples of agricultural robots are RoBoPlant, Blue River LettuceBot2, Energid Citrus Picking System, Agrobot E-Series, Agribotix, and vision Robotic [2].

21.3 Advantages

Traditionally, agriculture needs countless repetitions and a great deal of hard labour for activities such as planting, irrigating, fertilizing, monitoring, and harvesting a wide area of crops. The flexibility to carry out a wide range of jobs and applications in any setting is one of the significant advantages of robots. They are more precise and consistent than humans.

Since robots can accomplish tasks more quickly, they help increase production and profit significantly. Robots can work around the clock and can do tasks with great ease and fast pace, which people find time-consuming. They also make zero or very few mistakes compared to humans and save much time. Robots eliminate the jobs which are dangerous and threatening to human beings.

Elimination of labour - The farmer is freed from the laborious milking operation and its associated schedule, allowing him to devote more time to animal supervision, feeding, and other activities. Robotic cameras and sensors can identify weeds, pests, parasites, sickness, and other sources of stress. Sensors are often selective, spraying only the afflicted area. They can be relatively compact, allowing for data collection close to the crop and mechanized weeding, mowing, spraying, and fertilizations. Soil testing offers information about the soil. When that information is processed as intelligence, a report is generated and given to artificial intelligence via robotics, providing us with faultless results that may increase the quality. The utilization of fertilizers, pesticides, insecticides, herbicides, and water use might all be reduced dramatically. On the farm, the robots can reduce pesticide usage on a large scale. Robots can effectively replace human operators while providing a reasonable investment return. Robots can manufacture high-quality goods at a cheaper cost of production. In several industries, robots are more efficient and can quickly navigate past obstacles such as trees, rocks, ponds, and other obstacles. It brings a new age in cattle grazing, farming, and agriculture. Technicians who can fix robots may be able to work with them. The robot gantry might be used to spray fertilizer or liquids and an autonomous system of self-control that adjusts to changing climate conditions. Because the robots do not feel sick or tired, they do not need to vacation. Although it is a one-time expenditure, the cost of farming will significantly reduce.

21.4 Disadvantages

Unlike human beings, robots need a huge investment. Robots are expensive to build or purchase. Therefore, it is costlier to implement. To keep them functioning, they require proper maintenance. Also, time management and fully skilled labour are required to operate these agricultural robots. Another disadvantage faced is the lack of a continuous power supply. Robots require power, while in India, the agricultural region has a power outage of more than 65 percent. Since a single robot can do multiple tasks more efficiently compared to human beings, this, in turn, leads to farmers losing their jobs. Agricultural robotics research & development is costly. Due to its high price and maintenance, it is unaffordable and inaccessible to poor farmers.

21.5 Applications

21.5.1 Robot Drone Tractors

The production of various robots, such as introducing the first ever robot drone tractor, is altering how we farm. The robot will decide where to plant when to harvest, and how to determine the ideal path for crisscrossing the countryside. These are presently in use in the United States. Venerable American farm equipment manufacturer John Deere and French agricultural robot start-up Naio debuted their latest innovations at the Consumer Electronics show in Las Vegas.

21.5.2 Flying Robots to Spread Fertilizer

The robot can fly autonomously and apply fertilizer on its own. Thanks to camera equipment and an automated fertilization system on the front. It was created by the Chinese Academy of Sciences Shenyang Institute of

Automation's national important robotics laboratory. On July 25, 2011, a flying robot in Ili, a Kazak autonomous prefecture in Northwest China's Xinjiang Uygur autonomous province, analyzed the growing status of crops above farmlands. Drones' ability to move around quickly and navigate to their intended destinations is one of the most important uses of drones in agriculture. Drones with this capability may spray fertilizers and pesticides on crops to nourish them and give them the necessary nutrients. Crops can be healthy and thrive with such supplements. Drone controllers have complete control over the drone spraying nutrients to keep bugs, parasites, and worms at bay and extend crop life.

21.5.3 Fruit Picking Robots

The study is ongoing because the robots must be properly built to avoid bruising the fruit while harvesting. Suction grippers, which are utilized in automated fruit-picking machines such as those made by ACRO, are one alternative. Robotic citrus fruit pickers have received the most attention thus far, but cherry pickers are also being studied. Vision Robotics, in particular, has developed several robots capable of taking over the job. Researchers in the United Kingdom are developing an intelligent harvesting machine using newly discovered image technologies. The device can examine the leafy layers of a crop, identify the different components, and provide exact size determination via microwave measurement. A system like this might reduce crop waste like cauliflower and help U.K. farmers deal with an impending labour crisis caused by a decrease in migrant workers. Strawberries are fragile fruits that must be picked with care. Dogtooth Technologies, a start-up, created a robotic arm capable of collecting delicate crops like strawberries. F.F. Robotics is a firm that designs and manufactures a robot to help farmers save money by replacing harvesting labour. FFRobot is the name of the robot. Picking different sorts of fruits with the grabbing hand is simple. As a result, the robot may be used throughout various harvest periods.

21.5.4 Robot Cattle Grazing and Automatic Milking

It is a method of milking dairy cattle, especially dairy cows, without using human labour. Automatic milking systems (AMS), also known as voluntary milking systems (VMS), were developed in the late twentieth century. They have been on the market since the early 1990s. A sort of agricultural robot is at the heart of such systems, which allows total automation of the milking process. As a result, robotic milking is another name for automated milking. Computers and sophisticated herd management software are used in most systems. They provide us with fresh meals daily. The dairy sector has been transformed by robotic milking equipment. Gascoigne Melotte used a robotic milking box to milk the first dairy cow in 1986 at the experimental farm de Waiboerhoeve in Lelystad, the Netherlands, more or less without typical human intervention, according to U.S. Patent 4010714A. [7].

21.6 Automation in Agriculture

Agricultural jobs, such as cultivation, inspection, spraying, pruning, and harvesting, have become more automated to solve labour shortages. The phrase "automated agriculture" refers to any agricultural machinery or technology that eliminates the need for manual labour. Agricultural automation focuses on autonomous vehicle applications such as robots and tractors, which are utilized to reduce the harsh, dangerous, and lengthy working conditions that farmers face while providing a precise and efficient operation and control system [4]. Furthermore, quantity and quality of output must be maintained to ensure the product is of excellent grade and appropriate for human consumption. Thus, creating a good agricultural automation system is currently a focus of contemporary agricultural research to ensure food security's long-term sustainability [5, 6].

The organization, planning, and execution of automation and robotics in agriculture will alter as each agricultural firm has its unique set of goals to achieve. Following that, the evaluation will discuss some of the challenges and potential future advancements in agricultural automation and robotics.

Automating certain agricultural tasks has allowed farmers to regulate agricultural yield better while conserving effort and money. Farmers and academics have become interested in developing agricultural automation systems due to causes such as a labour shortage in agriculture, an ageing farmer population, and growing agricultural salaries. Autonomous robots and agricultural machineries such as tractors with cultivators, planters, multipacks, and chisel ploughs have been utilized and developed for agricultural automation.

21.6.1 Forestry

In agriculture and forestry, robotics has had a big impact. Farm owners know the need for autonomous vehicle routing to protect their soil's development zone. Produce detecting, understanding, and managing are now standard, while livestock activities are heavily instrumented and mechanized. Legged harvesters have yet to have much success in the forestry business; however, in the future, automation of cutting and moving and simultaneous localization and mapping procedures will alter the sector.

Some notable advancements in forestry harvesters walk on display, including equipment designed specifically for the rough terrain of Scandinavian woods. Traditional harvesters of the trim likewise have the highest level of automation, and tasks like delimbing must be completed quickly. Some of the stated localization and mapping procedures must be implemented before total autonomous harvesting is achievable.

A tractor may follow a row of plants and turn around at the end of a row, thanks to the combination of machine vision and satellite location (GPS). The timeline of a succession of projects is traced, culminating in the current product being sold. The location of macadamia nuts is another machine vision-based effort. It is vital to assign fallen nuts to the relevant tree before deciding which trees should be propagated. Produce colour sorting and grading requires more than just sensing; it also needs some produce handling, putting robots at their most cutting-edge.

The value of automated milking parlours has been demonstrated. Other initiatives mentioned here, such as sheep shearing robots and a robotic shearing robot slaughterhouse, have had a different level of success. Another project is given that physically divides sheep from goats using a using machine vision. A swinging fence is used to divide different species, so that wild species are not introduced and are kept out of the parched Australian outback's drinking holes.

Although robots are rapidly infiltrating these fields, they continue to provide a rich application project supply, some of which are sufficiently challenging to need the creation of new theoretical methodologies.

21.6.2 Animal Husbandry

Agriculture is a vital part of the global economy. Farmers must produce more food while maintaining good product quality to feed the world's rising population. Farmers can attain these aims with the aid of innovations like robots [3].

Agriculture robotization must consider several issues. Because agricultural organizations face various challenges, economic efficiency alone cannot support robotics adoption. The necessity to examine the appropriateness of adopting robots at the micro-level, considering numerous elements, necessitates the timely nature of the study subject since this issue still needs to be adequately researched. The study presents a technique for assessing the feasibility of agricultural organization robotization while considering the technical, technological, organizational, and social elements that characterize a given production process. The first stage of such an evaluation should examine the economic efficiency of robotics installation. The expert panel then calculates each recommended criteria's average and relative importance. The preliminary selection is completed by assigning a rating to each workplace's choice for robot implementation. In agricultural companies, the approach for measuring the efficacy of robots has been evaluated in animal husbandry workplaces.

One or more milking positions, robot arms to attach and detach the milking heads from the udders, a control computer, and several auxiliary pieces of types of equipment to clean and disinfect the udder and the milking head make up an AMS. The system can also be performed automatically to remove the initial milk pumps and, if necessary, filter removing milk from cows that have been given antibiotics and are therefore unfit for sale. The milking arm and its associated mechanism for precisely finding the space udders are unquestionably the most complex piece of engineering. Infrared, high-frequency sound, lasers, and television cameras are ways to obtain a 3-D picture. A single arm can also be used to service many milking stations, lowering the overall cost of the robot.

21.7 Precision Agriculture

Precision farming, often known as Precision Agriculture (P.A.), is a type of agricultural management approach that uses various techniques to control the variation in space and time involved with everything element of farming output. Its major purpose is to increase crop performance as well as the quality of the environment. P.A. is a farm management method that uses information technology (I.T.) to guarantee that crops and soil receive precisely what they require for optimal health and yield. Several writers have demonstrated the economic and environmental benefits when precision agricultural approaches are used. Despite this, academic surveys and professional assessments suggest that these technologies are still being adopted at a low rate.

Furthermore, rather than employing precision agriculture as a whole, most of the deployments documented use these approaches to address specific requirements or cover significant Spaces in farmers' expertise. Engineers have also worked extensively to supply the technology needed to adopt P.A. techniques, even though agronomists are leading the way in P.A. development. Engineering achievements for P.A. include sensors, controllers, and remote-sensing systems.

Vehicle guiding and auto-steer systems are the most extensively employed robotic technology in precision agriculture. The rationale is that the financial advantages may be obtained without requiring new components or decision support systems to be included. On the other hand, teams of agronomists and engineers utilize different technologies, particularly those linked to remote sensing, sensor development, and control systems.

21.8 Conclusion

Agricultural robots allow for fine-scale sensing and interaction with crops. As a result, high-throughput phenotyping for breeding new crop cultivars and ultraprecise farming, a critical technology for sustainably improving agricultural yield, is now possible. Agriculture robotics and automation will be critical in ensuring future food security. With today's advanced technology, supplied according to the defined system, and robot deployment, farmers have been able to finish agricultural tasks on schedule thanks to equipment. Cultivating, monitoring, dusting, and gathering activities will be carried out effectively with minimal operating expenses and human labour since the evolution of robotics systems in farming primarily focuses on imitating human labour behaviour in the fulfilment of agricultural operations.

Precision agriculture aims to use a variety of technologies to learn about crop spatial and temporal variability. Therefore, the present study sought to contribute to the condition of agricultural robot technology applications, transforming large farms into smart farms.

References

1. Vougioukas, S.G., Annual review of control, robotics, and autonomous systems agricultural robotics. *Annu. Rev. Control Robot. Auton. Syst.*, 2, 15–16, 2019, doi: 10.1146/annurev-control-053018.
2. Mahmud, M. S. A., Abidin, M. S. Z., Emmanuel, A. A., Hasan, H. S., Robotics and automation in agriculture: Present and future applications. *Appl. Model. Simul. (AMS)*, 4, 130–140, 2020.
3. Ren, G., Lin, T., Ying, Y., Chowdhary, G., Ting, K.C., Agricultural robotics research applicable to poultry production: A review. *Comput. Electron. Agric.*, 169, 105216, 2020.

4. Thomasson, J.A., Baillie, C.P., Antille, D.L., Lobsey, C.R., Mccarthy, C.L., Autonomous technologies in agricultural equipment: A review of the state of the art, [Online]. Available: www.asabe.org.
5. Oliveira, L.F.P., Silva, M.F., Moreira, A.P., Agricultural robotics: A state of the art survey, in: *Robots in Human Life-Proceedings of the 23rd International Conference on Climbing and Walking Robots and the Support Technologies for Mobile Machines, CLAWAR 2020*, pp. 279–286, 2020, DOI: 10.13180/clawar.2020.24-26.08.44.
6. Roldán, J.J., del Cerro, J., Garzón-Ramos, D., Garcia-Aunon, P., Garzón, M., De León, J., Barrientos, A., Robots in agriculture: State of art and practical experiences. *Service Robots*, 67–90, 2018.
7. Notsuki, I., Ueno, K., *System for managing milking-cows in stanchion stool.* U.S. Patent 4, 010, 714, issued March 8, 1977.

Index

Also of Interest

Check out these other related titles from Scrivener Publishing

ENERGY STORAGE TECHNOLOGIES IN GRID MODERNIZATION, Edited by Sandeep Dhundhara, Yajvender Pal Verma, and Ashwani Kumar, ISBN: 9781119872115. Written and edited by a team of experts, this exciting new volume discusses the various types of energy storage technologies, the applications of energy storage systems, the performance improvement of modern power systems, their role in the real-time operation of power markets, and the operational issues of modern power systems, including renewable-based generating sources.

SMART GRIDS FOR SMART CITIES VOLUME 1, Edited by O.V. Gnana Swathika, K. Karthikeyan, and Sanjeevikumar Padmanaban, ISBN: 9781119872078. Written and edited by a team of experts in the field, this first volume in a two-volume set focuses on an interdisciplinary perspective on the financial, environmental, and other benefits of smart grid technologies and solutions for smart cities.

SMART GRIDS FOR SMART CITIES VOLUME 2: Real-Time Applications in Smart Cities, Edited by O.V. Gnana Swathika, K. Karthikeyan, and Sanjeevikumar Padmanaban, ISBN: 9781394215874. Written and edited by a team of experts in the field, this second volume in a two-volume set focuses on an interdisciplinary perspective on the financial, environmental, and other benefits of smart grid technologies and solutions for smart cities.

SMART GRIDS AND INTERNET OF THINGS, Edited by Sanjeevikumar Padmanaban, Jens Bo Holm-Nielsen, Rajesh Kumar Dhanaraj, Malathy Sathyamoorthy, and Balamurugan Balusamy, ISBN: 9781119812449. Written and edited by a team of international professionals, this groundbreaking new volume covers the latest technologies in automation, tracking, energy distribution and consumption of Internet of Things (IoT) devices with smart grids.

SMART GRIDS AND GREEN ENERGY SYSTEMS, Edited by A. Chitra, V. Indragandhi and W. Razia Sultana, ISBN: 9781119872030. Presenting the concepts and advances of smart grids within the context of "green" energy systems, this volume, written and edited by a global team of experts, goes into the practical applications that can be utilized across multiple disciplines and industries, for both the engineer and the student.

SMART GRIDS AND MICROGRIDS: Concepts and Applications, Edited by P. Prajof, S. Mohan Krishna, J. L. Febin Daya, Umashankar Subramaniam, and P. V. Brijesh, ISBN: 9781119760559. Written and edited by a team of experts in the field, this is the most comprehensive and up to date study of smart grids and microgrids for engineers, scientists, students, and other professionals.

MICROGRID TECHNOLOGIES, Edited by C. Sharmeela, P. Sivaraman, P. Sanjeevikumar, and Jens Bo Holm-Nielsen, ISBN 9781119710790. Covering the concepts and fundamentals of microgrid technologies, this volume, written and edited by a global team of experts, also goes into the practical applications that can be utilized across multiple industries, for both the engineer and the student.

INTEGRATION OF RENEWABLE ENERGY SOURCES WITH SMART GRIDS, Edited by A. Mahaboob Subahani, M. Kathiresh and G. R. Kanagachidambaresan, ISBN: 9781119750420. Provides comprehensive coverage of renewable energy and its integration with smart grid technologies.

Encyclopedia of Renewable Energy, by James G. Speight, ISBN 9781119363675. Written by a highly respected engineer and prolific author in the energy sector, this is the single most comprehensive, thorough, and up to date reference work on renewable energy.

Green Energy: Solar Energy, Photovoltaics, and Smart Cities, edited by Suman Lata Tripathi and Sanjeevikumar Padmanaban, ISBN 9781119760764. Covering the concepts and fundamentals of green energy, this volume, written and edited by a global team of experts, also goes into the practical applications that can be utilized across multiple industries, for both the engineer and the student.

Energy Storage, edited by Umakanta Sahoo, ISBN 9781119555513. Written and edited by a team of well-known and respected experts in the field, this new volume on energy storage presents the state-of-the-art developments and challenges in the field of renewable energy systems for sustainability and scalability for engineers, researchers, academicians, industry.

Energy Storage 2nd Edition, by Ralph Zito and Haleh Ardebili, ISBN 9781119083597. A revision of the groundbreaking study of methods for storing energy on a massive scale to be used in wind, solar, and other renewable energy systems.

Hybrid Renewable Energy Systems, edited by Umakanta Sahoo, ISBN 9781119555575. Edited and written by some of the world's top experts in renewable energy, this is the most comprehensive and in-depth volume on hybrid renewable energy systems available, a must-have for any engineer, scientist, or student.

Progress in Solar Energy Technology and Applications, edited by Umakanta Sahoo, ISBN 9781119555605. This first volume in the new groundbreaking series, Advances in Renewable Energy, covers the latest concepts, trends, techniques, processes, and materials in solar energy, focusing on the state-of-the-art for the field and written by a group of world-renowned experts.

A Polygeneration Process Concept for Hybrid Solar and Biomass Power Plants: Simulation, Modeling, and Optimization, by Umakanta Sahoo, ISBN 9781119536093. This is the most comprehensive and in-depth study of the theory and practical applications of a new and groundbreaking method for the energy industry to "go green" with renewable and alternative energy sources.

DESIGN AND DEVELOPMENT OF EFFICIENT ENERGY SYSTEMS, edited by Suman Lata Tripathi, Dushyant Kumar Singh, Sanjeevikumar Padmanaban, and P. Raja, ISBN 9781119761631. Covering the concepts and fundamentals of efficient energy systems, this volume, written and edited by a global team of experts, also goes into the practical applications that can be utilized across multiple industries, for both the engineer and the student.

INTELLIGENT RENEWABLE ENERGY SYSTEMS: Integrating Artificial Intelligence Techniques and Optimization Algorithms, edited by Neeraj Priyadarshi, Akash Kumar Bhoi, Sanjeevikumar Padmanaban, S. Balamurugan, and Jens Bo Holm-Nielsen, ISBN 9781119786276. This collection of papers on artificial intelligence and other methods for improving renewable energy systems, written by industry experts, is a reflection of the state of the art, a must-have for engineers, maintenance personnel, students, and anyone else wanting to stay abreast with current energy systems concepts and technology.

SMART CHARGING SOLUTIONS FOR HYBRID AND ELECTRIC VEHICLES, edited by Sulabh Sachan, Sanjeevikumar Padmanaban, and Sanchari Deb, ISBN 9781119768951. Written and edited by a team of experts in the field, this is the most comprehensive and up to date study of smart charging solutions for hybrid and electric vehicles for engineers, scientists, students, and other professionals.

Printed in the USA/Agawam, MA
September 18, 2026

910661.117